Political Opportunities for Climate Policy

Analysis of climate change policies has focused mainly on the prospects for international agreements or on how climate policies should be designed. Yet effective domestic climate policies are essential to any global solution, and we know too little about how and why such policies are adopted. *Political Opportunities for Climate Policy* examines in depth the causes of effective climate policies in the United States, using a statistical analysis of all fifty states and long-term case studies of California, New York, and the federal government. In this book, Roger Karapin analyzes 22 episodes in which policies were adopted, blocked, or reversed. He shows that actors and events have positively affected climate policy making, despite the constraints presented by political institutions and powerful fossil fuel industries. Climate policy advocates have succeeded when they mobilize vigorously and astutely during windows of opportunity – which opened when events converged to raise both the awareness of problems and the political commitment to address them.

ROGER KARAPIN is Professor of Political Science at Hunter College and the Graduate Center, City University of New York, and is Editor-in-Chief of *Polity*.

Political Opportunities for Climate Policy

California, New York, and the Federal Government

ROGER KARAPIN

Hunter College and the Graduate Center
City University of New York

CAMBRIDGE
UNIVERSITY PRESS

CAMBRIDGE
UNIVERSITY PRESS

32 Avenue of the Americas, New York NY 10013

Cambridge University Press is part of the University of Cambridge.

It furthers the University's mission by disseminating knowledge in the pursuit of education, learning, and research at the highest international levels of excellence.

www.cambridge.org
Information on this title: www.cambridge.org/9781107425545

First published 2016

A catalog record for this publication is available from the British Library.

Library of Congress Cataloging in Publication data
Names: Karapin, Roger.
Title: Political Opportunities for Climate Policy : California, New York, and the Federal Government / Roger Karapin, Hunter College.
Description: New York, NY : Cambridge University Press, 2016. | Includes bibliographical references and index.
Identifiers: LCCN 2015042185 | ISBN 9781107074392 (hardback)
Subjects: LCSH: Climatic changes–Political aspects–United States. | Climatic changes–Political aspects–California. | Climatic changes–Political aspects–New York (State) | Energy policy–United States. | Climatic changes–International cooperation.
Classification: LCC QC903 .K2045 2016 | DDC 363.738/745610973–dc23 LC record available at http://lccn.loc.gov/2015042185

ISBN 978-1-107-07439-2 Hardback
ISBN 978-1-107-42554-5 Paperback

To those who work for effective climate policies

Contents

Figures

Tables

Preface and Acknowledgments

This book argues that actors and events have positively affected U.S. climate policy, despite institutional and other structural constraints on it. Administrative capacities, political culture, fossil fuel endowments, and air pollution problems strongly affected what could be achieved, but they did not determine the outcomes. Environmental organizations, environmentally minded elected officials, segments of business, and other members of broad advocacy coalitions also influenced the adoption of policies. They have done by mobilizing vigorously and astutely during windows of opportunity that opened in the framework of structural factors – when events converged to raise both the awareness of problems and the political commitment to address them. Timing matters.

These actors have had success not only in leading states such as California and New York, but also sometimes at the federal level. Climate policy making in the United States extends back at least to the 1970s, in the form of energy efficiency and renewable energy policies, and forward to emissions trading in the Northeast and the wide-ranging regulatory measures of California's Air Resources Board and the federal Environmental Protection Agency since the mid-2000s. These four decades of activity present a rich mixture of successes and failures, from which we can learn about the conditions under which actors can make a difference. After climate policy advocates made a strategic retreat to the state level in the first decade of this century and then the Obama administration put a renewed focus on federal climate policy, this is a good time to re-examine climate policy making by both the federal government and the leading states.

I think this book will make some fresh contributions to an increasingly studied field. While most work in political science focuses on the international level of climate policy, this book examines state and national policies. Even after the landmark agreement in Paris in December 2015, the domestic climate policies of the major economic powers – adopted largely through domestic processes – will continue to be the core and basis of any global solution. While most writing about climate policies argues for why they should be adopted and how they should be designed, this book instead looks at what makes them more likely to be enacted. There are many good policy proposals available, but our understanding of how to get them adopted lags far behind our knowledge of how to craft them.

While most analyses of domestic climate policy making have focused either on structural factors or on actions and events, implicitly debating their relative contributions, this book includes both viewpoints and explicitly sets them against each other. It draws out a synthesis that acknowledges the importance of structural constraints and drivers, and the partial dependence of future paths on past ones – while also delineating the ways that problems and politics have converged to create windows of opportunity and that actors have stepped forward to use them. It seeks to avoid the pitfalls of structural determinism and intentionalism, by recognizing that actors can make a difference but only when conditions are right. Finally, while most observers despair that U.S. climate policies are too weak to prevent dangerous anthropogenic interference with the climate, this book is comparatively upbeat. Many major and minor successes of the past have been overlooked, and we will have chances to repeat and build on them. If we can do so, we will at least be able to *reduce* global warming and its effects compared to business as usual – the path of inaction.

* * *

This book originated in my search for what was working relatively well in climate policy and policy making. That led me initially to Germany, to a cross-national view of the problem, and to a focus on the effects of policies on greenhouse gas emissions. As a newcomer to the study of environmental politics, I received early support and encouragement from Martin Jänicke and Miranda Schreurs at the Free University of Berlin, and from other comparativists whom I met in Germany, especially Ian Bailey, Andreas Duit, Detlef Jahn, Helmut Weidner, and Rüdiger Wurzel. Miranda gave me many opportunities to present my work in progress, and Martin Jänicke reminded me to avoid too much pessimism when studying environmental policy outcomes. When I turned to cases in the United States, I benefited from the encouragement of Christopher Borick, Robert Denning, Kathryn Harrison, Daniel Mazmanian, and Barry Rabe. Dan also suggested path dependence as part of the explanation of the California case and is kind enough to consider me a Californian by birthright despite my long absence from the Golden State.

Of course, in writing this book, I also benefited from more specific help. Owen Temby helped me with sources on the New York case. In the statistical analysis of Chapter 5, I received advice from Charles Tien and Zachary Shirkey, and especially from Detlef Jahn, who read that chapter multiple times and answered countless questions. Chapter 5 also benefited from comments by Todd Vachon and Jeffrey Ladewig, and by Lyle Scruggs and other members of the Political Economy Workshop at the University of Connecticut at Storrs. Maxwell

Boykoff, Robert Brulle, Robert Lowry, Daniel Matisoff, and Clive Thomas generously provided access to their data, as did staff at the World Resources Institute (Thomas Damassa), the Center for Climate and Energy Solutions (Scott Burgess), and the California Air Resources Board (Anny Huang). I also received helpful comments from audiences at the Free University of Berlin (especially Katja Biedenkopf, Detlef Jahn, Kristine Kern, Miranda Schreurs, and Helmut Weidner), Rutgers University at Newark (Gabriela Kütting and Frank Fischer), and the Midwestern Political Science Association (Andrea Olive, Leigh Raymond, Cindy Simon Rosenthal, and Owen Temby). Sheldon Kaminiecki, Michael Kraft, and three anonymous reviewers for Cambridge University Press also gave helpful comments on early stages of this work. To all, I give my thanks, while of course accepting responsibility for the remaining flaws.

Institutional support is always helpful. At Hunter College, President Jennifer Raab, Provost Vita Rabinowitz, and Dean Andrew Polsky helped make it possible for me to take a year's leave from teaching on a Scholar Incentive Award. Charles Tien and Carolyn Somerville, as chairs of the political science department, helped to create a good environment for research and writing. A PSC-CUNY Research Award (number TRADB-45–78) funded a course release and the preparation of the index, and Bob Buckley helped with the paperwork. At Cambridge University Press, Robert Dreesen supported this project from the outset, and his confidence in it helped me to develop it into the book it needed to become. Robert Swanson created the index, and Bindu Vinod oversaw the copyediting and proofreading.

Last but not least, my wife Silke Springorum inspired me with her local educational and organizing work on environmental problems and solutions. Our children, Anna and Lukas Karapin-Springorum, energized me with their curiosity and desire for a recognizable world to live in. All three patiently listened to me talk about the book at the dinner table and in the car. Thank you for being who you are and for being so big in my life.

Roger Karapin

Berlin and Montclair
July and December 2015

Abbreviations

AB 32	Assembly Bill 32, the Global Warming Solutions Act
ACEEE	American Council for an Energy-Efficient Economy
AFL-CIO	American Federation of Labor-Congress of Industrial Organizations
BTU	British Thermal Unit
CAFE	Corporate Average Fuel Economy
CARB	California Air Resources Board
CCES	Center for Climate and Energy Solutions
CEC	California Energy Commission
CO_2eq	CO_2 equivalent
CPUC	California Public Utility Commission
EDF	Environmental Defense Fund
EPA	United States Environmental Protection Agency
IPCC	Intergovernmental Panel on Climate Change
KWh	Kilowatthour
mpg	Miles per gallon
MT	Megatons
NRDC	Natural Resources Defense Council
NYSERDA	New York State Energy Research and Development Authority
OPEC	Organization of Petroleum Exporting Countries
PURPA	Public Utility Regulatory Policies Act
RGGI	Regional Greenhouse Gas Initiative
RPS	Renewable portfolio standard
UCS	Union of Concerned Scientists

1

Introduction and Overview

Human-caused climate change has been an issue of public and elite concern for over twenty-five years. It is a growing and increasingly visible problem, with little remaining uncertainty about the links between greenhouse gas emissions and global warming, and with increasing understanding that climate change is likely to severely threaten the living conditions of all societies. Addressing the problem requires government policy to reduce net emissions, but so far progress has been meager compared with the scope of the challenge. Uncertainties about the costs of climate change and the costs of various solutions, as well as collective action problems, have hampered progress. A solution requires countries to adopt politically or economically costly policies that may be ineffective if others do not adopt similar measures.

Since the climate change problem concerns a global commons, the atmosphere, much attention has focused on efforts to create binding international agreements. The 1992 United Nations Framework Convention on Climate Change led, five years later, to the Kyoto Protocol, which went into force in 2005 and required its participants to cut their collective emissions by a modest 5% over 1990–2010. But the United States did not implement its commitments under the Protocol, and the agreement included no commitments for large developing countries such as China, India, and Brazil. Efforts to extend the Kyoto Protocol through 2020 have made only slow progress; by the end of 2015, the Doha Amendment concerning a second, 2013–20 commitment period had been ratified by only 3 of the 32 countries that have commitments under it.[1] While the December 2015 international agreement in Paris was a major breakthrough, it rests on voluntary commitments by the 195 countries that participated in it. This means that the development of domestic climate policies, especially by large emitting countries, will be crucial for any global solution to the problem.

The good news is that many countries have not been waiting for forceful international action during the last twenty-five years. Even in the absence of strong, binding international agreements, many industrialized countries have pursued emissions reductions through formal reduction targets that are backed by substantial national climate policies.[2] Fifteen countries have adopted carbon or other broad energy taxes, 31 participate in the European Union's Emissions Trading System, several others have national trading systems, and all members of the European Union have binding national targets for renewable energy, which many of them have implemented with vigorous national policies.[3] Collective action problems do not make climate policies impossible, in part because such policies bring with them perceived co-benefits for air pollution reduction, energy security, economic development, and job creation. *Climate policy*, that is, government policy to mitigate climate change by reducing greenhouse gas emissions, is intertwined with these other policy areas.

National climate policies vary in stringency and effectiveness, with relatively effective policies found in countries such as Germany, Sweden, Denmark, and the United Kingdom. By 2012, those countries, Belgium, and the European Union as a whole each had reduced their emissions by more than 15% compared with 1990.[4*] Altogether, nine western countries and the European Union as a whole met their Kyoto Protocol commitments to reduce emissions, although their achievements were aided by the 2009 global recession. On the whole, there is a fairly strong relationship between the strength of national climate policies and reductions in greenhouse gas emissions.[5*] Effective domestic policies also are necessary for progress on international climate agreements, because they motivate governments to take international leadership, give them credibility in negotiations, and demonstrate that mitigation policies are consistent with economic prosperity.

At the same time, many industrialized democratic countries, including Australia, Canada, and Norway, did little to try to meet their Kyoto targets and have seen their emissions rise since 1990. The most prominent of these climate policy laggards has been the United States, which is still the world's second largest emitter of greenhouse gases, after China.[6*] The U.S. federal government's response has been seen as notably weak and ineffective, especially before 2009, whether viewed in its own terms or compared with other countries.[7] In 2001, the U.S. government rejected its Kyoto target of a 7% reduction over 1990–2010, and it has not adopted any target by law.[8*] Its explicit national climate policies have been largely limited to voluntary measures. There is no federal energy tax, carbon tax, emissions trading system, renewable portfolio standard, or legally binding climate action plan. Partly as a result, U.S. greenhouse gas emissions increased 9.5% from 1990 to 2010.[9*]

The weakness of U.S. federal policy is often attributed to structural barriers, including the separation of powers, a pluralist interest group system, a lack of green parties, and strong fossil fuel industries. While these barriers have limited federal climate policy, observers have exaggerated its weakness. In reality, several long-standing federal policies have been relatively effective at reducing or preventing increases in greenhouse gas emissions, including fuel economy regulations, standards and voluntary programs to promote energy efficiency, and the regulation of replacements for ozone-depleting substances. Recently, in 2009 and 2011, the federal government adopted relatively ambitious policies to reduce greenhouse gas emissions from motor vehicles, and it issued draft rules for reducing emissions from existing power plants in 2014; rules on emissions from landfills, aircraft, and the oil and natural gas industry are also in the works. Partly as a result of federal policies, total greenhouse gas emissions peaked in 2007 and declined 10% over the next six years. As I will argue in Chapter 9, U.S. federal climate policy making since the 1970s presents a rich mixture of failures and successes, the study of which can help us to understand how relatively strong policies can be adopted despite the structural constraints found in the United States.

WHY STUDY STATE-LEVEL CLIMATE POLICIES?

But there are very good reasons to focus on the state level, too. The most obvious is that California in the late 2000s adopted comprehensive climate policies in order to achieve legally binding reductions in greenhouse gas emissions, which go far beyond federal policy and are comparable to those of West European countries.[10] Yet the phenomenon of state-level climate policy goes well beyond California and extends much further back in time than its 2006 Global Warming Solutions Act. Within the U.S. federal system, states have much legal authority, administrative capacity, and autonomy from the federal government in energy and environmental policy, both historically and at present.[11]

In the absence of strong federal action, many state governments have adopted climate policies since the 1990s. Major state climate policies include emissions targets and climate action plans, support for energy efficiency improvements and renewable energy development, standards for energy efficiency in appliances and buildings and for vehicle emissions, and emissions trading.[12] Studies show these policies to be effective in reducing greenhouse gas emissions, as discussed in Chapter 2. Since the 1990s, about half of the U.S. states have adopted significant climate policies in these areas.[13] About a dozen states, mostly clustered in the Northeast and on the West Coast, have adopted what I call "relatively strong climate policies." That is, they have pursued a broad range of policies and have relatively ambitious goals, stringent standards, and high funding levels. Generally, they also started earlier than other states and hence have had more time for implementation.[14]

Subnational climate policy is likely to remain important regardless of federal developments. For example, U.S. state governments are likely to play large roles in implementing the federal rules on existing power plant emissions that are being developed by the Environmental Protection Agency (EPA). In addition, more than 140 U.S. municipalities had climate action plans completed or in process in 2009, and more than 1,000 mayors had signed the U.S. Conference of Mayors Climate Protection Agreement, promising to cut their cities' emissions by 7% over 1990–2010.[15*] States with strong climate policies have counterparts in other industrialized democratic countries, as regional governments have adopted binding climate policies in Canada (Quebec and British Columbia), Australia (New South Wales), and the United Kingdom (Scotland).[16*] Collaborations among large regional groupings of U.S. states also emerged in the 2000s and usually included Canadian provinces as well, in the New England Governors' and Eastern Canadian Premiers' Climate Change Action Plan (founded in 2001), the Regional Greenhouse Gas Initiative (RGGI) among Northeastern states (2005), and the Western Climate Initiative led by California (2007).[17]

Especially within federal systems such as the United States, subnational governments are part of a system of multi-level governance in climate policy, which includes international, national, state or regional, and local actors.[18] The system is not organized in a top-down fashion, but rather, each of these policy-making levels has a degree of autonomy from the others, they all potentially influence each other, and hence each can make a significant contribution to the development of climate policy. Multi-level governance sometimes involves mutual reinforcement between levels of government, when different levels take turns driving more stringent environmental policy. For example, California and the U.S. federal government made air pollution standards stricter in this way, and the European Union and some of its member states have had similar effects on climate policy in Europe.[19]

Within the U.S., state governments can serve as valuable testing grounds for demonstrating the technical and economic feasibility of climate policies, proving their political acceptability, and building political commitment for larger-scale action. Strong environmental policies may diffuse from leading states to others and to the federal government, in what David Vogel calls the "California effect." This concept was developed to explain the perhaps surprising diffusion of stricter environmental standards among economically competing nations engaged in liberalizing trade. But it is also quite apt for explaining the spread of climate policies within the United States, given the leading role that California has played in this area. In the California effect, the diffusion of regulatory standards occurs at least partly at the impetus of businesses that benefit from them. This is because those businesses have already adapted to higher standards in the leading states (which tend to be powerful and wealthy ones), and they seek a competitive advantage and want to avoid the costs of meeting multiple standards in different states.[20]

Furthermore, legal support for the California effect has been written into federal air pollution laws concerning automobile emissions standards since the late 1960s. These permit California to adopt standards stricter than federal standards, allow other states to copy them, and hence provide for upward pressure on industry practices and on federal standards. A similar process occurs with federal appliance and equipment standards, where the law allows states to raise certain standards in advance of the federal government.

California and other leading states already have spurred stronger climate policy at the federal level – for example, standards for appliances, motor vehicle emissions, and power plant emissions – and they probably will continue to do so. While subnational climate policies cannot be sufficient to address the global dimensions of the climate change problem, state-level action in the U.S. is crucial as a potential driver of federal climate policies and hence, at least indirectly, of international ones.

Multi-level governance is complex. One way to sharpen our understanding of it is to focus on the policy initiatives of leading countries, such as Germany, and of large, influential subnational governments, such as California. We need to understand why these initiatives emerge and develop, as well as how they affect other jurisdictions. Much of the literature on environmental policy leaders considers how policies diffuse from one jurisdiction to another, including horizontal diffusion across units and vertical diffusion from international regimes, the European Union, and the U.S. federal government.[21] While diffusion is an important process, it is important only because some jurisdictions, such as California and New York State, adopt unusually strong policies in the first place.

In some ways, the high-water mark for climate policy in the United States seemed to pass in the late 2000s, when federal legislation was defeated, two out of three regional accords among state governments collapsed, and party polarization on climate change, among many other issues, intensified. Many states that joined regional cap-and-trade agreements in the West and Midwest in the late 2000s had abandoned them by 2011, and officials in New Jersey, which was the first state to adopt an emissions target, lost interest in its earlier commitments and left the RGGI.[22] The vast majority of Republican governors and congressional representatives now deny the reality of human-caused climate change.

However, the current situation is actually mixed, not completely unfavorable for climate policy making, and the different components of the mixture can be usefully analyzed. The federal government has adopted some strong policies since 2009, including regulations on vehicle fuel efficiency, emerging rules on power plant emissions, and an international commitment to reduce emissions by 17% from 2005 to 2020. That goal now seems attainable, and indeed, the target recently was extended to a 26% to 28% reduction by 2025. Moreover, some states are still forging ahead. California has been implementing its 2006 legislative targets on schedule, including an emissions trading system that now covers 85% of the state's emissions, and Governor Jerry Brown has made climate policy a priority of his administration. Under Governor Andrew Cuomo, New York adopted power plant standards in 2011 and got the agreement of the other RGGI states to lower that system's emissions cap in 2014.

At the same time, there are limits to what even the leading states can do without the support of the federal government.[23] In some areas, stringent standards in some states may lead to a shifting of greenhouse gas emissions to more lenient states ("leakage"), as might have been the case with the Pavley vehicle emissions rules had the federal government not adopted its own version of them.[24] Furthermore, if there is federal inaction or hostility to climate policy, and if other states do not follow the leading states, even states such as California will be limited in how aggressively they can pursue emissions reductions – if these have short-term economic costs that place them at competitive disadvantages with other states. On the other hand, if the federal government adopts policies such as greenhouse gas emissions trading or a renewable portfolio standard, potential conflicts with state policies, including the issue of federal preemption, would need to be addressed.[25]

WHY CALIFORNIA, NEW YORK, AND THE FEDERAL GOVERNMENT?

Among the states with relatively strong climate policies, I selected California and New York for three reasons. First, compared with other U.S. states and indeed with many other industrialized countries, California and New York have advanced climate policies, and they clearly are among the top five or ten states in comparative measures such as those published by the Center for Climate and Energy Solutions (CCES). Their key policy instruments include *explicit climate policies* adopted since the 2000s, which intentionally aim to reduce greenhouse gas emissions, as well as energy efficiency and renewable energy policies that were adopted beginning in the 1970s. I term the latter *implicit climate policies*, since they have had the effect of reducing or curbing greenhouse gas emissions without this being part of the initial intention or justification for them. Because of their importance for reducing emissions, I include implicit climate policies prominently in this book. I also include early air pollution policies in California because they laid the legal and administrative groundwork for the later adoption of explicit climate policies in that state and, indirectly, elsewhere in the country.

California is pre-eminent among the states engaged in climate policy and hence is a crucial case for understanding why U.S. states sometimes adopt relatively strong climate policies. The range, ambitiousness, longevity, and continuity of its climate policies are comparable to many West European countries.[26] Crucially, it is the only state with legally binding greenhouse gas emissions targets that cover most of its economy.[27] Major climate policies adopted in California include greenhouse gas emissions standards for cars and trucks (2002), an ambitious renewable portfolio standard (2002, 2011), the Global Warming Solutions Act (2006), and a low-carbon fuel standard (2007). The state has set a target of reducing its emissions 15% from 2006 to

TABLE 1.1 *Key Data on the Three Cases, 2010*

	U.S. national	California	New York
Population (millions)	309.3	37.3	19.4
Economy			
Gross economic product (GDP or GSP in billions)	$14,860	$1,950	$1,200
Manufacturing share of total	12%	11%	6%
Agriculture share of total	1%	2%	0.2%
Economic product per capita	$48,045	$52,322	$61,755
Government			
Government spending (billions)	$3,456	$210	$138
As share of economic activity	23%	11%	12%
Energy			
Total production (million BTUs per capita)	241	68	43
Renewable energy production, share of total primary energy production	11%	28%	43%
Renewable energy production, share of total primary energy consumption	8%	9%	10%
In million BTUs per capita	317	210	186
Energy imports (share of consumption)	24%	68%	77%
Emissions			
Greenhouse gas emissions, total (million metric tons CO_2eq)	6,875	450	215
In tons per capita	22.2	12.0	11.1

Sources: U.S. Census, U.S. Bureau of Economic Analysis, U.S. Energy Information Agency, U.S. EPA, CARB, and NYSERDA; greenhouse gas emissions are excluding land use changes.

2020, despite expectations of rapid population growth; the planned reduction is 29% when compared with a business-as-usual scenario.[28] Legislation requires the California Air Resources Board (CARB) to achieve the target, which it is doing through a variety of regulatory programs and a cap-and-trade program.

Second, California and New York are also important cases in their own rights, as shown in Table 1.1. They are large states, with economies and greenhouse gas emissions that are significant in relation to the United States and other countries. They also have diverse economies and energy supplies, and hence are more comparable to other states and to the U.S. as a whole than are small states such as Vermont or Oregon. If California were an independent country, its GDP of $2 trillion in 2010 would have made it the 9th largest economy in the world, roughly the size of Italy or India.[29] Its population of

37 million would place it 35th, about the size of Poland or Canada.[30] Its greenhouse gas emissions would make it the world's 19th largest emitter, similar to France or South Africa.[31] New York State is also a significant case on these dimensions. Worldwide, it would rank 13th in GDP, 60th in population, and 36th in emissions if it were an independent country.

Third, both these states have been leaders in climate policy. They have adopted policies in advance of, and have influenced, other states and the federal government. California also has entered into cooperative relationships on climate and energy policy with other countries, including agreements with the United Kingdom and Mexico, informal cooperation with the European Union, and participation in United Nations climate conferences.[32] Both states have influenced federal appliance standards, and New York influenced a federal emissions trading system for sulfur dioxide, which was an important precursor for greenhouse gas emissions trading in the United States and the European Union.[33] New York was a leading force in RGGI, California was in the Western Climate Initiative, and California's path-breaking regulation of greenhouse gas emissions from motor vehicles was endorsed by 17 other states and ultimately a version of it was implemented nationally by the federal government. Since the 1970s, federal air pollution laws have given California a special role in influencing the emissions standards of other states and the federal government, because California has the legal right to apply for waivers from federal air pollution standards and other states are permitted to adopt the California rules.[34] Finally, both states continue to aggressively implement the commitments they made during the 2000s, even though some other states have not.

Besides analyzing California and New York State, the book also includes a case study of federal climate policy making, for two reasons. The latter's failures, especially from 1993 to 2005, provide a contrast to the adoption of strong policies in California and New York; this increases the amount of variance to be explained and hence the analytical leverage of the research design. The factors that explain the strength of climate policy in California and New York should be relatively lacking in federal policy making, and vice versa. Also, federal and state climate policy making are closely related to each other. Federal inaction often has created scope for state policy making, which in turn has influenced the adoption of federal policies, while at other times, such as the 1970s, federal initiatives have spurred state action.

QUESTIONS AND THEORIES

This book asks why relatively strong climate policies have been adopted at some times and in some places in the United States, and not at other times and in other places. In developing answers, I aim to contribute to scholarly literatures and political discussions about the political possibilities for effective climate policies in the U.S. at both state and federal levels. Hence, I will use two contrasting theoretical frameworks, based on structures and processes.

These comprise the most common explanations in academic research on climate and other environmental policy, and they also bear directly on a practical question: Under what conditions do climate policy advocates, such as environmental organizations and elected officials, have the opportunity to take effective action?

Scholarly work on climate policy contains a debate, often an implicit one, between those who argue for political institutions and socioeconomic structures as causes and those who focus on political processes. In the first view, if a state or nation has strong policies, this is due to built-in advantages that other states or nations simply do not enjoy. Hence, the structural theory argues basically that the United States has weak climate policies because of a set of structural barriers – including the separation of powers, strong fossil fuel industries with ready access to government through a pluralist interest group system, electoral rules that prevent the rise of a strong ecological party, and a political culture hostile to government regulation.[35] These factors are unlikely to change in the foreseeable future, and the implication is that the United States will not adopt strong climate policies unless its institutions or basic economic structures change.

In the second view, strong policies result from short-term political interactions, processes, and events. These processes can be subsumed under Kingdon's windows of opportunity theory,[36] which argues that major environmental policies can be adopted if policy entrepreneurs take advantage of opportunities that arise when *policy windows* open. Such windows appear when there is a convergence between events in the "problem stream" (e.g., natural disasters, reactions to previous policies) and events in the "political stream" (e.g., election results, interest group mobilization). *Problem events* are those that raise the perceived importance of climate change issues, helping them to gain access to the crowded policy-making agenda, while the relevant *political events* are those that marshal political commitment and political resources to a cause. Since climate policies provide co-benefits for air pollution reduction, economic development, and energy security, changes in any of these problem areas can help to create opportunities for climate policy making. Once a window has opened, the adoption of policies also depends on whether an advocacy coalition mobilizes more vigorously and astutely than the opposing coalition.[37] From the standpoint of process theories, it seems that favorable processes could happen anywhere, or at least in many places other than the leading states and countries, and that they could be influenced by individual political actors or coalitions of them.

Thus, the debate that the book addresses is not merely an academic one. What is at issue is whether a laggard country like the United States can overcome the barriers to climate policy presented by its institutional and socioeconomic settings, and how it might do so, through action by the states, the federal government, or both.

I will use these theories to address several questions about climate policies in the U.S. Why have some states adopted stronger policies than others, even though they all share the political-institutional barriers to policy innovation that have helped to limit federal climate policy? In particular, why have California and, to a lesser degree New York, adopted strong climate policies? Did the successes of these two states depend on built-in advantages in socio-economic structures and political institutions, or do their causes suggest that other states or even the federal government could adopt strong policies under favorable circumstances? Indeed, why has the federal government already adopted some effective climate policies, despite the institutional and other structural barriers? More generally, how much "room to maneuver"[38] do climate policy advocates have, given the structural constraints?

My aim in contrasting the structural and political-process theories is to assess the contributions that each makes, rather than to uphold one and reject the other. This leads me to examine how structures and processes combine to produce policies, and also to use a non-deterministic theory of path dependence to bridge the structural and process theories and to help explain the cases. In path dependence, past crises and conflicts create new structural conditions, such as administrative capacity or a sizeable clean energy sector, which then affect policy making in later decades. Thus, the book's intended theoretical contributions are threefold: to go beyond structural determinism; to place process theories in a context that includes structural factors; and to relate the two approaches through a theory of critical junctures and path dependence.

To test and develop theory, this book combines a cross-sectional statistical analysis of all fifty states with comparative, longitudinal case studies of California, New York, and the federal government. I used the two kinds of evidence to conduct largely independent tests of the structural and process theories; the last section of Chapter 3 provides some details. The case studies take a long-term historical approach, going back to the 1940s (California), 1960s (New York State), or 1970s (federal government). In my efforts to make the theoretical claims falsifiable, the case studies include longitudinal data on public and political agendas, as well as detailed process tracing of policy-making processes.

The case studies are embedded in a comparative research design that is a blend of universalizing comparisons, in which a variety of successful cases are investigated in order to try to identify their common causes, and variation-seeking comparison, in which instances of failure are used as contrasts to the successes.[39] The three case studies are differentiated into 22 subcases, episodes in which bursts of climate policy making occurred or were possible. Although most of the subcases represent the adoption of significant climate policies, 7 of them involve the blockage of attempted reforms or the reversal of previously adopted policies.

To carry out this research design, I used a mixture of sources and methods, including a variety of cross-sectional data for the statistical analysis, and newspapers and government documents to measure agendas longitudinally. For the qualitative, historical case studies, I drew on a wide range of narrative sources, including newspaper articles, reports by nonprofit organizations and interest groups, government data and documents, and secondary literature; relevant information was updated in December 2015. In using all this evidence, my aim has been to probe the structural and opportunity theories, and to demonstrate the utility and limitations of each, in an even-handed way. While doing so, I provide synthetic explanations of the most important climate policy successes and failures in the United States.

OVERVIEW AND PLAN OF THE BOOK

Overall, this book argues that climate policy making is affected by certain structural conditions, the appearance and use of windows of opportunity, and the path dependent creation of administrative and legal capacities. Major climate policy reforms, both in the states and at the federal level, are possible if the right circumstances converge to overcome the barriers posed by the United States' political institutions and certain socioeconomic structures. These circumstances include a fortuitous confluence of natural disasters or human-made crises, favorable election results, committed political leadership, and mobilization by environmental organizations and their allies. These processes have been most effective when they combined with certain favorable structural conditions, such as large air pollution problems, relatively weak fossil fuel industries, and the prior creation of government agencies possessing scientific expertise and independent funding sources. These arguments are developed in the remaining chapters as follows.

Chapter 2 describes climate policies in the United States – at the federal level, in the states, and specifically in California and New York – with a focus on their effects on reducing greenhouse gas emissions. It distinguishes between implicit and explicit climate policies and provides a classification of the most effective climate policies adopted by the federal government and the states. The relationship between federal and state climate policies and the legal scope for state action in light of potential federal preemption are also examined. A detailed comparison of California and New York climate policies shows that while both have strong policies, California's are clearly more comprehensive and stringent. Finally, this chapter analyzes the effectiveness of federal and California climate policies. It estimates that their climate policies will have reduced greenhouse gas emissions by about 29% and 38%, respectively, over 1990–2020, when compared with no-policies, business-as-usual scenarios.

Chapter 3 provides the theoretical foundations of the book, first by detailing the arguments for the effects of political and socioeconomic structures on climate and other environmental policies and then by describing political and economic process theories. I subsume these processes under Kingdon's windows of opportunity theory, in which major policy reform depends on the convergence of events that heighten problem awareness and events that spur political commitment. I supplement this approach with elements of Sabatier's advocacy coalition framework, to analyze mobilization by coalitions of proponents and opponents of policy reform. Finally, I offer a theory of path dependence to bridge the structural and process theories, and describe the methods of theory testing used in the empirical chapters.

Chapter 4 tests the main structural theories against the federal, California, and New York cases. While structural factors help to explain the weakness of national climate policy, their effectiveness as barriers also depends on political processes, such as election results producing divided government, party polarization, interest group mobilization, and issue framing. Moreover, to a large degree, the structural factors fail to explain the two state cases. In particular, both states have political-institutional profiles that should hinder the development of strong climate policy. However, some structural factors are supported in both cases: high administrative capacities, a liberal (pro-intervention) political culture, and high incomes and education. Yet these structural factors would predict stronger policy in New York than California, and hence fail to explain the exceptional policies of the latter.

Chapter 5 presents a statistical analysis of climate policies in the fifty U.S. states, using as the dependent variable a climate policy index drawn from data assembled by the CCES. I used a variety of independent variables representing the main structural and process explanations in order to develop structural, process, and combined structural-process models. The results support several aspects of the structural theory, though not most of the political-institutional factors. Administrative capacity, liberal citizen ideology, an air pollution index, and fossil fuel production per capita were found to be significant explanatory factors. Since these structural factors are not completely fixed, I examine their recent trends in order to judge their likely effects on the prospects for climate policy over the next decade. The statistical analysis also supports process explanations, since government ideology and environmental organization membership were significant variables. The chapter uses a path analysis to examine the mechanisms through which these variables produce climate policy. It also analyzes the relative sizes of the independent variables' marginal effects and tests the robustness of the findings in a variety of ways. The conclusions discuss how the statistically significant variables were incorporated into the case study chapters.

Chapters 6 through 9 present qualitative case studies of climate policy making by California, New York State, and the federal government. Each case is analyzed into a number of episodes in which climate policies were, or might have been, adopted. I then show how the problem stream and the political streams converged or failed to do so at those times; instances of convergence set the stage for interest group contestation over potential policy reforms. Each chapter also shows how problem events concerning energy and global warming drive changes in the public agenda and in decision-making agendas – measured through the content analysis of news sources and government documents – which in turn help to produce policy-making episodes.

The first two case study chapters examine the California case, which is divided into ten episodes that mark key developments in climate policy making. Chapter 6 treats seven policy-making episodes from the 1940s to the 1980s, in which the state government adopted implicit climate policies or crucial institutional and legal precursors of climate policies. The first three of these episodes involved the expansion of air pollution policy from the 1940s to the 1960s. Next, I analyze the origins of the California Environmental Quality Review Act of 1970, which made new power plant siting much more difficult, and the Warren-Alquist Act of 1974, which created the California Energy Commission. The last two episodes in this chapter concern the energy efficiency and renewable policies adopted by the Brown administration in the late 1970s and the reversal of the latter policies in the early 1980s under the Deukmejian administration. Intense smog problems and elite and protest mobilization led to the creation of CARB and California's legal right to exceed federal air pollution standards, both of which were crucial for later climate policy making in the state. In the 1970s, the energy supply crises, legislative leadership, environmental mobilization, and the election of Jerry Brown as governor converged to produce innovative and relatively ambitious energy efficiency and renewable energy policies. These included appliance and building standards and renewable energy subsidies that helped to make California the wind power capital of the world. However, the combination of falling fossil-fuel prices and the election of a conservative Republican governor led to the undoing of the state's renewable energy policies in the early 1980s.

Chapter 7 examines two episodes (2002 and 2005–08) that saw the adoption of the first explicit climate policies in California, and one episode in which an attempted policy reversal failed (2010). Among the major policies adopted were the Pavley bill's limits on greenhouse gas emissions in transportation and the Global Warming Solution Act (Assembly Bill 32, or "AB 32"), which gave CARB authority to ensure that statewide greenhouse gas reduction targets are met. Problem events (such as Bush's rejection of the Kyoto Protocol, scientific reports about climate change impacts, droughts, and wildfires) converged with the elections of Fran Pavley and Arnold Schwarzenegger and their political leadership to make these reforms possible. The mobilization of a broad advocacy coalition of environmental, business, health, and governmental organizations, and their framing of the legislation in terms of co-benefits for energy security and public health, also were important for legislative passage. Finally, CARB played a key role in climate policy making and implementation, as did California's legal right to adopt air pollution standards stricter than the federal government's; both show the importance of path dependence.

Chapter 8 examines four main episodes of climate policy making in New York State: power plant siting in the early 1970s; energy conservation policy in the late 1970s; the setting of greenhouse gas emissions targets, the adoption of a renewable portfolio standard, and the creation of RGGI under Governor George Pataki in the early 2000s; and the acceleration of energy efficiency and renewable energy policies under Pataki's Democratic successors after 2007. Smog problems, environmental mobilization, the energy crisis, and the election of Hugh Carey as governor helped to produce the first two sets of reforms. Pataki's election, his political ambitions, and Bush's rejection of Kyoto helped to produce the third. The Democrats' rise to power in Albany, together with high fossil fuel prices and a nationwide surge in interest in global warming, helped to produce the fourth.

Chapter 9 analyzes eight episodes of climate policy making by the federal government from the early 1970s to the present: energy conservation and renewable energy policies under Jimmy Carter in the late 1970s; policy reversal under Ronald Reagan in the early 1980s; the emergence of the global warming issue and explicit climate policy in the early 1990s during the George H. W. Bush administration; two episodes of congressional-presidential stalemate and modest progress under Bill Clinton in the 1990s; further stalemate under George W. Bush in the early 2000s; and two episodes of legislative failure and regulatory successes during the Obama administration. Among these, several cases of major policy reforms were driven by a convergence of problems and politics, such as the energy crisis and the 1974–76 elections of congressional Democrats and Jimmy Carter, the 1988 congressional hearings on global warming and George H. W. Bush's election, and Obama's election just after the global warming issue peaked in the late 2000s. Yet, policy reforms were also weakened or blocked by a mistiming between those streams of events, such as the Republican takeover of Congress before the negotiation of the Kyoto Protocol. Strategic errors made by climate policy advocates also played a role in these failures.

Chapter 10 assesses the structural, policy windows, and path dependency theories in light of a comparative analysis of the 22 episodes analyzed in the case chapters, as well as the statistical analysis of all 50 states. It discusses the crucial roles of elections, environmental organizations, and focusing events, and it proposes modifications to the policy windows theory concerning the interdependence of the problem stream and the political stream, and the duration of policy windows. The chapter also examines the roles of advocacy coalitions and details five mechanisms of path dependence that operated in these cases. It concludes with a discussion of the prospects for further climate policy developments in the states and by the federal government, grounded in the theoretical conclusions and the foregoing case studies. Leading states such as California and New York are poised to continue to consolidate and expand their climate policies, and even at the federal level it is possible that favorable conditions could converge to produce major policy reforms, despite the evident structural constraints.

2

Climate Policies in the United States

TYPES OF CLIMATE POLICY

Government policy to address anthropogenic climate change is complex on several dimensions. Mitigation policies, which aim to reduce net emissions of greenhouse gases, are distinct from adaptation policies. Adaptation to climate change has been gaining rapidly in attention, including in the United States and other industrialized countries, as the effects of anthropogenic climate change become more apparent.[1] Adaptation policy includes a wide range of measures to address anticipated threats to property, transportation and energy infrastructure, public health, agriculture, forestry, tourism, and biodiversity due to increases in sea level, heat waves, drought, wildfires, storms, flooding, disease agents, and air pollution.[2]

By contrast, climate change mitigation policies, the subject of this book, aim to minimize human-caused climate change. This can be done by reducing emissions of carbon dioxide, methane, nitrous oxide, the F-gases (HFCs, PFCs, and SF_6), and soot produced by human activity. It also can be done by increasing the storage of carbon by vegetation and soil, or by at least reducing the rate at which human activity is reducing the storage capacities of those natural sinks. In the United States, carbon dioxide from fossil fuel combustion made up 77% of greenhouse gas emissions in 2013, while the growth of forests makes land-use changes a net contributor to carbon storage. Methane (10% of U.S. emissions) is produced primarily by livestock farming, natural gas production, landfills, and coal mining, while nitrous oxide (5% of emissions) is emitted mainly by fertilizer use. The F-gases (3% of emissions) come primarily from industrial processes and manufactured goods, especially in refrigeration, foams, and aerosols, where they are substitutes for ozone-depleting chemicals.[3]

Because of the great variety of emissions sources, climate change mitigation policies potentially include many different policy areas. When all greenhouse

TABLE 2.1 *Greenhouse Gas Emissions in the Three Cases, 2010*

	U.S. national	California	New York
Greenhouse gas emissions, total (millon tons CO_2eq)	6,875	450	215
in tons per capita	22.2	12.0	11.1
% From electricity sector	33	20	22
% Transportation	27	38	33
% Industry	19	20	5
% Agriculture	9	8	2
% Residential	5	7	15
% Commercial	5	3	12

Sources: Calculated by the author from U.S. EPA, CARB, and NYSERDA data; greenhouse gas emissions are excluding land use changes; sectoral percentages may not total 100 because small categories are omitted here.

gases are included, the main sources of emissions in the United States come from electricity generation, motor vehicles, and industrial combustion and processes, and secondarily from agriculture and residential and commercial space heating, as shown in Table 2.1. Climate policies in the United States have largely, but not completely, followed the major sources of emissions. They have focused on some of the main areas of fossil fuel combustion – electricity generation and consumption, transportation fuel use, and energy consumption in buildings. To a lesser degree, they have targeted industrial processes, waste management, agriculture, and forestry.

This book takes a broad approach to understanding climate change mitigation policy. It includes policies that have had the effect of reducing greenhouse gas emissions without intending to do so (*implicit climate policies*), as well as *explicit climate policies*, which deliberately target emissions and can reasonably be expected to reduce them in comparison to business-as-usual scenarios.[4] Sometimes, climate policies are implicit rather than explicit because of when they were adopted. Before the late 1980s, government was not seen as responsible for addressing greenhouse gas emissions, because policy makers did not yet widely recognize them as problematic.

Yet in retrospect, a number of energy policies since the 1970s have had significant effects on reducing U.S. greenhouse gas emissions. They have done so mainly by curbing the construction of new coal power plants,[5*] encouraging energy efficiency and other conservation measures, and promoting the development of renewable energy sources as well as nuclear energy.[6*] After the passage of the California Global Warming Solutions Act in 2006, Arthur Rosenfeld, an energy efficiency expert and member of the state's Energy Commission, was asked when to expect statewide reductions in greenhouse gas emissions to become apparent. "Around 1975," he said, referring to the effects of the appliance and building efficiency standards adopted by California in the mid-1970s.[7] Besides those policies, important implicit climate policies treated in this book include many renewable energy policies and vehicle fuel economy standards that were initiated in the 1970s. These implicit climate policies, together with rising energy prices, helped lead to a 30% drop in the energy intensity of the U.S. economy from the early 1970s to the mid-1980s, a faster rate of decline than before or after this period.[8]

Alternatively, climate policies may be adopted without being explicitly directed at climate change because such policies have co-benefits for air pollution control, energy diversification and security, economic development, employment, and energy efficiency.[9] Policy makers and other proponents of emissions-reducing policies may be motivated exclusively by those reasons, or they may opportunistically frame policies in such terms in order to maximize potential support. For example, the renewable portfolio standards adopted in Texas (1999) and California (2002) were justified by reference to energy security and not climate change.

While climate policies vary according to the sectors they target and their degree of explicit focus on climate change, they also vary according to the degree of government coercion employed.[10*] Four basic types of policy instrument can be distinguished, although actual policies often combine elements from more than one type. Traditional *regulatory policies* set a mandatory standard or ban or require an activity, with penalties for non-compliance. *Financial incentives* reward certain activities with subsidies, tax credits, grants, or loans. *Market-based instruments* with mandatory participation, such as taxes or emissions trading, send price signals, such as from a carbon price, to a wide range of market participants. Finally, purely *voluntary measures* provide information, or technical or organizational assistance, in order to encourage certain actions. These policy types are shown in Table 2.2, with examples of policies adopted by the federal or state governments in six economic sectors, as well as cross-sectoral measures.

Therefore, in each sector, there are a number of policy measures in use or under consideration, which potentially complement or compete with each other. A large normative literature on climate policies argues for particular measures or design features and against others, usually on the basis of their efficiency, equity, or effectiveness. My approach is different. It is to focus on *explaining* those policies that have been most effective in reducing greenhouse gas emissions, with their effects, wherever possible, estimated by post-hoc analyses. Details on these estimates are in the last section of this chapter.

This means that I adopt an agnostic position on policy debates concerning climate policy instruments, such as on the merits of emissions trading vs. carbon taxes, feed-in tariffs vs. renewable portfolio standards vs. research and development funding, or fuel economy standards vs. zero-emission vehicle mandates. This is partly because for each of these instruments, effectiveness can range widely, depending on the details of policy design and implementation,

TABLE 2.2 *Examples of Climate Policies in the United States*

Sector/policy type	Traditional (command-and-control) regulations	Financial incentives	Market-based instruments	Voluntary programs
Cross-sectoral	Mandatory greenhouse gas emissions registries Binding emissions reduction targets		Cap and trading of greenhouse gas emissions Carbon taxes	Voluntary emissions registries Energy plans Non-binding emissions reduction targets Climate action plans
Electricity	Power plant licensing Appliance efficiency standards Utility rate structures and mandatory conservation programs Net metering Mandatory green pricing options Renewable portfolio and energy efficiency resource standards Power plant emissions standards	Decoupling of revenues from sales Tax credits and feed-in tariffs for renewable energy Public benefits funds (used for energy efficiency, renewable energy) Government purchasing of renewable source electricity Research and development for renewable energy, carbon capture and storage	Renewable energy credit tracking systems	Energy Star labeling program for appliances and equipment Demand-side management programs
Buildings	Residential and commercial building codes			Energy Star program for buildings

	LEED and other efficiency standards for government buildings			
Transportation	CAFE fuel efficiency rules Zero-emission and low-emission vehicle mandates Pavley and EPA greenhouse gas emissions standards Biofuel mandates Low-carbon fuel standard Greenhouse gas emission requirements in transportation planning	State purchasing of alternative fuel vehicles Tax credits for such vehicles Public transit subsidies: Research and development funding for hydrogen and electric vehicles, biofuels	Gasoline and diesel taxes	Carpooling programs Tire inflation programs
Industry	EPA's Significant New Alternatives Policy program on substitutes for ozone-depleting substances (with a large voluntary aspect)			Energy Star program for industry Natural Gas Star Program
Agriculture and forestry	Sustainable forest management policies	Conservation Reserve Program		Manure digester programs
Waste management	Landfill regulations on methane			Voluntary recycling programs

Source: Author's compilation.

and those cannot be known until the legislative or implementation process has completed. Another reason is that there is much similarity across policy instruments in the types of factors that make policy adoption more or less likely. My approach is also pragmatic. For example, although carbon taxes might or might not be preferable to emissions trading, the latter has gotten much more political traction in the United States to date. Since this is a retrospective analysis, which seeks to use past experience to assess the present political opportunities for climate policies, this book will have much more to say about emissions trading than carbon taxes.

Furthermore, I will focus on explaining *major policy changes*, which are those which reasonably can be expected to produce relatively large reductions in greenhouse gas emissions. As I sorted through the huge array of climate policies adopted in the United States,[11] I identified policies which target relatively large emissions sectors, for which outcomes studies find relatively large effects compared with "business-as-usual" scenarios, and which involve relatively high regulatory standards or spending levels compared with the range of options that have been adopted in other U.S. jurisdictions. Hence, major policy change is a relative term and does not necessarily mean radical change or changes involving the imposition of significant direct costs on economic actors.[12] Specific justifications for the policies selected here can be found in the section on climate policy outcomes, especially in Tables 2.4 and 2.5, at the end of this chapter. For the comparative analysis of policy-making episodes (summarized in Chapter 10), major change will be contrasted with episodes in which little or no policy change occurred or in which policies were reversed. Major policy changes are often, though not always, those with the most intense political conflicts.

In the next section, I will describe federal and state-level climate policies in the United States, both implicit and explicit, and give an overview of state-level policy adoption. The chapter then details the climate policies adopted in California and New York, followed by a comparison of those two states and a discussion of how they have influenced the climate policies of other states and the federal government. The chapter concludes with analyses of greenhouse gas emission trends in the United States, California, and New York, and of the emissions reduction effects of the major climate policies adopted by the federal government and California.

U.S. FEDERAL AND STATE-LEVEL CLIMATE POLICIES

The United States has a mostly well deserved reputation as a climate policy laggard in cross-national perspective, since it lacks binding national targets for greenhouse gas reductions and mandatory policies to achieve them. Partly as a result, the U.S. performed far worse that its Kyoto target of a 7% reduction over 1990–2010, during which its greenhouse gas emissions rose by 10%.[13]

However, a closer examination of the most important areas of U.S. federal climate policy shows a mixed picture, with strengths as well as weaknesses.

The most important climate policies adopted by the federal government, California, and New York State are summarized in Table 2.3.[14*] Collectively, these policies had major impacts on emissions. By 2011, the most effective federal policies had reduced greenhouse gas emissions by an amount equal to 17% of 1990 emissions, compared with no-policies scenarios; the effects are summarized in Table 2.4, in the last section of this chapter. The comparable percentage for Germany's major policies ranges from 22% to 31%, depending on whether the restructuring of the East German economy in the course of unification is included.[15] Partly as a result of federal policies, U.S. emissions declined 10% from their peak in 2007 to 2013, as shown in Figure 2.2, which is also in the last section of this chapter.

Before the 1970s and to a large extent since then, U.S. federal energy policy has aimed to promote the general economic interest in large, inexpensive energy supplies, as well as the specific interests of the major energy industries. The oil depletion allowance and inexpensive leases for exploration on federal lands allowed the oil and gas industries to prosper, while legal liability limitations and other subsidies promoted nuclear power. Inexpensive energy made suburban sprawl possible, limited the economic feasibility of public transit, and encouraged energy inefficiency in building construction and motor vehicles.[16] Hence, the long legacy of energy policy in the United States has helped to make it the industrialized democracy with the second highest per capita greenhouse gas emissions, at 20.7 tons per person in 2012.[17*] U.S. energy policy continues to be strongly oriented toward ensuring low-cost supply. However, the de facto set of policies that affect energy and greenhouse gas emissions have become more complex during the last four decades. Policy goals that have emerged relatively recently often conflict with the goal of low-cost energy provision: air pollution control (since the late 1960s); energy conservation and renewable energy promotion (since the 1970s); and greenhouse gas emissions limitations (since the early 1990s).

Targets and Cross-Sectoral Policies

The United States lacks binding national targets and major cross-sectoral climate policy instruments. The Kyoto Protocol was never submitted to the Senate for ratification, and since the U.S.'s 7% reduction target was not legally binding, it was not implemented through, or used to guide, federal climate policies. The United States lacks a national emissions trading system, unlike the European Union and a few other industrialized countries.[18*] It also has no carbon or broad-based energy tax, unlike eleven other industrialized countries, including the United Kingdom and Japan.[19] However, the federal government does prepare detailed inventories of U.S. emissions and climate action plans according to its obligations under the 1992 United Nations Framework

TABLE 2.3 *Major Climate Policies Adopted by the Federal Government, California, and New York State*

Sector or policy type	Federal government	California	New York State
Cross-sectoral	Non-binding targets (1998 Kyoto signing, 2009 and 2015 Obama pledges)	Binding targets, with CARB required to implement (2006) Cap and trading of emissions, economy-wide (2013, 2015)	Non-binding targets (2002, 2009) Cap and trading of emissions in electricity sector (2009, 2013)
Energy efficiency in electricity and buildings; other utility policies	Appliance and equipment standards (1987+) Energy Star programs (1992+)	Appliance standards (1976+) Building codes (1978+) Mandatory utility programs (1979) Decoupling (1978, 1982) and decoupling-plus (2007) Public goods charge (1996) Energy efficiency targets for utilities (2004) Electricity performance standard (2006) Green buildings strategy (2007)	Appliance standards (1976+) Building codes (1979+) Mandatory utility programs (1977) Decoupling (early 1990s) System benefits charge (1998) Energy efficiency resource standard (2008)
Renewable energy	PURPA regulations (1978) Production Tax Credit (1992) Green Power Partnership (2000)	Tax credits (1978–86, 2001–05 only)	Generous feed-in tariffs (1981–92 only)

		Generous feed-in-tariffs (1979–1985 only) Research and development spending (1970s) Renewable portfolio standards (2002, 2006, 2011)	Research and development spending (early 2000s) Renewable portfolio standards (2004, 2010)
Transportation	CAFE standards (1978+) Greenhouse gas emission standards for motor vehicles (2009, 2011)	Zero-emission vehicle program (1990) Greenhouse gas emission standards for vehicles (2002) Low-carbon fuel standard (2007) Greenhouse gas emissions targets in transportation planning (2008)	Zero-emission vehicle program (early 1990s) Greenhouse gas emission standards for vehicles (2005)
Industry	SNAP program on substitutes for ozone-depleting substances (1990)	High-GWP programs (2009+)	
Waste management	Landfill air regulations (1996)	Mandatory commercial recycling (2011)	

Source: author's compilation.

Convention on Climate Change. Recently, the federal government has adopted greenhouse gas targets through executive branch action. In December 2009, President Obama announced a goal of reducing emissions by 17% from 2005 to 2020, which would amount to a reduction of about 3% below the 1990 baseline.[20] In 2015, the federal government submitted to the United Nations a new commitment of a 26% reduction over 2005–25. However, these targets are not binding under domestic law.[21]

In response to federal inaction in the decade after Kyoto, many state governments began to adopt greenhouse gas emissions targets, which made them prominent actors in explicit climate policy. In 1998, New Jersey set a target of a 3.5% reduction from 1990 to 2005, and twenty states now have adopted targets by executive order, administrative order, or legislation.[22*] Most statewide targets call for reducing emissions to or slightly below the 1990 baseline by 2020 and for much steeper reductions, of 50% to 80% below the baseline, by 2050. In 18 states, targets are supported by emissions reporting requirements for large emitters such as power plants and industrial sources, and by climate action plans, which 34 states have adopted in order to identify or recommend possible policies to reduce emissions.[23] While statewide targets and plans may influence energy or transportation planning and may help to justify new initiatives in those sectors, they are not legally binding except in California.

Other states took steps toward giving their targets some teeth by creating, or trying to create, regional cap-and-trade systems. RGGI was formed in 2005 and went into effect in 2009; nine Northeastern states are now members.[24] RGGI is an emissions trading system that covers only carbon dioxide emissions from the electric power sector, which makes up about 20% of total greenhouse gas emissions in the participating states.[25] Most permits have been auctioned, with revenues mainly funding energy efficiency and renewable energy programs. The system currently calls for a cap that will decline by 2.5% per year during 2014–20 (see the section on New York's cross-sectoral policies later in this chapter). In addition, California adopted emissions trading in 2013, and the governor of Washington State proposed a similar program to its legislature in December 2014.

Other states began to create cross-sectoral regional trading systems, but have so far failed to develop them. The Western Climate Initiative began in 2007 as an attempt to create a multi-sector, U.S.-Canada emissions trading system, with a 2020 cap that would be 15% below 2005 emissions. It involved seven states and four Canadian provinces at its peak, but all the U.S. states except California left the initiative in 2011, due to the recession and the election of Republican governors in some of the member states.[26] In the Midwest, six states joined Manitoba in the Midwestern Regional Greenhouse Gas Reduction Accord in 2007, which issued a model rule in 2010, outlining a multi-sector cap-and-trade system to reduce emissions by 20% from 2005 to 2020 and by 80% in 2050. However, the plan was not pursued further, partly because all the governors who initiated it were out of office by 2011.[27]

Energy Efficiency Policies

By comparison, federal and state-level energy efficiency policies are relatively strong in the U.S. Driven by the efforts of leading states, the federal government has adopted national efficiency standards for appliances since 1987, and such standards now cover 29 residential appliances, 26 kinds of commercial and industrial equipment, and 16 types of lighting.[28] California and New York adopted the country's first standards in 1976, and at least 14 other states have at various times adopted standards that went beyond existing federal standards, often spurring the latter to catch up. The Department of Energy in the Obama administration adopted 17 appliance and equipment standards between 2010 and 2014; it plans to finalize 25 more standards by 2016.[29]

In addition, the federal government's Energy Star program sets voluntary standards and provides information to purchasers through labeling. The program was initiated in 1992 and soon expanded to commercial buildings, industry, natural gas, power transmission equipment, and consumer products; it operates through a partnership among the EPA, the Department of Energy, manufacturers, retailers, and builders.[30] Furthermore, the federal Green Power Partnership encourages voluntary green-power purchases by large organizations, while the Combined Heat and Power Partnership promotes cogeneration; both were launched around 2000.

Energy efficiency policy extends to the buildings sector. Federal law in 1992 required states to update commercial building codes and to consider updating residential codes by 1994 and again when the Energy Department determines a revision is appropriate. It did so in 2002, 2008, and 2011, although compliance is very uneven; only 15 states had up-to-date commercial codes and 7 had no statewide commercial or residential codes at all in 2014.[31] However, over 30 states had codes that met at least the standards of the 2009 International Energy Conservation Code. After 2001, many state governments also adopted programs to increase energy efficiency in state buildings, with 24 states requiring a LEED (Leadership in Energy and Environmental Design) standard in new buildings.[32]

The federal government adopted requirements on utilities to take energy-saving measures in 1992, requiring them to undertake integrated resource planning (including renewable energy), make demand-side investments at the same marginal cost as new generation, and adopt rate structures that encourage efficiency.[33] Some states led the federal government by more than ten years in those areas and in others related to utility regulation. State utility commissions in the 1970s began to alter rate structures to reduce incentives for consumption.[34] Moreover, 11 states have adopted policies to decouple gas and electric utilities' profits from revenues, thus removing the incentive for utilities to encourage consumption in order to maximize their profits; another 12 states have adopted decoupling for either electric or gas utilities, though not for both.

Finally, beginning in 1996, states established public benefits funds through small taxes on utility bills, in order to finance utilities' energy efficiency and renewable energy programs; 27 states now have such funds or similar mechanisms for reimbursing utilities from ratepayers.[35] More recently, mostly after 2005, 24 states have adopted energy efficiency resource standards or targets, which require utilities to reduce energy use progressively over a future period.[36] Utilities' energy efficiency programs are estimated to cost about three cents per Kilowatthour (KWh), about one-third to one-half the cost of energy production from renewables, fossil fuels, or nuclear energy.[37]

Renewable Energy Policies

Federal renewable energy policies also have been significant, though not as strong as energy efficiency policies. Federal research-and-development spending on new renewable energy sources rose in the 1970s, but it fell sharply in the early 1980s and has been low since then.[38] There is no national renewable portfolio standard, despite efforts by a bipartisan group of senators to create one in 2010.[39] The federal PURPA law (the Public Utility Regulatory Policies Act) in 1978 allowed states to require utilities to give generous feed-in tariffs to independent power producers, and several states did so, although this provision of the federal law was repealed in 2005. The most important federal source of support for renewable energy is the Production Tax Credit, which was adopted in 1992 and now provides a subsidy of 2.3 cents per KWh for wind, geothermal, and closed-loop biomass. However, the credit expired briefly in 1999, 2002, and 2004, and its renewal has been in doubt at other times, which has created uncertainty for investors and a boom-bust cycle in new installations. Nonetheless, this federal tax credit has been a necessary ingredient in the rise of wind power, along with state policies.[40] Wind power made up 4.1% of U.S. electricity generation in 2013, placing the United States in 12th place in generation share among industrialized countries; it was 2nd in the world in total installed capacity, behind only China.[41] However, other "new renewables" made up only 2.5% of U.S. electricity generation, including 1.6% for biomass and 0.4% for solar energy, in 2014.[42]

State governments also have been very active in promoting renewable energy. Beginning in the 1970s, research and development funding, high feed-in tariffs (permitted under the federal PURPA law), and tax credits offered by some state governments were important supports to the wind, solar, geothermal, and biomass industries.[43] While many of these policies lapsed or were sharply curtailed when fossil fuel prices dropped in the 1980s, some states returned to supporting renewable energy with new policies beginning in the late 1990s. Renewable portfolio standards (RPSs) require utilities (often only large, privately owned utilities) to generate or sell a gradually increasing percentage of electricity from renewable sources. Iowa adopted the first RPS in 1983 and was followed by several states in the late 1990s.[44] Now 27 states have RPSs, and 3 others have alternative energy portfolio standards that include renewable energy requirements. RPSs are significant sources of support for renewable energy, subsidizing the price of wind energy by an amount ranging from 0.5 to 5.5 cents per KWh, depending on the state.[45] In addition, green-power choice policies, adopted in 12 states, require utilities to give customers the option of choosing a renewable energy provider while paying a price premium, a subsidy that is worth 1.5 cents per KWh on average and that affects about 30% of wind-power generation.[46] Finally, net metering laws, which require utilities to purchase excess electricity produced by retail customers, usually at retail rates, have spread since the 1980s, and are now on the books in 45 states.

Transportation Policies

In transportation, by contrast, the United States' policies have been relatively weak, though with major improvement recently. Federal road fuel taxes are very low, at 18 cents per gallon, and about 85% of the revenues are dedicated to highway construction.[47] States also levy gasoline and diesel taxes on top of the federal tax, but the total average road fuel tax is still only about fifty cents per gallon, compared with about three dollars per gallon in the European Union. Low fuel taxes translate into low retail prices, encouraging consumption; gasoline and diesel prices in the United States have ranged from about two to four dollars per gallon in recent years, which is less than half the price in the average country in the European Union.[48] Although a federal Renewable Fuel Standard was adopted in 2005, it resulted mainly in the blending into gasoline of corn ethanol, which has life cycle greenhouse gas emissions similar to those of gasoline.[49] The use of cellulosic ethanol would reduce emissions, but the EPA recently abandoned ambitious targets for this fuel source.[50]

The main federal policy for improving efficiency in transportation has been the Corporate Average Fuel Economy (CAFE) program, created by 1975 legislation. However, before 2010, CAFE standards for passenger cars were very low (nominally 27.5 mpg) and had not been raised since 1985. Moreover, a separate, even lower standard for light trucks (about 21 mpg) allowed sport utility and other large vehicles to grow from 2% of the U.S. market in 1975 to 25% in 2002.[51] (By comparison, the share of new cars sold in the European Union that had four-wheel drive peaked at 9% in 2007.[52]) Hence the total fuel economy of the U.S. vehicle fleet leveled off from 1982 to 2005,[53*] despite the opportunities afforded by technical improvements in the industry, which raised fuel economy in Germany by 20% during 1991–2005. Therefore, U.S. fuel economy is only about half that of European countries, in a pattern directly related to fuel prices.[54]

Against a background of stagnation in federal fuel economy regulations for automobiles, state governments became active in trying to force technological advances that would reduce greenhouse gas emissions in transportation.

California initiated its zero-emission vehicle program in 1990, and other states in the Northeast soon adopted it. The program was a mandate on manufacturers to achieve increasing sales targets for electric or hydrogen fuel cell vehicles, and it initially required that 10% of cars for sale in 2003 be zero-emission vehicles.[55] However, under pressure from the car industry, including a successful lawsuit which argued that the rules were preempted by the federal law concerning fuel economy, CARB modified the California program repeatedly. It now permits manufacturers to earn credits also by selling hybrid electric, natural gas-powered, or conventional vehicles with zero evaporative emissions.[56] California has gained waivers for its program from the EPA, and nine other states have adopted its mandates on manufacturers.

Moreover, California's foray into the regulation of greenhouse gas emissions from motor vehicles, in the 2002 Pavley bill, has had far-reaching consequences. While CAFE has been a federal policy that preempts state action on fuel efficiency, the Pavley rules evaded this restriction by regulating greenhouse gases rather than fuel economy as such. After California passed the Pavley bill in 2002, 17 other states announced that they intended to adopt the California rules once the EPA granted the necessary waiver.[57] Although the EPA in 2007, during the Bush administration, initially denied the waiver request, two years later, soon after Obama took office, the agency approved it.

Furthermore, California's actions led to the start of federal regulation of greenhouse gas emissions by vehicles. Agreements between the Obama administration and major carmakers in 2009 and 2011 established greenhouse gas emissions standards for cars and light trucks, which nationalized the Pavley approach to the regulation of motor vehicles emissions. The federal rules effectively required an increase in fuel economy in cars from 28 mpg in 2011 to 35.5 mpg by 2017 and to 54.5 mpg by 2025; greenhouse gas emissions by new vehicles will be cut in half over the entire period.[58*] The United States' standards for future model years are now approaching those of the European Union, with 163 gCO_2/mi required for cars and light trucks in the United States by 2025. That is the equivalent of 101 gCO_2/km in the U.S., compared with 95 grams/km required for cars in the European Union four years earlier (by 2021).[59]

Moreover, California's early actions also carved out a potential role for California in future enhancements of greenhouse gas standards for vehicles. Despite the initiation of federal regulation in this area, California retained its own vehicle emissions program for greenhouse gases and the right to adopt tougher standards in the future. After the EPA granted a waiver for California's program, the 2009 agreement affirmed the state's right to apply for future waivers in order to adopt greenhouse gas emissions standards stricter than those of the federal government.[60] This has given CARB leverage, which it used during the 2011 negotiations, when it threatened to adopt its own standards if it considered the federal rules to be insufficiently strict.[61]

Finally, low-carbon fuel standards are an emerging area of state climate policy making. Under the Global Warming Solutions Act, CARB adopted rules in 2010 that require the carbon intensity of transportation fuels sold in California to decline 10% by 2020.[62] Thirteen other states are currently working on similar standards, including eleven that are trying to develop a joint program for the Northeastern and mid-Atlantic states.

Industrial and Power Plant Emissions Policies

Although industrial carbon dioxide emissions are not addressed by current federal policy, the government has curbed the growth of an important source of F-gases through its implementation of the Montreal Protocol on ozone-depleting substances. The 1990 amendments to the Clean Air Act established the EPA's Significant New Alternatives Policy program, which regulates chemicals used in industrial processes, air conditioning, fire suppression, and other areas. Under this program, the EPA, in cooperation with industry, evaluates substitutes for ozone-depleting substances for their global warming potential, among other criteria, bans some, and restricts the use of others to certain applications. In addition, regulations on solid-waste landfills, also required by the 1990 amendments and adopted in 1996, require large landfill operators to collect and combust methane; this is another important source of reductions in greenhouse gas emissions.

Finally, another explicit climate policy that first emerged at the state level is the direct regulation of carbon dioxide emissions from power plants. Mostly after 2006, six states (California, New York, Washington, Oregon, Montana, and Illinois) adopted CO_2 emissions standards, which have the effect of preventing new coal plant construction unless carbon storage technology is used. Then, the federal EPA became active following the 2007 Supreme Court ruling in *Massachusetts v. EPA*, which held that this agency has the authority to regulate carbon dioxide under the Clean Air Act, without new legislation. At the start of the Obama administration, in December 2009, the EPA issued an "endangerment finding," concluding that six greenhouse gases posed a threat to public health and welfare.[63] On this basis, the recent revisions to the CAFE regulations, described in the previous section, have been expressed in terms of a carbon dioxide emissions standard, and the EPA also began rule making on power plant emissions. It completed draft rules for new and existing power plants in June 2014 and issued final rules in August 2015, although the rules are likely to be challenged in federal court.[64]

State-Federal Relations on Climate Policy

Since the U.S. is a federal system, the relations between federal and state actions are important for climate policy.[65] State and federal climate policies, for the most part, have been complementary.[66] Historically, state governments have had large roles in utility regulation and public health,[67] which evolved into significant roles in air pollution and energy policy beginning in the 1960s and

1970s. In many non-regulatory climate policy areas where the federal government has become involved – such as tax credits for renewable energy, research and development spending, and road fuel taxes – federal and state policies remain mostly complementary or additive. In others, such as renewable portfolio standards, emissions trading, or low-carbon fuel standards, federal inaction has left much scope for state-level policy making.

Furthermore, in many climate policy areas, state governments have led in innovations that have influenced the federal government and other states. Leading states have driven federal appliance standards since 1987 and the EPA's regulation of greenhouse gas emissions by vehicles and power plants since the agency's 2009 endangerment finding. This is similar to what leading states have done in other environmental policy areas, such as criteria air pollution standards and NO_x emissions trading.[68]

Federal climate policies could potentially preempt state policies,[69*] but so far this has not occurred to a great extent. The main exception is the CAFE program during 1978–2009. In this area, what turned out to be weak federal standards were adopted at an early date and prevented states from adopting stricter standards later on. In other areas of climate policy, preemption of state climate policy could occur if industry were to press for uniform national standards that are lower than those in leading states, for example in emissions trading, renewable portfolio standards, or low-carbon fuel standards.[70] Such attempts were made with air pollution standards in the late 1960s and with appliance standards in the 1980s, but they were largely unsuccessful. A more recent attempt at preempting state-level emissions trading was made in the Waxman–Markey bill in the late 2000s, but the legislation failed in the Senate. Finally, the power plant regulations now under development by the EPA would not preempt state policies. Rather, they foresee highly flexible implementation by state governments, with a wide range of existing policy instruments permitted, including increased efficiency in fossil fuel plants, shifting to generation by combined cycle natural gas plants, increased use of renewable sources and nuclear power, and increased energy efficiency.[71]

The scope of the preemption issue for state-level climate policies so far has remained limited partly because exemption provisions were written into federal motor-vehicle air pollution laws and, to an extent, energy efficiency legislation. The Clean Air Act of 1970 gives California the right to apply for a waiver from federal requirements in order to implement its own stricter air pollution standards for motor vehicles, if they are needed to address the state's "compelling and extraordinary conditions." Until it rejected the Pavley regulations in 2007, the EPA had routinely approved all of California's waiver requests, at least in part.[72] Moreover, in 1977, the Clean Air Act was amended to give other states the right to follow California's lead, and in 1990, these rights were expanded to off-road mobile air pollution sources (see also Chapter 6).[73] The Pavley bill was written to regulate greenhouse gas emissions rather than fuel economy in order to be covered by California's right to an exemption from federal air pollution standards under the Clean Air Act, and to avoid claims that the federal law on fuel economy preempted California from adopting such regulations. The strategy ultimately succeeded.[74*]

Finally, energy efficiency legislation generally provides for federal preemption of state standards if a federal standard is in place. But the laws also permit state governments to retain or adopt their own standards on appliances for which federal standards have not been adopted or until federal standards go into force.[75] In 2015, states had standards in place for 11 residential, 4 commercial, and 3 lighting appliance categories, including such important items as television sets, battery chargers, pool pumps, commercial refrigeration equipment, and hot food holding cabinets.[76] For the vast majority of these, no federal standard is planned before 2020. In short, despite federal preemption, states still play an important role in leading the federal government in appliance standards.

Variation in State Climate Policies

State governments vary greatly in the climate policies that they have adopted. Twelve states, located in the Northeast and on the West Coast, have adopted more than 19 out of a maximum 25 policies in an index compiled by the Center for Climate and Energy Solutions.[77*] As shown in Figure 2.1, the top five states include California and New York, as well as Massachusetts, Connecticut, and Vermont, which have at least 23 policies out of 25 possible.

At the same time, many state governments do not adopt a large number of climate policies, and those they do adopt are likely to be relatively low-impact measures found in nearly all states, such as renewable energy certificate

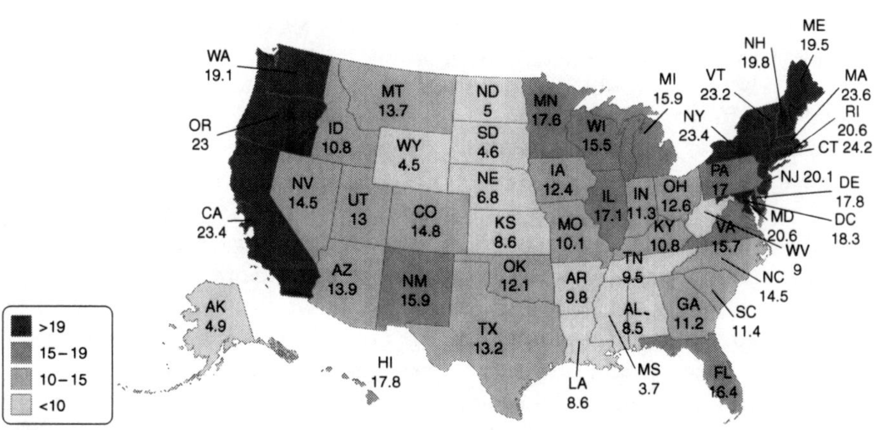

FIGURE 2.1 State-Level Climate Policy Index, 2014
Source: CCES data, modified by the author as described in Chapter 5.

tracking systems or biofuel mandates. Twelve states, concentrated in the South and Great Plains, have fewer than ten climate policies; the five lowest ranked states are Mississippi, Wyoming, Alaska, South Dakota, and North Dakota, all with five policies or fewer. In fact, while some states have been embracing climate policy, others have been explicitly rejecting it. Sixteen states adopted anti-Kyoto resolutions, and some passed a legislative ban on state agencies engaging in actions to reduce greenhouse gas emissions.[78] Between these two extremes, there are about two dozen states with middling climate policies, which have adopted between 10 and 19 measures. Chapter 5 will attempt to explain this variation in the degree to which state governments have adopted climate policies. But next, I will examine in more detail the climate policies of two leading states, which were summarized in Table 2.3, earlier in this chapter.

CALIFORNIA AND NEW YORK STATE CLIMATE POLICIES

Targets and Cross-Sectoral Policies

California

Ambitious greenhouse gas reduction targets were adopted in California by executive order in 2005 and by the legislature in 2006, the latter legally requiring and authorizing a state agency (CARB) to achieve the targets. The Global Warming Solutions Act caps emissions of the Kyoto-regulated greenhouse gases from all significant sources, including electricity imports, with a target of no growth from 1990 to 2020. The target is more ambitious than it appears, since it requires about a 13% cut from 2004 emissions, a 29% reduction from the business-as-usual scenario, and a 32% cut in per-capita emissions over 1990–2020, because of projected population growth.[79] Moreover, the governor's executive order also included an 80% reduction target by 2050 and required the California Environmental Protection Agency to report every two years on the impacts of climate change in California and on progress being made to reach the targets.

To implement the law, CARB drew up a scoping plan in 2008 that detailed the means to achieve the 2020 reduction target. The plan included a cap-and-trade program projected to achieve about 20% of the state's total emissions reductions. But most reductions were expected to come from a number of existing and anticipated regulatory policies, which are summarized in Table 2.5, in the last section of this chapter. These include the Pavley rules (adopted in 2002 and approved by the federal EPA in 2009), the California Solar Initiative (adopted in 2006), and a low-carbon fuel standard (adopted in 2007). Additional policies intended to contribute to reductions consist of an RPS with an accelerated target of 33% by 2020 (adopted in 2011), new and existing energy efficiency measures in electricity and buildings, and increases in commercial recycling (made mandatory in 2011).

Although California's climate policies rely much more on regulation than on emissions trading, most of the regulatory measures are designed not as traditional command-and-control policies. Rather, policies with flexibility mechanisms for compliance are responsible for three-quarters of the emissions reductions.[80] The policies typically set performance standards that private actors can meet in a variety of ways, and they include other flexibility features, such as the trading of renewable energy credits under the RPS and the averaging and banking provisions in the Pavley rules and the low-carbon fuel standard.

The cap-and-trade program began operating in 2013, initially covering in-state electricity generators, electricity importers, and large industrial facilities (those emitting over 25,000 tons CO_2eq/year), which account for 35% of statewide emissions. In 2015, CARB added distributors of transportation fuel, natural gas, and other fuels to the program. About 350 businesses with 600 facilities are now included, representing 85% of the state's greenhouse gas emissions; only agriculture, forestry, and waste management are excluded. The cap was set to decline 2% a year in 2013 and 2014, and 3% a year during 2015–20.[81] Allowances have been distributed both for free and through auctions, with free allocations initially equal to 90% of previous emissions and then declining. CARB auctioned off the first quarterly batch of allowances in November 2012, when allowances to emit 23 million tons of greenhouse gases were sold for slightly over the minimum price of $10 per ton. Auction revenues, which were about $780 million in the first year, are to be spent on greenhouse gas reduction measures. One-quarter of the revenue is set aside for programs benefiting low-income communities, and the state legislature recently committed another one-quarter of the funds to helping build a high-speed rail system.[82] Flexibility mechanisms include three-year compliance periods and the availability of offset credits. However, the latter are limited to 8% of total allowances for each regulated entity and to CARB-approved, U.S.-based projects in forestry, urban forestry, manure digestion, and the destruction of ozone-depleting substances.

New York

New York officially adopted a greenhouse gas emissions target in June 2002, when Governor Pataki released a comprehensive energy plan drawn up by the New York State Energy Research and Development Agency (NYSERDA). The target was to reduce emissions 5% over the 1990–2010 period and by a total of 10% in 2020.[83] Since emissions had already risen since 1990, this meant a 19% reduction would be needed during 2000-2020. The inclusion of this target was prompted by the Greenhouse Gas Task Force, which Pataki had created in June 2001 and which released its own final report in 2003. Hence, although the target was not authorized by legislation or executive order, it had the endorsement of the governor and a variety of state agencies involved in the task force, including the Department of Environmental Conservation, the Public Service Commission, and NYSERDA.

When Democrats took over the governorship in 2007, they adopted new, accelerated goals for greenhouse gas emissions, as well as for energy efficiency and renewable energy. In August 2009, Governor Paterson issued Executive Order 24, which set a target of an 80% reduction in emissions over 1990-2050 and created a Climate Action Council. The council drew up a draft climate action plan in November 2010, outlining how to achieve the 80% goal. Its report focused on climate change and mitigation strategies, including an emissions inventory.

Under Governor Pataki, New York State was the leading force in establishing RGGI, the first greenhouse gas emissions trading system in the United States, which currently includes nine Northeastern states. RGGI, which is limited to carbon dioxide emissions from electricity, is the first emissions trading system in the world to rely mainly on the auctioning of emissions allowances.[84] It was built on the experience with earlier trading systems for sulfur dioxide and nitrogen oxide emissions.[85] The former was established at the urging of New York and other states through the 1990 Clean Air Act Amendments, and was credited with a 50% reduction in sulfur dioxide emissions by 2007, with low compliance costs.[86]

RGGI originated in April 2003, when Pataki's task force's recommended that New York establish an emissions trading system and Pataki invited the governors of Northeastern and Mid-Atlantic states to join in a regional system. The ten participating states developed rules through multiple rounds of negotiation, finalized an interstate agreement in December 2005, and then adopted implementing policies by legislation or executive actions.[87*] RGGI held the first quarterly auction of allowances in September 2008, and compliance began in January 2009. Although allowances and auctions are regional, each state has its own carbon budget, rules concerning the proportion of allowances to be auctioned, and rules about the use of its share of the auction funds.

RGGI includes all electricity generation from fossil fuel plants larger than 25 Megawatts, which accounted for about 190 million short tons (172 Megatons, MT) of carbon dioxide at the start of the program. The program includes 95% of all carbon dioxide emissions from electricity in the region and currently about 20% of total greenhouse gas emissions in New York State.[88] The total number of allowances was set to be stable from 2009 to 2014 and then to decline by 2.5% in each of the next four years, in order to force an overall emissions reductions of 10% in the electricity sector by 2018.[89] Offsets for other projects, such as landfill methane destruction, SF_6 production avoidance, or carbon dioxide sequestration through deforestation, are limited to 3.3% of total annual allowances.[90]

Initially, states planned to allocate most allowances for free, but in December 2006, Pataki announced that New York would specify that all its allowances be auctioned.[91] This move was largely followed by the other states, so that RGGI has auctioned about 90% of all allowances, generating nearly $1.5

billion in the first four years of operation.[92] Through September 2013, over 600 million allowances had been auctioned at prices of about $2 to $3 per ton, a very low price that would have been even lower were it not for a reserve price of about $2 set by RGGI. New Jersey left RGGI in 2011, after Chris Christie, a Republican, was elected governor, but the other nine states have remained, partly because of the revenues provided by auctioning. According to RGGI, over 80% of auction proceeds have gone to end-use energy efficiency programs, renewable energy, and aid for low-income households.[93] However, in 2010, some funds were used as general revenues by New York ($90 million, half its proceeds), New Jersey ($65 million), and New Hampshire.[94]

RGGI's strengths include the auctioning feature, which has probably made it durable even in the face of recession, and its wide geographic coverage. Its weaknesses include its limitation to the electricity sector and what proved to be an unambitious rate of decline in the emissions cap. Due largely to declining natural gas prices and fuel switching away from coal, emissions from the power sector in the RGGI states unexpectedly declined by nearly half during 2005–13, far exceeding the planned future decline in the cap. Hence in February 2013, RGGI proposed a new model rule that would reduce the 2014 cap from the previously agreed 165 million short tons to a new level of 91 million short tons. The new cap would decline 2.5% annually for the next four years, although RGGI would allow generators to use excess allowances that they had accumulated from previous years.[95] This rule, which had been approved by all nine RGGI states in 2014, is intended to lock in the currently lower emissions levels by the power sector even if rising natural gas prices and economic growth were to put pressure on emissions later this decade.

Energy Efficiency Policies

California

California's greenhouse gas reduction targets are supported by a wide range of pre-existing and new sectoral policies. Of these, the energy efficiency and renewable energy policies date to the 1970s. The state legislature created the California Energy Commission (CEC) in 1975, with a broad mandate to oversee power plant siting, produce energy forecasts and plans, adopt energy standards for appliances and buildings, and conduct research and development on energy-saving technologies.[96] The agency sponsors academic research on which its efficiency standards are based. It issued its first mandatory appliance standards in 1976, for refrigerators, freezers, air conditioners, and gas appliances, and its commercial and residential building codes were implemented in 1978. California has continued to take a leading role in both areas up to the present, and many of the mandatory federal appliance standards adopted in 1988 and 1992 were based on California's.[97] Recently, California has had standards for 13 different appliances or pieces of equipment that exceeded federal standards.[98]

California also has the "most aggressive and best enforced building code[s]" in the United States.[99] Furthermore, the CEC adopted an ambitious Zero Net Energy Building plan in 2007 as part of a broader green building strategy. The goal is to require all new residential buildings, beginning in 2020, to produce at least as much energy as they consume; all new commercial buildings are to meet the standard by 2030.[100] The CEC has begun implementing the goals for residential buildings through incremental, triennial tightening of the state's building codes and through a 2013 action plan developed with builders, utilities, local governments, and nonprofit organizations. A 2012 executive order also requires that all new state buildings designed after 2025 be zero net energy and that state agencies try to bring half of existing square footage up to that standard by that date.[101]

The California Public Utilities Commission (CPUC) also became active in promoting energy efficiency in the late 1970s. It uncoupled utilities' profits from sales of natural gas and electricity in 1978 and 1982, respectively, and it began to fine utilities in 1979 for their failure to pursue energy conservation and promote alternative energy sources. The state's three largest private utilities began to cooperate in implementing energy efficiency programs funded by ratepayers, such as financial assistance with insulation or more efficient appliances, and by the early 1990s they were participating in biennial energy planning with the CEC and CPUC.[102] In 1996, to counteract some of the anticipated effects of electricity market deregulation, California by law required the private utilities to collect a public goods charge from ratepayers to fund renewable energy and energy efficiency programs. The charge, at almost one half of a cent per KWh, was the highest among all states in 2007, when it generated $350 million in revenue.[103]

After the California electricity crisis in 2000–01, energy efficiency gained renewed importance in the state. The 2003 California Energy Action Plan, adopted by the CEC and CPUC, called for increased energy efficiency and renewable energy as the best approaches to avoid price spikes and power outages in the future.[104] The CPUC gave utilities mandatory electricity savings targets in 2004, which were increased in 2009. In 2007, the agency adopted "decoupling plus," a policy that gives utilities economic incentives to reduce electricity consumption by allowing them to split the savings with their customers.[105] The CPUC also created the California Institute for Climate Solutions in 2008, in order to conduct research and development on ways to reduce greenhouse gas emissions. The institute was generously funded at $60 million per year for the next ten years, with the money coming from a surcharge on utility fees.[106] California also adopted an electricity performance standard by law in 2006, requiring new plants or import agreements to not exceed the carbon emissions rate of a combined cycle natural gas plant. This effectively bars new long-term contracts to purchase coal-fired electricity.[107]

New York

New York's overall emissions reductions target is also supported by energy efficiency policies that have been ongoing since the 1970s, affecting mainly electricity and heating fuel consumption. In 1976, the state legislature created the New York State Energy Office, which was qualified to receive federal subsidies for conservation programs under federal legislation.[108*] New York State's 1977 Home Insulation and Energy Conservation Act required all eleven major public utilities to provide low-cost energy audits to residential customers, and although initially they resisted, ultimately they did pursue conservation. In addition, NYSERDA, created by law in 1975, has carried out research and development of energy efficiency technologies. This agency began creating some of the nation's first appliance efficiency standards in 1976, helping to spur federal legislation along with California. All these efforts were guided by a 1977 energy conservation plan that called for a 7% cut in energy consumption in the state.

In terms of energy efficiency policies in place in 1979, New York was one of a dozen states that were leaders within the United States It had lighting and heating standards for new buildings, loans for weatherization, and regulatory bans on declining block rates for large customers and master metering for apartment buildings.[109] Energy efficiency efforts accelerated in the 1980s, as utilities became concerned about whether new power plants could be built in the face of environmental opposition. Moreover, after the federal government reached a multistate legal settlement with oil companies that were accused of overcharging for crude oil in the 1970s, New York received $335 million from the settlement by the end of the 1980s. The state used the money to improve energy efficiency in buildings, equipment, lighting, and industrial processes, partly in efforts to meet clean air standards.[110] Progress continued in the early 1990s, when regulators put in place mechanisms to decouple utility revenues from sales, as well as incentives and penalties for performance relative to energy efficiency goals. New York adopted a system benefits charge in 1998, implemented by NYSERDA, to fund the efficiency and demand reduction programs previously paid for by the utilities.[111]

Finally, in April 2007, energy efficiency gained renewed focus when Governor Spitzer announced a goal of reducing electricity consumption by 15% below the business-as-usual level in 2015.[112] This would entail a small absolute cut in consumption and would require a variety of updated or new programs for about 80% of the reductions.[113] In June 2008, the Public Service Commission established an energy efficiency portfolio standard to help implement this goal, with three-year interim targets, incentives for utilities, and implementation through NYSERDA and the utilities, using funds from the system benefits charge. Two years later, the legislature passed two laws to establish energy efficiency standards for six types of appliances and to provide loans to consumers to improve energy efficiency in natural gas use, to be repaid through their utility bills.[114]

Building codes comprise another important climate policy area in New York State. Fuel combustion to heat residential and commercial buildings is responsible for about one-quarter of New York's greenhouse gas emissions, on top of the emissions from the buildings' electricity consumption.[115] New York has been a leader in promoting energy efficiency in buildings since the 1970s, with the State Energy Office aggressively developing and adopting codes for all new buildings, beginning with the 1979 Energy Conservation Construction Code and continuing in recent years.[116] In 1999, Governor Pataki and the state legislature agreed to update New York's construction code with a model energy code, supported by a federal grant and implemented in 2002. The code was updated in 2010, based on the latest International Energy Conservation Code and the latest ASHRAE[117*] code, with enhancements for New York State; the state also has been active in the International Code Council since 1998.[118]

In addition, in 2000, Pataki announced his support for a green building tax credit program, which the legislature passed that year. It provided tax breaks for construction features that increase energy efficiency or use environmentally friendly materials.[119] In June 2001, Pataki issued Executive Order 111, directing state agencies to reduce energy consumption in all their buildings by 35% over the 1990–2010 period; however, they had achieved only a 13% reduction in use per square foot by the 2009–10 fiscal year.[120] Efforts in this area continued in the next decade, as a 2012 executive order issued by Governor Andrew Cuomo established a target of a 20% reduction in energy use per square foot in state buildings over 2010–20.

Renewable Energy Policies

California

California's renewable energy policies have made it a national and sometimes international leader since the late 1970s. In response to the energy crises of the 1970s, California offered generous tax credits for solar and wind power, set ambitious targets for renewable-source electricity, and guaranteed generous prices for it. A 1977 law provided that solar energy be used in new state buildings where feasible, and the next year a 25% tax credit was established for wind and solar power projects; for projects under $12,000, the state permitted homeowners to deduct 55% of the costs from their income taxes, up to a maximum $3,000 tax credit.[121] The next year, another law set ambitious targets for renewable-source electricity: 1% by 1987 and 10% by 2000. The CEC also began mapping wind resources in 1977, providing crucial data that were not available in other states.[122] In 1980, the CPUC started a demonstration project that installed solar hot water systems in 160,000 houses.[123]

In addition, several large fines by the CPUC spurred the state's three large private utilities to negotiate with independent power producers and the commission, which led to contracts with high purchase prices for wind power in the early 1980s.[124] The CPUC, under the authority of the federal PURPA law, locked in high prices for renewable electricity by issuing standard ten-year contracts with an average price of seven cents per KWh.[125] This led to a boom in wind power, so that in 1985, 87% of the world's wind power capacity was located in California.[126]

Although most of these supportive policies were discontinued in the mid-1980s,[127] new policies to promote renewable energy were adopted beginning in the 1990s. The legislature passed a net metering law for homeowners in 1995, which was extended to small and large commercial installations in 1998 and 2001, respectively.[128] The 1996 electricity deregulation law provided for $135 million per year in renewable energy funding during 1998–2001, and a subsequent law extended this funding stream through 2011, after which it was replaced by an even larger solar program.[129]

Finally, California adopted strong renewable portfolio standards beginning in 2002, when it set a target of 20% of the electricity sold by the large private utilities to come from new renewable sources (not including large hydroelectric facilities) by 2017. In 2006, the legislature moved up the target to 2010, and in 2011 it expanded the program to include municipal utilities and increased the target to 33% by 2020.[130] A 2006 law also created the large California Solar Initiative, a "million roofs" program, which provides rebates, financial incentives, and grants for existing and new homes. Implemented by the CPUC, CEC, and publicly owned utilities, the goal is to create 2,000 to 3,000 Megawatts in new electricity generation capacity and to save 585 million therms in natural gas over 25 years. The law committed $3.3 billion to support residential installations over 2007–16, an average of $330 million per year, with the funds coming from a charge on the utilities' customers.[131] When it was created, the initiative was the second largest solar program in the world, after Germany's.[132] As a result of the policies adopted beginning in 2002, new renewable energy sources' share of net electricity generation in California increased from 14% to 21% during 2003–13.[133]

New York

New York has large hydropower resources, but it developed very little new renewable energy before the 2000s, other than a temporary increase in the use of biomass (mainly wood) during the 1970s and 1980s.[134] The state's "six-cent law," adopted in 1981 and rescinded in 1992, gave independent power producers one of the most generous feed-in tariffs in the country, which led to a boom in cogeneration. But in the absence of tax credits, it was not adequate to spur the long-term development of new renewable-source electricity.[135] However, since 2000, policy changes have helped New York to become a regional leader in wind power. In the early 2000s, NYSERDA provided research and development funding for three small wind farms in upstate New York, and later pledged $20 million for four additional wind projects. Governor Pataki issued an executive order in June 2001 requiring state-owned facilities to buy renewable-source electricity in the amount of 10% by 2005 and 20% by 2010.[136] By the end of 2012, 1,638 Megawatts of wind power facilities had been installed, generating about 2% of the state's electricity demand.[137]

In 2004, the Public Service Commission adopted a renewable portfolio standard to raise the share of renewable energy sold in the state from 19% to 25% in 2013; in 2010, the commission raised the goal to 30% by 2015. A surcharge on customers' bills funds the RPS.[138] The target is less ambitious than it appears because it includes large hydroelectric power, mostly from facilities owned by the public New York Power Authority near Niagara Falls and on the St. Lawrence River, as well as small hydropower, but most generation from waste is excluded. Net metering, to facilitate small-scale photovoltaic solar installations, was adopted in 1997 and expanded in 2008 and 2010 to include larger installations.[139] Finally, the NY-Sun Initiative, announced by Governor Andrew Cuomo in 2012, aims to rapidly increase consumer solar installations, with $800 million in investment planned by NYSERDA and two utilities, and supporting legislation was passed to give homeowners tax credits and property-tax exemptions.[140]

Transportation Policies

California

The transportation sector is a crucial area for California climate policy, representing about 40% of the state's greenhouse gas emissions as well as its major source of criteria air pollutants such as ground-level ozone. State policy makers have tried to spur the development of electric and hydrogen fuel cell cars through mandates concerning the percentage of new cars offered for sale in the state. Based on a 1988 law, CARB began developing requirements for zero-emission as well as for low-emission vehicles in 1990. The strategy of trying to force the technological development of electric and hydrogen vehicles, though aimed at criteria pollutants, offers potential reductions in greenhouse gas emissions if the fuel mix used to produce the electricity or hydrogen has a relatively low carbon intensity compared with gasoline. This is the case in California, where natural gas and renewables are important sources of electricity generation.[141*] The initial rules were the subject of litigation, and CARB relaxed them in a settlement with carmakers, but regulations adopted in 2012 will increase the requirements starting in 2018.[142] CARB estimates that complying with the new rules will result in about 6% of new car sales in the state being zero-emission vehicles in 2025, with another 9% being plug-in hybrid vehicles.

A policy change with a larger impact was the path-breaking "Pavley bill," which was the first major law explicitly aiming to regulate greenhouse gas emissions in the United States. The 2002 law directed CARB to develop standards for new motor vehicles in order to achieve the maximum feasible cost-effective reductions in greenhouse gas emissions. In 2004, CARB adopted regulations requiring greenhouse gas emissions per mile driven to fall by 37% for cars and small light trucks and by 24% for large light trucks during 2009–16, reaching 205 and 332 grams CO_2eq/mile in 2016, respectively.[143] In 2011, CARB finalized rules that further tightened the standards for the 2017–25 model years, by another 19% for cars and light trucks.[144]

CARB did not require carmakers to meet these standards through any particular technologies; they could do so with alternative fuel vehicles, changes to air conditioning systems, or increases in fuel economy for conventional vehicles. If they pursued the latter exclusively, the rules translated into an increase in fuel economy for cars from 26 to 40 mpg. Compliance with the regulations was expected to add between $1,100 and $1,900 to the price of a new car in 2016, which CARB estimated would be recouped through reduced fuel consumption.[145] The Pavley regulations were endorsed by 17 other states that intended to adopt them, but were blocked by carmakers' lawsuits and by the EPA in 2007, before they were permitted by an EPA ruling in 2009 and effectively implemented nationwide (with modifications) by the Obama administration later that year; see this chapter's earlier section on U.S. transportation policies.

California also adopted a low-carbon fuel standard for road fuels by an executive order issued in 2007, which CARB implemented beginning in 2010. It requires refineries and distributors to reduce the carbon intensity of transportation fuels 10% by 2020.[146] Reductions are to be gradual over 2010–20 and are calculated on a "well to wheels," life cycle basis. The reductions can be achieved in a variety of ways, which are accounted for through a credit market run by CARB. Compliance methods include refinery efficiency improvements, adding liquefied petroleum gas, ethanol, or compressed natural gas to the fuel mix, or buying credits from alternative fuel suppliers or from electric utilities that supply low-carbon electricity to electric vehicles.[147] Corn ethanol and oil companies successfully challenged the program in federal court, with a district court ruling in 2011 that the program discriminated against out-of-state commerce; however, that ruling was overturned by a federal appeals court in 2013.[148] A practical difficulty for the program will be obtaining enough low-carbon biofuels, since corn-based ethanol has life cycle emissions similar to gasoline, and the production of other biofuels, such as sugarcane ethanol from Brazil or cellulosic ethanol, is small compared to the size of the California market.

Finally, climate policy has also affected California land-use planning, because of the latter's effects on transportation emissions. In 2006–07, both the Center for Biological Diversity and Attorney General Jerry Brown engaged in successful litigation concerning the impacts of new housing developments on greenhouse gas emissions. The next year, the legislature followed up with a statewide policy targeting the greenhouse gas impacts of development. Senate Bill 375 directed CARB to set regional greenhouse gas emissions reduction targets within the state, totaling 5 $MTCO_2$eq by 2020. The law encourages transportation, housing, and land-use planning bodies to orient their plans to the targets by creating Sustainable Communities Strategies. However, compliance is uncertain because lobbying by city and county governments weakened the bill.[149*] Metropolitan planning organizations need only to include targets in their plans in order to get CARB approval, they can opt out if target attainment is not feasible for any of a variety of reasons, and they suffer no funding consequences for non-compliance. Their main incentives to comply lie in gaining access to streamlined environmental impact procedures under the state's environmental review law.[150]

New York

Despite New York's advantages in urban density and mass transit, transportation accounted for 31% of the state's greenhouse gas emissions in 2011. For the most part, New York's policies on motor vehicle pollution and efficiency have followed California's. The state adopted California's "Pavley" greenhouse gas emission standards for motor vehicles in 2005, and two years later it joined the *Massachusetts v. EPA* lawsuit to try to force the federal agency to regulate greenhouse gas emissions by vehicles.[151] New York also adopted California's zero-emission vehicle standard in 1990 and modified it in 2012 to reflect the changes to the California program. However, New York has not yet adopted a low-carbon fuel standard, and although it began working with ten other Northeastern states on a regional standard in 2009, they still have not adopted one.[152]

New York did strike out on its own with a program for research and development on electric and alternative fuel buses. This was funded by NYSERDA, the Metropolitan Transit Authority's purchases, and later by $28 million from a 1996 environmental bond act. The program led the Metropolitan Transit Authority in New York City and transit authorities in other cities worldwide to adopt hybrid electric buses built in upstate New York.[153] New York State also has initiated smart-growth programs since 2007 in efforts to curb greenhouse gas emissions. A 2010 law requires evaluations of public infrastructure projects according to smart-growth criteria, such as improving existing infrastructure, building in municipal centers and in-fill sites, and promoting compact development.[154] NYSERDA's Cleaner, Greener Communities Program, adopted in the Andrew Cuomo administration, provides $100 million in grants for public-private partnerships to conduct regional planning projects to reduce transportation and other energy use.

CALIFORNIA-NEW YORK COMPARISON

A comparison of California and New York climate policies, which were summarized earlier in this chapter in Table 2.3, shows two main features to be explained. In the first place, both states have adopted strong policies in all the major areas described here, and both are clearly among the states with the strongest policies in the United States. In the climate index published by the Center for Climate and Energy Solutions from 2008 to 2014, California and New York were almost always in the top five states, with their average ranks being 2nd and 4th, respectively. Second, California's policies are clearly stronger than New York's in the areas of enforceable targets, emissions trading, and renewable energy, and somewhat stronger on energy efficiency and transportation policies. While both are national leaders in climate policy, California is pre-eminent.

In the setting of emissions targets, both states have ambitious targets for 2050 (an 80% reduction), and roughly similar targets for 2020. New York's target for that year is nominally more ambitious than California's (a 10% reduction vs. stabilization relative to 1990). However, given differences in past and projected population growth since 1990, California's target actually is more ambitious than New York's on a per-capita emissions basis.[155*] But what is more important is that California is the only state in the country with legally binding emissions targets; its legislation requires CARB to implement regulations to achieve them. By contrast, there is no state agency in New York with the legal authority or obligation to act on statewide targets. Moreover, given a combination of slow population growth and the recent decline in New York State emissions, which is at least partly due to lower natural gas prices and fuel switching, New York's targets do not provide pressure for policy measures to reduce emissions further. There are also signs that the New York State's commitment to its reduction goals is somewhat weak. NYSERDA has carried out statewide emissions inventories infrequently, and the climate action plan still has not been finalized, several years after its 2012 due date.

In emissions trading, New York State would appear to have some advantages on California. New York initiated emissions trading four years earlier than California, it successfully led a regional effort that contrasts with the failure so far of the Western Climate Initiative, and RGGI auctions 90% of its allowances, compared to only 10% at the outset of California's program. Auctions have the advantage of creating an independent funding source for energy efficiency and renewable energy programs and an interest in maintaining the trading system on the part of the constituencies that benefit from those programs – utilities, their customers, and state officials. Auctioning allowances also reduces the risk of overallocations and windfall profits.[156] However, what is of greater importance is that California's emissions trading system covers 85% of the state's greenhouse gas emissions, while RGGI covers only 20% of New York's.[157] Moreover, California's cap began declining by 2% per year immediately after the system's start in 2013 (and will decline 3% per year after 2015), while RGGI was designed to produce emissions cuts only starting in 2015, six years after its inception. This eliminated the advantage that New York could have gotten from its earlier start.

The two states have broadly similar policies for increasing energy efficiency in electricity and the heating of buildings, although New York has lagged

California at times. Both were leaders in adopting appliance standards and building codes in the 1970s, but California adopted rate decoupling more than a decade before New York, and its "decoupling plus" policy is not matched in the latter state. In 2006, the American Council for an Energy-Efficient Economy rated New York well behind California in its scorecard, which is based on a wide range of energy efficiency policies. California was tied for 1st place, with 75% of the maximum score, while New York had 57% of the maximum score and was in 7th place. But New York has improved its relative standing, and by 2013, it was in 3rd place nationally, with 76% of the maximum score and just behind California's 82% score. In terms of utility spending on energy efficiency, California also has scored higher than New York State in most years since 2004, but New York had caught up by 2011.[158]

California has led the nation in policies to support the development of new renewable energy sources, and although New York has become more active since 2000, its policies are still weaker. In the 1970s and 1980s, both states offered generous feed-in tariffs (six to seven cents per KWh), but California also offered a 25% tax credit and technical assistance to wind power producers, a combination of policies that was more effective than New York's six-cent law alone.[159] This helped California to become the global leader in wind power.

More recently, California's renewable portfolio standard calls for a steeper rate, and a longer period, of increase in the sale of new renewables than New York's does. California's planned increase is 1.1% per year (rising from 14% of electricity from new renewables in 2003 to 33% in 2020), while New York's is only 0.7% per year (rising from negligible amounts in 2004 to 8% in 2015).[160*] Comparative studies of RPS stringency also place California well ahead of New York. One study gives California a 74% score and New York only a 45%, ranking them 8th and 22nd, respectively, out of 33 state programs.[161] Furthermore, New York is lagging well behind its target for new renewables; in 2014, with only one year to go, only 56% of the 2015 target had been achieved.[162] California, by contrast, is only somewhat behind schedule to achieve its goal of 33% by 2020. About 23% of electricity sold by investor owned utilities in 2013 was from new renewables.[163] This is a rise of 0.9% per year, compared with the 1.1% rate needed to attain the target, and the recent rapid growth of California's solar program might help to close this gap.

Finally, in transportation policies, New York has copied California's zero-emission vehicle program and the Pavley rules, as did many other states, mainly in the Northeast and on the West Coast. These states all lent their weight to California's efforts to influence national vehicle markets and federal policy. However, New York has still not adopted a low-carbon fuel standard for transportation fuels, pending discussions with other states on a regional policy. Those fuels also remain outside RGGI, in contrast to California, where they were included in the emissions trading system in 2015.

INFLUENCE ON OTHER STATES AND THE FEDERAL GOVERNMENT

California and New York State have been climate policy leaders in two senses. They have adopted innovative policies early, at times when other states and the federal government have not, and they have influenced those other actors to adopt policies similar to theirs.

New York has played a key role in emissions trading beyond its borders. Its pioneering law on SO_x emissions from power plants in 1984 helped lead to Title IV of the 1990 Clean Air Act amendments. Those created a national cap-and-trade system for SO_x emissions, which was credited with greatly reducing emissions at low cost. The Environmental Defense Fund helped to make this policy instrument a "living legend" well known to climate policy proponents federally and in other states. This spurred interest in creating similar systems for other pollutants, including greenhouse gases.[164] New York also played a major role in the Ozone Transport Commission, which worked with the EPA to create an NO_x emissions trading system in 1999.[165] Then, in the early 2000s, New York spearheaded the formation of RGGI, gaining the participation of nine other Northeastern states to create the first greenhouse gas emissions trading system in the United States and the first in the world to rely mainly on auctioned allowances.

California began emissions trading a few years later than New York, just as it had with statewide emissions reduction targets, but its creation of the first cross-sector, comprehensive trading system inspired efforts to create similar systems at the federal level and among the states in the Western Climate Initiative. Although those efforts have failed so far, in 2013 California, Washington, Oregon, and British Columbia reached an agreement to cooperate on targets for emissions and carbon intensity in fuels, and the next year, California and Quebec began to link their emissions trading system.[166] Governor Jay Inslee of Washington State also has proposed instituting a statewide, multi-sector emissions trading system like California's, although the Washington legislature rejected his plan in 2015.[167*]

In appliance standards, both states took pioneering roles beginning in 1976. After New York and five other states had adopted California's standards in the 1980s, manufacturers sought uniform national standards, leading to the National Appliance Energy Conservation Act in 1987 and to later federal legislation.[168] The federal appliance and equipment standards adopted since 2010 also followed the leads of California, New York, and other states. The now-widespread strategy of using increased energy efficiency as a means to avoid power plant construction, including a quantitative measure of avoided marginal cost, first was developed by an academic team in California and then was adopted by the CEC.[169] California officials developed policies to decouple utilities' revenues from sales, which they applied to natural gas utilities in 1978 and to electric utilities in 1982; similar policies are now used in 23 states.

California also has been influential in renewable energy and transportation emissions. In the early 1980s, the state became an important test market for wind power, and it has done the same for solar energy since the mid-2000s. The existence of a large state with highly supportive policies allowed the renewable energy industries to try out technologies and prove their economic feasibility, and it helped advocates and regulators to develop effective policies and build political constituencies.[170] California's zero-emission vehicle policy is currently followed by nine other states in what is yet one more iteration of a pattern of states adopting California emission standards. In 2013, New York and six other Northeastern states joined California in an agreement to coordinate their zero-emission vehicle policies in order to try to get 3.3 million innovative vehicles, mainly plug-in hybrids and electric vehicles, on the road by 2025.[171]

Finally, the Pavley law uses California's rights under the Clean Air Act to limit greenhouse gas emissions from motor vehicles, an approach that 17 other states had endorsed by 2007. Altogether, the Pavley states, which included Florida and Pennsylvania, represented more than 40% of the U.S. population and nearly 50% of new vehicle sales.[172] Although the EPA rejected California's waiver request in 2007, the economic weight of the states endorsing the Pavley rules probably influenced the Supreme Court's decision in *Massachusetts v. EPA*, which permitted and encouraged the EPA to regulate carbon dioxide emissions in transportation under the Clean Air Act. The economic importance of the coalition of states that were following California also made the carmakers willing to reach deals with the Obama administration to nationalize the Pavley approach in 2009 and, two years later, to extend and accelerate the emissions reductions rules for an additional eight years, through 2025.

THE EFFECTIVENESS OF CLIMATE POLICIES IN THE UNITED STATES

How effective have these climate policies been at reducing greenhouse gas emissions? One way to begin to address this question is to examine aggregate trends in emissions, although by itself this approach has severe limitations. It cannot take into account what would have happened in the absence of policies (a business-as-usual scenario) and cannot assess the effectiveness of particular policies. It also cannot distinguish between the results of policies and other factors, such as changes in economic activity, demographic change, and fuel switching caused by market-driven changes in relative fuel prices. Nonetheless, aggregate emissions are what matters for global warming, and they are critical benchmarks for policy makers in general and for emissions trading systems in particular, so they provide an important frame of reference and a useful starting point for assessing policy outcomes.

Figure 2.2[173*] presents the greenhouse gas emissions trends for the United States, California, and New York from 1990 to the most recent year of

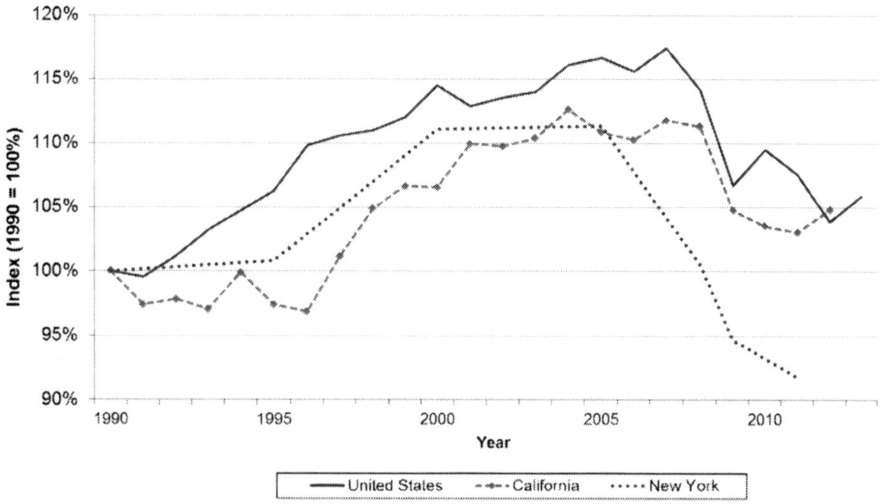

FIGURE 2.2 Greenhouse Gas Emissions Indices for the United States, California, and New York, 1990 to 2011–13
Sources: see note 173.

available data (which ranged from 2011 to 2013). All three cases show a broadly similar pattern. Emissions increased in an uneven way after 1990, rose most steeply in the late 1990s, and reached a peak in the mid-2000s. This was followed by a decline that was sharpest during the 2008–09 recession and largest in New York State. So far, there have been no appreciable increases in emissions during the economic recovery that began in 2009. These data from official emissions inventories are production-based, which means that they count only emissions produced within a territory. Since they do not include emissions due to manufactured goods imported to the United States from other countries, they underestimate total greenhouse gas emissions. However, this does not bias the findings about recent trends, because consumption-based estimates also indicate that U.S. emissions have declined since about 2007.[174*]

Therefore, the trends in the official, production-based inventories can be used to make comparisons among the three cases. Because of annual fluctuations,[175*] I have used the average of emissions in 2008–12 (or 2008–11 for New York) as the endpoint, with 1990 as the baseline. By this measure, New York's recent emissions were 4.9% lower than in 1990, while California's were still 5.5% higher and the United States' still 9.5% higher than in 1990. These trends are affected by population growth, which is much higher in California and nationally than in New York. On average, California's emissions index (relative to the 1990 baseline) was 5.1 percentage points lower than the United States' during 1991–2012, although this narrowed to an average 2.9 percentage point difference during the last five years.

To understand the effects of climate policies, studies that compare emissions to business-as-usual (or "no policies") scenarios are needed. Fortunately, reasonably good estimates for the effects of federal and California climate policies are available. They show significant reductions in greenhouse gas emissions for the major policies described in this chapter. For the U.S. federal government case, Table 2.4[176*] shows that the largest effects have come from mandatory appliance standards, the voluntary Energy Star programs to increase energy efficiency, the CAFE standards for motor vehicle efficiency, the Significant

TABLE 2.4 *Estimated Annual Effects of U.S. Federal Climate Policies on Greenhouse Gas Emissions in 2011 and 2020*

Policy measure	Estimated effects in 2011 (in $MTCO_2eq$)	Estimated effects in 2020 (in $MTCO_2eq$)
Electricity sector		
Mandatory energy efficiency standards for appliances and equipment	179.5	291.2
Energy Star for products	99.7	141.2
Energy Star for commercial and residential buildings	89.3	97.3
Green Power Partnership; Combined Heat and Power Partnership	29.6	73.3
Wind power policies	33.1	95.2
Power plant greenhouse gas emissions standards (proposed 2014)	–	68
Transportation		
Fuel economy standards (CAFE), 1978–2004	183.5	183.5
Fuel economy standards (CAFE), 2005–11	35.0	79.3
Motor vehicle greenhouse gas emission standards, 2012–25 (light and heavy duty)	–	273.7
Waste and industry		
Landfill air regulations	146.4	183.1
Significant New Alternatives Policy program (F-gases)	206.9	311.1
Total effects of major policies	1,003.0	1,796.9
Total effects as % of 1990 greenhouse gas emissions	**15.9%**	**28.5%**

Sources: see note 176.

New Alternatives Policy program regulating substitutes for ozone-depleting substances, landfill air regulations, and wind power subsidies. By 2011, the estimated effects of these major[177*] policies alone totaled about 1,000 MT, which was 16% of the United States' total emissions in 1990. In other words, without these policies, U.S. emissions in 2011 would have been 23% higher than the 1990 baseline; actually they were "only" 7.5% above the baseline in 2011. In 2020, the effects of major climate policies, including the proposed rules for existing power plants, are projected to rise to about 1,800 MT, which is 29% of the 1990 baseline. These estimates, like those I am about to present for California, are conservative, since they exclude many smaller measures that also have had effects.

Table 2.5 depicts the effects of the most significant[178*] California climate policies, in absolute terms and as a percent of the 1990 baseline. Compared with the United States as a whole, the effects relative to the baseline are smaller for the policies adopted before 2006 and larger for those going into effect during the present decade. The energy efficiency and renewable energy policies adopted in California from the 1970s through the early 2000s reduced emissions in 2008 by about 33 MT, which is only 8% of the 1990 baseline.

TABLE 2.5 *Estimated Annual Effects of California Climate Policies on Greenhouse Gas Emissions in 2008 and 2020*

Policy measure	Estimated effects in 2008 (in $MTCO_2eq$)	Estimated effects in 2020 (in $MTCO_2eq$)
Electricity and buildings sectors		
Utility energy efficiency programs, building and appliance standards, combined heat and power programs	21.2	50.7
Green buildings program	–	4.5
Renewable energy policies	12	32.5
Transportation		
Pavley standards for motor vehicles	–	29.9
Low-carbon fuel standard	–	15.2
Waste, industry, and cross-sectoral		
High-GWP programs	–	5.4
Recycling (mandatory commercial)	–	5.0
Cap-and-trade program	–	23.0
Total effects of major policies	33.2	**166.2**
Total effects as % of 1990 greenhouse gas emissions	7.6%	38.0%

Sources: see note 177.

However, the acceleration of those policies, the addition of the Pavley rules and the low-carbon fuel standard in transportation, and the start of emissions trading (all of which were adopted in the late 2000s) will greatly increase the impact of California's climate policies. Altogether, the state's policies are projected to produce an estimated 166 MT of emissions reductions by 2020, which is 38% of the state's 1990 baseline.

Other states' climate policies also have been effective in reducing greenhouse gas emissions. Comprehensive estimates of the emissions reductions due to policies adopted by New York and the other states are not available. However, Byrne et al. estimate that just two sets of common state-level policies, on energy efficiency and RPSs, prevented 669 MT in emissions in 2010 and will avoid 1,822 MT by 2020;[179] these amounts are 11% and 29% of the United States' 1990 baseline emissions, respectively. Those reductions are compared to business as usual, and go beyond the reductions expected from technological innovations independent of policies. Lutsey and Sperling provide a much lower estimate, which, however, is of net reductions from current emissions rather than from business-as-usual scenarios.[180] They conclude that if 17 states with reduction targets and climate action plans – which include 45% of the U.S. population – could achieve cuts of 728 MT by 2020 if they meet their targets; that is about 9% of the United States' 1990 baseline emissions.

Plainly, the United States has adopted significant climate policies, both at the federal level and in many states. These policies have also varied greatly across places and time. That raises questions about their causes, which the theories in the next chapter address.

3

Theories of Climate Policy

Climate change policy is a very complex area, and hence a great variety of theories can help us to understand why strong climate policies are sometimes but not usually adopted. Relevant literatures include those on environmental policy and outcomes (especially concerning air pollution), on energy policy (especially concerning energy efficiency and renewable energy), and on explicit climate change mitigation policy. Of these literatures, those on environmental policy and environmental outcomes are the best developed, offering a comprehensive set of causal factors, while the literature on explicit climate policy making is most directly pertinent to the book's topic. Therefore, in the following I will draw mainly on the first two bodies of writing and will include the latter wherever possible. Given the context of multi-level governance in which U.S. climate policy unfolds, literatures comparing national policies (and positions in international climate negotiations), on U.S. federal policies, and on U.S. state-level policies all pertain to the topic of this book, and will be included in this chapter.

These literatures propose an enormous number of causal factors. But running through all of them is an implicit debate between structural explanations, which are based on stable political institutions and socioeconomic features, and process explanations, which are based on rapidly changing events and processes, including the actions of individuals, groups, and coalitions. Even though it is usually implicit in these literatures, I think the structure-process (or structure-agency) debate is worth explicating. Doing so helps to address core questions of this book: Are the strong policies that have been adopted in the leading states only possible there, because of their structural advantages? Or, alternatively, could they be adopted elsewhere, such as in other states or by the federal government, through some combination of intentional attempts by actors and fortunate events and timing? The overarching question here is to what extent political actors are constrained by structures and to what extent they can achieve their goals despite those constraints. In addressing this question through a combination of structural and process explanations, I aim to steer clear of the pitfalls of both structural determinism and intentionalism.[1]

Most theorizing about climate and environmental policies tends to emphasize either structures or processes, but not both.[2*] Cross-national work on climate policy as well as large-N studies of state-level climate, environmental, and energy policies explain policy change mainly in terms of political institutions and fixed socioeconomic features. For example, cross-national studies argue that climate policies result from weak fossil fuel industries, parliamentary systems, and neocorporatism, while writing on state-level environmental policy has emphasized problem severity, wealth, and governmental capacity. This work implies that jurisdictions without the right structural conditions are unable or very unlikely to adopt strong climate policies. By contrast, the literatures on U.S. federal environmental policies and climate policies tend to focus on processes, such as focusing events, leadership, framing, and public opinion, and hence they imply that the requisite processes can arise in many places and times, if not in every one of them.

After discussing structural and process theories in the next two sections, I then will offer a comprehensive, process-oriented theoretical framework. For this, I will draw on windows of opportunity theory[3] and advocacy coalition formation theory,[4] into which the process variables advanced by other authors can be incorporated. Next, I will relate this framework to the structural theories, in part through a theory of partial (non-deterministic) path dependence.[5]

The overall approach that I offer in this chapter and this book is intended to build on the strengths of existing work and to redress some of its weaknesses. I distinguish sharply between structures and processes in order to try to identify the conditions under which political actors may be able to influence major reforms. By structures, I mean features that are completely or nearly stable for at least one or two decades, and hence are stable during the shorter period in which any particular climate policy would be adopted or rejected. In contrast, processes are features that often change, sometimes under the influence of actors and coalitions of actors (both advocates and opponents) during the time frame of policy consideration. I will aim to identify both the structures and the processes that contribute to major policy change, and will address whether the processes that operate are determined by structural factors or are largely independent of them.

Drawing a sharp analytical line between structures and processes will help to delineate a third, intermediate type of factor – structural conditions that are created by processes and then operate over periods of decades to affect later policy making, such as administrative capacity or legal authority. They will be the focus in my discussion of path dependence theory.

STRUCTURAL THEORIES

Scholars have identified a number of *political institutions* and *socioeconomic structures* as factors that promote environmental, and specifically climate, policy making. While they do not reach a consensus, the following factors are prominent in the various literatures and have substantial empirical support. Therefore, they will be tested through an examination of how they affect climate policy making by the federal government, California, and New York in Chapter 4, and in the statistical analysis in Chapter 5; they will also reappear in the conclusions of the case studies, in Chapters 6 through 9.

The first political institutional feature that is held to make climate policies more likely is the concentration of governmental authority in the executive branch. Concentrated authority makes strong government intervention more likely by reducing the number of veto points at which the opponents of climate policy can block policy making, and by increasing coordination among governmental actors and the continuity of policy making and implementation.[6] Concentrated authority is found in parliamentary rather than presidential systems and in countries lacking strong independent judiciaries. For example, Dolsak's analysis of 91 country cases concludes that those with parliamentary systems had stronger commitments to climate policy than those with presidential systems.[7] Studies of industrialized democracies also find that those with concentrated governmental authority had better environmental outcomes.[8]

On this dimension, the United States clearly has fragmented, not concentrated governmental authority. Studies find that the weakness of U.S. climate policy compared to that of many European countries is partly due to the multiple veto points in the U.S. system. These include the strong legislative roles of both houses of Congress and their committee chairs, as well as the Senate's procedural rules giving strong rights to individual senators and the minority party.[9] Similarly, much research on U.S. environmental policy making finds that its basic institutions inhibit policy making.[10] The U.S.'s "institutional pluralism" also includes the dispersal of policy implementation between multiple administrative agencies and the system of federalism, where state governments often implement EPA regulations.[11]

However, the opposite argument also has been made about the effects of U.S. institutions on environmental policy. The separation of powers and federalism present multiple pathways that reformers can try to use, and different institutions may compete to increase standards. The separation of powers also may protect policies from being undermined when backlashes against them develop.[12] Blockage of reform efforts in one branch of government may lead to initiatives by the other branches, and blockage at the federal level may lead to action by the states. In particular, federalism in the United States provides opportunities for state governments to adopt climate policies. U.S. constitutional doctrine give states authority to act if the federal government does not, and state governments historically have had jurisdiction over areas related to climate policy, such as electricity generation, land use, and agriculture.[13] Although federalism helps to fragment government institutions, it is ambivalent in its effects, since it also gives scope for state governments to innovate, and their innovations may lead to stronger policy by the federal government.[14]

The second institutional explanation concerns the nature of the interest group system, with neocorporatist systems held to promote climate policy while pluralist systems hinder it.[15] Neocorporatism can be defined as a pattern of formal consultation, negotiation, and society-wide deal-making between government and a small number of monopolistic, well-organized interest associations representing broadly defined societal groups.[16*] It includes a high degree of cooperation, in contrast to pluralism, which is characterized by adversarial relations between government and a large number of sometimes competing interest groups that often use lobbying and litigation to press their particularistic interests.

Proponents of neocorporatism argue that it can help to overcome opposition in a variety of ways. Similar to the institutional concentration of governmental authority, neocorporatist systems reduce the number of veto points at which powerful producer interests can block climate policies, by allowing government actors to concentrate on negotiating with a small number of interest groups.[17] Neocorporatist systems can compensate losers, communicate the benefits of environmental policies to members, and create consensus. Hence, they reduce free riding, promote the internalization of environmental externalities, reduce the risk of backlash against environmental policies, and yield more cooperation in policy implementation. By contrast, in a pluralist system such as the United States, the economic actors that would be harmed by environmental policies tend to lobby successfully for preferential treatment, allowing them to impose external costs on others. Indeed, many cross-national studies show a correlation between neocorporatism and environmental outcomes, with the Nordic countries, the Netherlands, and Germany scoring relatively high on both.[18] In the United States, by contrast, a pluralist system has allowed anti-regulatory business interests to hamper environmental policy, especially since the Reagan administration.[19]

Third, high green party vote shares and participation in government are favorable conditions for climate policy making.[20] Green parties include explicitly ecological parties and left-socialist parties that have adopted strong environmental programs.[21] Through their strong identification with environmental issues, green parties help to mobilize environmental public opinion, and through their vote shares, they demonstrate that environmentalism is a political force that all parties need to reckon with. Several studies find that green party strength is correlated with stronger environmental performance,[22] although others find only unstable or statistically insignificant effects.[23]

Green party strength varies over time, and hence to a large degree it is a process variable, but the weakness of green parties in the United States is closely tied to electoral institutions. Because green parties are small parties, they require a degree of proportionality in the electoral system in order to gain significant political strength. Hence, electoral-system proportionality is favorable for environmental outcomes.[24] In the few industrialized democracies that have strictly majoritarian electoral systems – the United Kingdom, Australia, Canada, and the United States – green parties have not won more than one seat in the lower chamber of the national legislature or more than 3% of the presidential vote in any national election since 1990.[25*] Although minor parties that have large, spatially concentrated pockets of support, such as regional parties, can do well in such systems, green parties do not benefit from this effect.[26]

Therefore, although national green party strength is a process variable in most countries,[27*] in the United States the extreme weakness of this political force is due to a very stable institutional feature, namely its electoral system, which is anchored in the Constitution, law, and custom. Green parties are virtually irrelevant in U.S. national politics largely due to the country's first-past-the-post, single-member-district rules for congressional elections and the winner-take-all (statewide or district-level) rules that state legislatures use for allocating electoral college votes in presidential elections.

Fourth, much work on comparative and U.S. environmental policies has argued for the importance of strong governmental capacities. Cross-national work holds that large capacities, in terms of revenue, spending, and administrative personnel, are necessary for effective government intervention in the environment.[28] Centralized bureaucracies and professionalized legislatures also have been advanced as important causes of environmental policy in the U.S. states, since these have the ability to gather information, develop policy, and implement it.[29] While Rabe points to the importance of the capacities of state government agencies, he notes that these have been growing in some states, as agencies have diversified and expanded.[30] Hence, administrative capacities are not necessarily fixed features, and their growth also requires explanation.

Finally, climate policies are more likely if the dominant ideology or political culture favors government intervention.[31] Ideology can influence policy making through the views of elites, citizens, or both. Hence, cross-national studies have argued that a political culture which favors market forces over government intervention is an obstacle to climate policy making.[32] Citizens in Europe are also more accepting of the value of the precautionary principle and the concept of sustainability than those in the United States.[33] Indeed, studies of U.S. states find that those with predominantly liberal citizens (i.e., who favor government intervention) have better outcomes in air and water pollution and are more likely to adopt renewable portfolio standards and other climate policies than states with conservative citizens.[34]

Scholars also posit several socioeconomic structures that are generally favorable to climate and other environmental policies. First, high incomes give countries more economic capacity to adjust to the costs of domestic climate policies, and they also increase citizens' demands for environmental policies.[35] Cross-national studies of industrialized democracies have found an association between income and environmental outcomes, at least until very high levels of income are reached.[36] Similarly, high education levels increase public awareness and understanding of environmental issues, openness to new ideas, and scientific capacities for addressing problems.[37] Studies of environmental policies in the United States also find that richer states have tended to adopt innovations earlier than others.[38] Finally, income and education underlie the theory of postmaterialism, which holds that citizens socialized into rich, highly educated societies in the absence of war have developed preferences for non-material goods such as environmental protection, as well as the skills to act politically to attain them.[39]

Second, the composition of the economy affects the chances for strong climate policy. A service-based economy is beneficial for climate policy since service-sector firms have lower costs of adjustment and are less exposed to international competition than are manufacturing firms.[40] Also, green or clean technology firms, which include those involved in energy efficiency or renewable energy, favor climate policy because they benefit directly from subsidies or regulations, the costs of which are borne by customers and competitors.[41] More broadly, the agricultural, real estate, insurance, recreation, and tourism sectors of the economy favor climate policy because of their exposure to the risks of climate change and their relatively low energy consumption.[42]

Conversely, the fossil fuel, automobile, and energy-intensive manufacturing sectors of the economy (such as steel, aluminum, and cement) will work against climate policy because they bear high costs from policies that seek to reduce carbon dioxide emissions through increased energy costs or regulations.[43] Fossil fuel companies, in particular, have lobbied intensely against climate policies, in international negotiations and at the national level.[44] Fossil fuel industries also have successfully influenced media coverage and public opinion by funding the research of climate change skeptics and publicizing their views in the United States.[45]

Third, a country's (or a state's) profile of energy production and consumption has additional effects on its propensity to adopt strong climate policies. If a country is a fossil fuel importer, it is more likely to favor climate policies, which tend to reduce fossil fuel consumption, since these policies also improve its balance of payments and reduce its dependence on foreign energy supplies. Hence, climate policies are in the national interest as perceived by policy makers and other elites. Conversely, countries with large domestic fossil fuel supplies will oppose climate policies since they lack those motivations, and, as already noted, they have powerful domestic interest groups favoring fossil fuel use.[46]

In addition, countries with domestic fossil fuel resources and industries tend to have economies that use energy inefficiently. Therefore, they perceive the costs of adjusting to a low-carbon energy system to be high, and will resist policies to reduce emissions.[47] This tendency exists even though inefficient economies have lower marginal costs for the initial measures involved in climate policy, the "low hanging fruit," such as energy efficiency standards.[48]

While an economy's energy intensity, its CO_2 intensity, or its energy prices are often offered as an explanation of climate policies,[49] these are intervening variables rather than independent explanatory factors. That is, they are partly driven by structural factors, such as domestic or in-state fossil fuel production[50] and partly by implicit climate policies, such as utilities' rate structures or renewable energy policies. Therefore, I will not use them as independent variables in the statistical analysis in Chapter 5.

Finally, the severity of problems faced by a country or state is held to be a factor in the adoption of environmental policies. Severe problems can motivate both citizens and elites to press for policies to alleviate the problems. Work on environmental policy in the U.S. states has showed that the adoption of ameliorative policies correlates with urbanization and with air and water pollution.[51] This is relevant for state-level climate policy, because improving air quality is a motive for many states to adopt policies curbing greenhouse gas emissions. Indeed, Matisoff finds that states with lower air quality are more likely to adopt renewable portfolio standards.[52] In climate change mitigation policy, problem severity is also measured by objective ecological vulnerability. Small island nations back international climate action because they are concerned about the effects of rising sea levels on their land and freshwater resources, while countries in sub-Saharan Africa are concerned about the effects of desertification.[53] In the United States, coastal states face risks from sea level rise and increased storm activity, and hence are more likely to favor climate policies.[54] Climate change also threatens forests in Maine and winter tourism in New England, vulnerabilities that can help to explain why certain states participated in the *Massachusetts v. EPA* lawsuit.[55]

In accordance with the structural theories, key comparative works explain the relative weakness of U.S. climate policies in terms of several largely immovable factors.[56] The separation of powers fragments authority between the executive branch, Congress, and the federal courts, while a pluralist interest group system gives the country's large fossil fuel industries ample footholds in policy-making institutions. Because of these structural conditions, business arguments about scientific uncertainty and costs have not been countered by a green party or a powerful environmental ministry, and public opinion does not support a strong regulatory role for the federal government. Therefore, the business opponents of climate policy have been able to block climate policy in Congress and the executive branch. However, this structural account cannot explain why relatively strong climate policies sometimes have been adopted by the U.S. federal government and by some states. For this, process theories are required.

PROCESS THEORIES

Scholars writing about U.S. federal climate policies and other environmental policies focus heavily on processes, which are also included in many cross-national and state-level works on climate policy. The main factors they treat are focusing events, political leadership, partisanship, the framing of the co-benefits of climate policies, first-mover advantages, interest group mobilization, public opinion, and economic processes. These will provide much of the conceptual framework for the case study analyses in Chapters 6 through 9.

First, scholars contributing to various literatures find that external *focusing events* can rapidly shift the content of the public agenda and encourage the simultaneous mobilization of diverse actors who might advocate for climate policy, thus helping to overcome free-riding problems.[57*] The latter are especially important barriers to policy adoption where the potential coalition includes many groups or groups with very large memberships, such as environmental organizations. Cross-national research on environmental policy innovations by pioneering countries identifies many focusing or triggering events that are relevant for environmental policy. These include fossil fuel price increases or supply disruptions, environmental disasters such as storms, fires, drought, or disease outbreaks, the publication of new scientific findings, and action by other governments at the national or international level.[58] Research on state-level climate policy produces a similar list of the types of events that can drive policy adoption.[59]

Second, *leadership by government officials* (elected or appointed) who are interested in or committed to climate policy can help to prioritize it on both the public agenda and the political decision-making agenda, and to assemble a broad coalition in favor of it. Cross-national studies emphasize the importance of political leadership that sees global warming as a pressing problem.[60] In the United States, President Bush's lack of a commitment to climate policy is clearly a major reason for the lack of progress on the issue in the 2000s. At the state level, governors, attorneys general, judges, and directors of state agencies at various times have been important advocates of climate policy.[61] Political ambition (the desire to run for re-election or for higher office) helps to make officials willing to confront powerful interests that are opposed to climate policy, in order to gain publicity and claim credit for innovations.[62] For example, a study of state attorneys general found that those who sought stronger consumer protections through multi-state litigation were likely to run for higher office.[63] In addition, governors often have played key roles in initiating stakeholder dialogues that built coalitions for climate policy.[64]

Third, aside from the influence of green parties, discussed in the previous section, the literature suggests that the major left- and right-wing parties, and hence the *election outcomes* that determine their control of major offices, affect the prospects for climate policy. Because of the ideological commitments of their activists and their electoral bases, left-wing parties in government are more likely to adopt climate and environmental policy, according to cross-national and state-level studies.[65] However, in the United States, differences within the parties may make elite ideology more important than partisanship. Calvert reaches this conclusion in explaining environmental policy making in the

Northeastern states, where Republicans have been more moderate than in other parts of the country.[66] Similarly, this book's case studies of California and New York State show that both Republican and Democratic officials (governors and legislative leaders) were key supporters of climate policy from the 1970s to the 2000s. A study of the adoption of renewable portfolio standards by state governments also shows a bipartisan pattern, even though nationally, Democrats are stronger advocates for renewable energy than are Republicans.[67]

In any case, partisanship and elite ideology on environmental policy appear to have been converging rapidly in the United States since the late 2000s. Polarization on climate policy has increased in public opinion about the causes of global warming and the desirability of renewable portfolio standards, clean coal policies, fuel economy standards, and a cap-and-trade program.[68] The Republican shift to the right on environmental policy has blocked action in Congress and forced a search for alternative policy-making paths.[69] Polarization has increased at the state level, too. Republicans increasingly oppose green energy laws in state legislatures,[70] and in 2014, they reversed policies that had created energy efficiency programs and standards in Indiana and Ohio.[71]

Fourth, appropriate *framing processes* promote climate policies by easing public acceptance and helping advocates to assemble broad coalitions in favor of them. For example, a Canadian study found that motivational frames were more effective than sacrifice frames in increasing public willingness to participate in grassroots advocacy.[72] Layzer argues that the framing of problems and solutions by advocates, in interaction with experts, media, and government officials, is of crucial importance in U.S. federal environmental policy making.[73] In the climate policy arena, energy security and green economy frames have been criticized for what is seen as their ineffectiveness in mobilizing public support over the last decade.[74] For their part, climate policy opponents have successfully deployed scientific uncertainty and economic cost frames to shift the debate onto terrain less favorable to climate policy making.[75]

At the state level, many studies of climate policy making also have found that appropriate framing eases acceptance. Through framing, advocates and political leaders may interpret a region as ecologically vulnerable or emphasize the co-benefits of climate policies for economic development, job creation, energy security, or air pollution control.[76] Indeed, climate policy goals often are secondary and are downplayed in legislative debates and texts, such as in the ambitious renewable portfolio standards adopted by Texas in 1999 and by California in 2002.[77] Some cross-national work also investigates how the linkage of climate policy to other issues on policy makers' agendas can facilitate or hinder reform.[78]

Fifth, scholars contributing to a variety of literatures have argued that governments pursue climate and other environmental policies because they confer *economic advantages* on jurisdictions that become leaders. Subsidies provide support for the development of new industries concerned with energy efficiency or renewable energy.[79] Furthermore, because stricter standards give first-mover advantages to certain domestic economic sectors, those sectors will press for such policies, and policy makers will see the policies as contributing to their jurisdiction's exports and overall economic development.[80] Jurisdictions that are environmental pioneers also gain first-mover advantages, because stricter environmental policies create demand for new products, turning the pioneers into environmental lead markets that become the sites of technological innovation.[81] In the domain of international trade, Vogel argues that stricter environmental standards confer a competitive advantage on firms that adapt in order to meet them. Moreover, having gained that advantage, they are likely to press for the adoption of higher standards in other parallel or higher jurisdictions – the "California effect."[82]

At the U.S. state level, leading states seek first-mover advantages over other states through renewable energy and energy efficiency policies, which promote their domestic clean technology sectors.[83] Besides helping domestic firms to develop new technologies, take advantage of new markets, and create jobs, officials in leading states gain expertise in making and implementing regulations and an opportunity to influence federal policy in ways that help their states.[84] Early adoption of policies also helps a state to demonstrate the economic feasibility of policies and technologies and to create and expand markets for them, especially if the state is a large one.[85] States might adopt climate policies before the federal government does because state officials are more aware of the possibilities for economic development[86] and because those potentials are concentrated in certain states that already have a critical mass of such firms and the research and development infrastructure that they require.

Sixth, *interest group mobilization* is widely seen as important for the fate of climate policy-making efforts.[87] Interest groups have the informational and organizational resources to monitor and intervene continuously in legislative, regulatory, and judicial processes. Strong environmental organizations are seen as important in cross-national studies of climate policy,[88] of U.S. federal environmental policy,[89] and of U.S. state-level air pollution and energy efficiency policies.[90]

However, business organizations are at least as important in promoting or resisting climate policy. Climate policy making in the United States has been hindered by a strong conservative counter-movement backed by mainstream business organizations.[91] However, in a cross-national study, Falkner argues that the climate policy positions of many sectors of business are not simply determined by their structural positions. There is much diversity in business responses between countries and regions and even within sectors, because businesses are influenced by more specific factors, including their membership in corporate networks and ability to apply technological innovations. This applies especially to firms in the many sectors that are in structurally ambivalent positions concerning climate policy. These include the natural gas industry (because of that fuel's relatively low carbon intensity) and electricity generating firms (which can use alternative energy sources). They also include most manufacturing industry (which can benefit from increased energy efficiency) and the freight, airline, and shipping industries (which are in proximity to retail customers and can be influenced by their preferences).[92] Similarly, Brewer argues that leaders as well as laggards can be found in many business sectors, even in energy-intensive manufacturing, the oil industry, and car manufacturing, because some firms see opportunities for long-term competitive advantages through innovation driven by regulation.[93] At the U.S. national level, other motivations for business are seen. Even some carbon-intensive industries supported emissions trading in 2009 because they wanted to avoid more costly regulation, shape the program to their advantage, gain regulatory certainty and uniformity, improve their images, or respond to public pressure.[94]

Seventh, scholars have emphasized the role of *public opinion*, including the salience of the climate change issue, as a cause of or constraint on policy.[95] Key policy makers have an eye on public opinion as they anticipate the next elections, and they use it to gauge whether supporting climate policy would earn them credit or blame. It is argued that some states lead in adopting climate policies because their publics are unusually aware of, and concerned about, the problem of climate change, as in California and New England.[96] At the same time, writing about public opinion and climate policy is ambivalent about the theoretical role that public opinion plays. Some argue that public opinion is driven by elites and the media, and hence is only an intervening variable within political processes.[97] Others imply that environmental public opinion is a largely autonomous force or responds directly to environmental problems.[98]

Finally, *economic processes* also are held to influence climate policy adoption. High economic growth tends to make strong policy more likely, by giving firms, governments, and households more resources to absorb any short-term costs. The priority given to environmental issues in contrast to economic concerns rises during periods of prosperity, and the reverse occurs during recession.[99] Therefore, the 2008–09 recession made it more difficult to pass federal climate legislation in the United States.[100] However, high unemployment has ambivalent effects. It reduces the perceived importance of environmental issues, but also provides opportunities for framing climate policies in terms of their co-benefits for job creation, especially in the areas of renewable energy and energy efficiency.[101]

Taken together, these processes suggest that the failures of U.S. climate policy making were due to contingent causes, rather than being inevitable results of the unfavorable structural conditions that hold in the United States. Focusing events, such as international negotiations or catastrophic storms, may have been too few in number or political actors may have failed to use them effectively. Certain presidents and congressional leaders clearly lacked adequate commitment to climate policy (e.g., George H. W. Bush, Newt Gingrich, and George W. Bush), and advocates may have chosen the wrong frames in their efforts to sell climate policy. Serious policy efforts may have been too late to capture first-mover advantages, and environmental organizations may have been too divided and elitist to counter the vigor with which the fossil fuel industry responded to the issue. Democrats and moderate Republicans did poorly in important elections, such as the 1994 congressional races and the 2000 presidential contest, which put foes of environmental policy in the driver seat. Public opinion became a drag on, rather than a driver of, climate policy due to the remarkable attack on climate science orchestrated by the conservative counter-movement. Just when national politicians were on the verge of passing significant climate legislation in 1991 and 2009, recessions undercut support. Similarly, these kinds of processes could help to explain the differences in state-level climate policies.

However, the process explanations also have some shortcomings as a basis for a theory of climate policy making. Works relying on them usually lack a broader theoretical framework, and they say little about structural conditions and whether they might affect or determine the processes that they treat.[102*] In the next section, I will offer a more comprehensive theoretical framework incorporating the aforementioned processes, which I will relate to structural conditions in the section on path dependence.

WINDOWS OF OPPORTUNITY AND ADVOCACY COALITION THEORIES

Agenda-setting theory posits that a major constraint on policy making is the limited space available on the *political* or *decision-making agenda*, that is, the continually evolving, brief list of issues that command policy makers' attention at a given time. These are the issues upon which they are actively considering taking action, and a great many issues compete for access to this agenda. According to John Kingdon's multiple-streams theory, an issue is most likely to reach the political agenda when three things occur at the same time: a problem is perceived as important and urgent by the public and elites; viable policy solutions are available; and political commitment to adopt a solution is high.[103]

The convergence of these three elements creates a brief *window of opportunity* or *policy window*, during which an issue has "gotten hot," policy makers are intensely focused on it, and significant policy change is possible, though not inevitable. If the advocates of policy change fail to take advantage of this opportunity, they will have missed their chance until the next window opens.[104] Partial convergence, when only two of the three streams come together, often occurs, but it does not lead to major policy changes. For example, an election may produce officials committed to a solution at a time when the problem does not seem pressing enough, or a problem may reach crisis proportions at a time when political leadership is unwilling to adopt any of the available, vetted solutions to it.[105] Similar to theories of critical junctures,[106] and of punctuated equilibrium,[107] the theory of multiple streams attempts to locate and explain moments of unusual openness, when policies that usually are rather stable may change rapidly.

Therefore, what is crucial in the first instance is the temporal convergence of intense problem perception, viable policy solutions, and strong political commitment. In turn, each of these factors is affected, both favorably and adversely, by streams of events and other processes, which are largely independent of each other. The problem stream and the political stream are largely responsible for shifting the political decision-making agenda and creating a policy window, while policy streams mainly affect the policy alternatives that will get serious consideration.[108] While analyzing the policy stream is important for understanding the specific content of policies, my aim is mainly to explain the overall strength of climate policies adopted, rather than the specific mix of instruments. Hence, I will focus on the problem stream and the political stream and will not systematically examine the origins of policy proposals.

Bracketing out the policy stream will help to simplify what is already a quite multi-faceted comparative analysis of cases in light of contrasting theories, and hence will make it possible to give structural factors adequate attention alongside the process factors. Moreover, I will use the windows of opportunity theory to explain leading state cases and the U.S. federal case (in Chapters 6 through 9), rather than cases of laggard states, where the availability of well-known, acceptable policy solutions is likely to be uneven and hence might be a constraining factor.[109*] Nonetheless, where instances of policy innovation arise in the case chapters, such as the regulation of greenhouse gas emissions from motor vehicles in California and the auctioning rather than giveaway of emissions permits in the RGGI program, I will describe the sources of those important policy ideas.

A *problem stream* is a series of events that affect the conditions which citizens, organized interests, and policy makers will be aware of and perceive as important and urgent.[110] High-visibility problem events often affect the decision-making agenda of government officials by first affecting the *public agenda*, that is, the ever-changing, brief list of issues to which the general public, or potential voters, are paying attention.[111] There are three kinds of events in the problem stream: dramatic information about problematic conditions, extraordinary focusing events, and strong feedback from existing policies.[112] Information about problems can come from routine indicators (such as the reporting of greenhouse gas emissions) or from studies commissioned on an ad hoc basis; the design and interpretation of indicators are important and can be contentious. For their part, *focusing events* are sudden, uncommon events that are harmful to geographically or otherwise concentrated interests and become known to the public and policy makers simultaneously.[113] The focusing events most relevant to the climate and energy policy areas are environmental disasters, such as hurricanes, wildfires, droughts, and oil spills, and energy supply disruptions, such as the 1970s oil crises or the 2000 California electricity crisis. Finally, *policy feedback* occurs when the implementation of existing policies provides information about problematic conditions or highlights policies that do not seem to be working. For example, in implicit climate policy, initial policies had little effect on common air pollution, which suggested that stronger measures were needed.

The recognition, importance, and urgency of a problem also depend on interpretation by policy entrepreneurs and other political actors.[114] Problems can be constructed through comparisons between observed (or predicted) conditions and accepted values, such as when the risks of property damage and loss of life due to climate change are considered. Alternatively, a jurisdiction's current performance can be compared to its own past performance or to what others are doing, such as when greenhouse gas emissions trends are noted or compared. The choice of categories into which to place observations of conditions can also strongly affect interpretation, such as when a hurricane is classified as a result of anthropogenic climate change rather than as a natural disaster.[115]

The *political stream* is just as important as the problem stream. The political stream consists of a series of events that affect the degree of *political commitment* – that is, the extent to which political actors who command large political resources commit them to advancing an issue and a policy course. Events in the political stream are those that affect the power resources and relations of important actors in the policy field. They concern who holds power, their ideological or value-based commitments, and the political constraints they face or anticipate. Therefore, the political stream is affected by elections and other events that result in changes in governmental leadership positions, shifts in public opinion, and mobilization by organized groups.[116] For example, climate policy is likely to rise on the political agenda if Democrats or moderate Republicans are elected, sympathetic elites are appointed to key administrative positions, advocates mount campaigns against global warming, and the public comes to see climate change as a serious problem and proposed solutions as reasonable. When political commitment begins to rise, it may continue to rise because of positive feedback among the actions of elected officials, news coverage, and public opinion.[117]

Although it is not part of Kingdon's theory, in state-level climate policy making it is evident that the problem stream and the political stream sometimes are both driven by actions of the federal government or international organizations (see also Chapter 10). For example, the 1970 Clean Air Act adopted modest air pollution standards that did not rise quickly. Therefore, given economic growth and an increase in polluting activities, the act's low standards were seen as helping to worsen the pollution problem that many states faced; hence, they affected the problem stream. At the same time, the Clean Air Act gave California the right to develop its own standards, which, after the 1977 amendments to the act, could be adopted by other states as well; since these features reduced the constraints on tougher state policies, they are part of the political stream.

In another example, the United Nations' processes affected both streams. Reports by the Intergovernmental Panel on Climate Change provided information and sparked media attention (affecting the problem stream); meanwhile, the adoption of the Kyoto Protocol increased political commitment to federal and state-level policy (affecting the political stream). Bush's rejection of the Protocol in 2001 blocked federal climate policy, and hence created conditions for greenhouse gas emissions to increase (problem stream); but it also created more legal space for state-level policy, created more political space for Republicans to move to his left on the issue, and spurred mobilization by environmental groups (political stream).

In Kingdon's theory, when problem perception and political commitment come together at the same time, their convergence creates a window of opportunity, assuming that at least one acceptable policy solution is already available. But before the window of opportunity is perceived and can be used, the problems, politics, and solutions need to be joined together in a single "package." The agents of this coupling process are *policy entrepreneurs* – policy advocates who possess the skills, relationships, and persistence to "attach problems to their solutions and find politicians receptive to their ideas."[118] Policy entrepreneurs can be elected officials, high-level bureaucrats, interest group representatives, academics, or journalists.[119] When elected officials serve as policy entrepreneurs, they are able to supply their own political commitment, as well.

While the theory of multiple streams is geared toward explaining how issues get onto the agenda and become ripe for action by policy makers, it has less to say about the outcomes of those decisions. To explain the latter, it is helpful to elaborate the political stream with certain concepts from the advocacy coalition framework, which is largely consistent with and complementary to multiple-streams theory.[120*] Among other things,[121*] advocacy coalition theory analyzes the changing balance of power between a coalition of advocates for a policy direction and a coalition of their opponents.[122] Both kinds of coalition can draw on specialists in a variety of governmental and private organizations, including government agencies, parties, parliaments, interest groups, non-profit organizations, social-movement organizations, research institutions, and media outlets, and at times they may also include top government officials.[123]

Status quo energy policies, which aim to ensure relatively low energy prices, are in the short-term interest of many actors, including industry, other businesses, and households, the latter two concerning emissions from buildings and transportation. Therefore, the adoption of climate policies that increase energy prices requires the assembly of a broad advocacy coalition in order to overcome opposition. In state-level and federal climate policy in the U.S., members of advocacy coalitions have been very diverse. They include environmental organizations, elected officials in legislatures and the executive branch (Democrats and moderate Republicans), officials in environmental and energy agencies, health advocates, business groups and labor unions, academics, and journalists.

The opposing coalitions have centered on fossil fuel industries, manufacturing, the car industry, most of the Republican Party, and moderate Democrats. Electric utilities and service-sector businesses sometimes have shifted from opposition to advocacy, or at least to neutrality, as have even some manufacturing and oil companies.[124]

Many events in the problem stream and the political stream influence the composition and political resources of advocacy and opposing coalitions, as well as the constraints on them and on their mobilization.[125*] These events include environmental and energy crises, rapid economic growth or recession, shifts in public opinion, changes in the governing coalition, or policies in other policy areas or jurisdictions. Focusing events signal increased opportunities, and hence encourage simultaneous mobilization by a range of advocates, helping to overcome the free riding problem. Election results that put advocates or sympathizers into high positions have similar effects. Rapid economic growth often worsens environmental problems and creates resources for addressing them, while recessions usually reduce the perceived urgency of environmental issues. Therefore, the problem stream and the political stream affect the balance of power between advocacy and opposing coalitions, and hence the authoritative decisions that officials are likely to reach.[126]

In sum, the policy windows and advocacy coalition theories together can explain the development of climate policy, as shown in Figure 3.1. When the problem stream and the political stream converge, so that problem perception and political commitment are high at the same time, this opens a policy window that policy entrepreneurs may shape and call attention to. The heightened

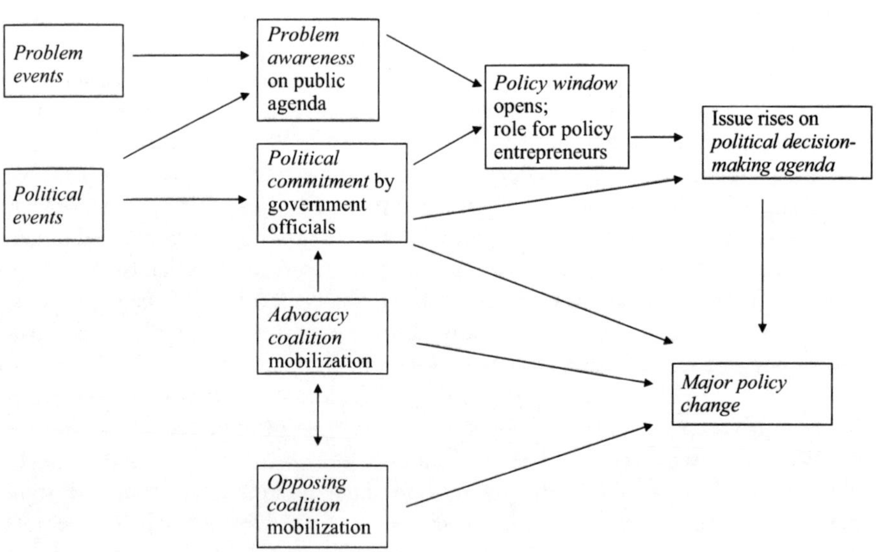

FIGURE 3.1 Policy Windows and Advocacy Coalition Theories

success chances encourage the mobilization of the advocacy coalition. If it is more powerful and mobilizes more energetically and astutely than the opposition coalition, a major policy change is likely to result.

The factors identified by the process explanations of climate policy making, discussed in the previous section, all fit into this theoretical framework. Focusing events and economic growth or recession are important parts of the problem stream, which affect problem awareness and salience. Political leadership by officials can provide both political commitment and policy entrepreneurship, which includes using framing processes to define problems and solutions, link them together, and build broad advocacy coalitions. Leaders' actions may be driven or constrained by public opinion, but also can shape it. Elections and the partisan control of elected office matter because party affiliation correlates strongly with political commitment to environmental issues and policy. But aside from partisanship, elections can result in turnover in the individuals holding key positions, and their individual leanings can sharply raise or lower political commitment. The first-mover advantages that are seen for certain economic sectors help to motivate them to join advocacy coalitions and to link climate policy to the problem of domestic economic development, which is usually highly salient. Finally, interest group mobilization, such as by environmental organizations and segments of business, is simply another term for the activity of advocacy and opposition coalitions.

PATH DEPENDENCE THEORY

My overall aim is to combine structural and process explanations of climate policy, in order to show the contributions and limitations of each. There are several ways in which I will relate structures and processes to policy making in the rest of the book. First, in Chapters 4 and 5, I will show that certain structural conditions promote strong climate policy, although others advanced in the theoretical literature are not supported. The case studies, in Chapters 6 through 9, amplify these findings while putting them in the context of processes; in the case studies, I sometimes also identify additional, ad hoc structural conditions that were favorable. Third, I will analyze the extent to which the problem stream and the political stream are affected by structural conditions, explicitly in Chapter 5 and more implicitly in the process tracing done in the case studies. I will argue that although structures influenced the problem stream and the political stream, the streams also were driven by contingent events – such as international events or the results of close elections – and hence were substantially independent of the structural factors.

In addition, there is a fourth, more fundamental way in which structures and processes combine to influence policy making: through critical junctures and (non-deterministic) *path dependence*.[127] Critical-junctures theory holds that systems and their policy outputs are usually stable (e.g., large fossil fuel industries, high energy consumption, and the policies and interest groups supporting them), but sometimes encounter crises (e.g., the energy crises of the 1970s). A crisis often affects multiple jurisdictions, which might respond differently because of different processes. For example, some states adopt strong energy efficiency and renewable energy policies, while others do not. In some jurisdictions, the responses create legacies – such as energy agencies and new economic sectors – that affect the system for several decades or longer.[128*]

In state-level energy and climate policy making, there have been a series of potential critical junctures, or turning points, that affected all or many states in the postwar period and that provoked varying responses from them.[129*] These junctures were driven by: urban air pollution problems from the 1940s through the 1960s; the eruption of local environmental conflicts and the rise of the national environmental movement in the late 1960s; the energy crises of the 1970s; the national and international rise of the global warming issue in the late 1980s and early '90s; the adoption of the Kyoto Protocol and its rejection by the Senate and the Bush administration in 1997–2001; and the rise in energy prices and resurgence of the climate change issue in 2005–09.

At these critical junctures, crises and conflicts led some states to respond with policy changes (legacies) that affected the chances for stronger climate policy for decades into the future. That is, the outcomes of earlier conflicts created new structural conditions that made later successes more likely. Five kinds of new structural conditions are relevant to climate policy. First, *specialized administrative agencies*, such as CARB and the CEC, have provided information about problems, expertise in interpreting them, proposals grounded in expert knowledge, neutral meeting grounds for stakeholders, and other implementation capacities. Second, *legal authority*, including potential exemptions from federal standards, have permitted the future incremental toughening or broadening of regulatory standards in California and other states.

Third, *new economic sectors*, such as renewable energy and clean technology, have grown with the help of government subsidies. They became important constituencies for the preservation and expansion of energy efficiency and renewable energy programs, as well as testimony that climate policies are economically feasible and could usher in a transition from fossil fuels to a green economy.[130] Fourth, certain environmental policies granted citizens or groups resources that helped them to influence official decision making, such as information, the right to comment on rule making and to litigate, and sites for coalition formation; these *political resources* helped them to overcome their opponents' advantages in other areas, such as structural economic power. Fifth, earlier environmental conflicts created new *environmental organizations*, strengthened old ones, and increased *public support* for environmental protection.[131*]

This is a non-deterministic theory of path dependence. Getting onto the right path in the 1960s and 1970s, as California and New York did, made the adoption of relatively strong climate policies more likely there in the 2000s. However, it did not make such policies inevitable in those states, and path

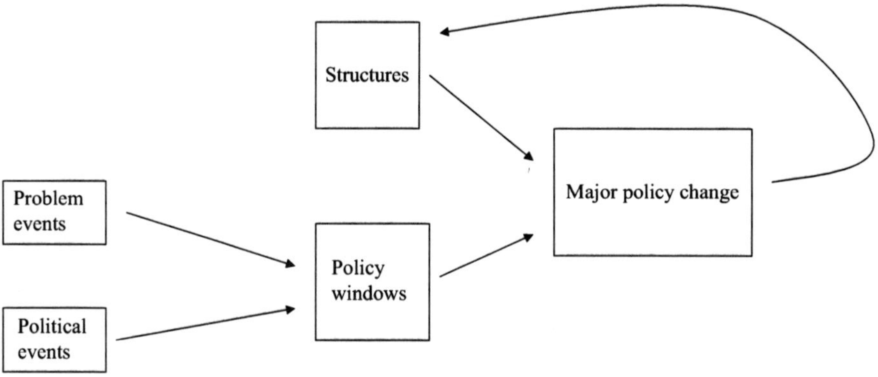

FIGURE 3.2 Structural, Policy Windows, and Path Dependence Theories

dependence does not mean that other states cannot catch up with them, at least to a large extent. Nor does it mean that the leading states' policies are irreversible. The decline of renewable energy policy in California after the early 1980s is a case in point.

In sum, path dependence is one of several ways in which structures and processes can relate to each other. Problem events and political processes create policy windows, during which structures (e.g., political institutions) and processes (e.g., coalition formation) together sometimes produce policy changes. In turn, those policies may have legacies that become new structural conditions, which then influence the next round of major policy making, as shown in Figure 3.2. That round becomes possible when the problem stream and the political stream again converge to produce policy windows.

In sum, through the process theory offered here, I aim to explain agenda setting (through stream convergence and windows of opportunity), to explain policy change and implementation (through the balance between advocacy and opposing coalitions), and to incorporate legacies from past crises and conflicts (through critical junctures and weak path dependence).

TESTING THE THEORIES

In writing this book, I worked recursively between theories and evidence. I first used and tested a range of structural and process theories in developing an explanation of the California case; the longitudinal exploration of that case also led me to an argument about path dependence. Then I developed a larger and more systematic set of contrasting theoretical claims, which I tested in the statistical analysis. I examined the significant variables from that analysis further in the California case study as well as in the New York State and national case studies. Throughout, I focused on explaining the adoption of major climate policies, and as I sought and found variation in outcomes,

I contrasted those cases to episodes in which little or no policy change occurred, or in which policies adopted earlier were reversed.

The order in which I test, illustrate, and elaborate upon the theories in the rest of the book will be somewhat different from that of the research process, however. In Chapter 4, I provide an in-depth analysis of how the main structural explanations help, or fail, to explain federal, California, and New York climate policies. In Chapter 5, I offer a statistical analysis of climate policies in all fifty states, in which I test the ability of a large number of structural and process factors to explain why states vary so much in the strength of their climate policies.

In the three case studies (Chapters 6 through 9), I define a total of 22 episodes of potential policy change (15 major changes, 5 failures to adopt major changes, and 2 reversals). This collection of subcases permits both universalizing comparisons (seeking common causes for all the major policy adoptions), and variation-finding comparisons (contrasting the major reforms with the cases of failure and reversal).[132] In these case studies, I also check for those factors found to be significant in the statistical analysis, using process tracing where possible to see whether and how they contributed to the policy changes. Certain structural conditions make it possible, or more likely, for key processes to unfold; for example, high levels of air pollution helped to produce environmental protests.

But the main task in the case study chapters is to test the windows of opportunity and path dependence theories, applying evidence in a way that makes their theoretical claims falsifiable. First, I use data on public agendas and political agendas to test several hypotheses. One is that public attention on energy and climate problems tends to cluster in response to problem events. Another is that peaks of attention to an issue on the public agenda will precede or occur simultaneously with peaks in the political agenda, if political commitment to the issue is also high. The last is that peaks in the political agenda, in turn, will precede or occur at about the same time as major policy changes, as suggested by Figure 3.1 earlier in this chapter. I measured public attention through counts of newspaper articles and public opinion measures of salience, and measured the political agenda by analyzing legislative hearings, the introduction of legislation, and the content of gubernatorial or presidential speeches.[133*] Second, I carried out process tracing[134] to check for the presence, temporal proximity, and relative importance of the main variables in the windows of opportunity theory outlined in this chapter: problem events, political events, policy entrepreneurs and framing, advocacy and opposing coalitions, policy adoption, and, in instances where it was contentious, policy implementation.

For example, to trace the effects of problem events, I used several kinds of evidence in the case studies. Objective measures of problems – such as smog, oil prices and shortages, and blackouts – were taken from secondary sources, government data, and news reports. I linked them temporally to the public agenda, as measured in newspapers, and to the political agenda, as measured in legislative bills and hearings and executive branch speeches. I used a range of sources to confirm the importance of those problems in policy debates and adoption, including the justifications used by supporters in public debates and in the texts of bills and executive orders, as well as public opinion polls.

In the narrative parts of each case study, I looked not merely for temporal proximity, but also for mechanisms. On crucial or controversial points, I sought multiple sources. I also have tried to avoid the kind of circular reasoning that would infer the presence of a process from the existence of the expected outcome. For example, evidence that key government officials had environmental commitments comes not simply from their actions in office, but from the actions and positions they took prior to taking office and from circumstantial and direct evidence about any strategically motivated shifts while in office. Similarly, the roles of policy entrepreneurs are indicated not simply by their support for a policy change, but also by their public statements mentioning and connecting problems and solutions.

4

The Structural Theory Applied to the Cases

The previous chapter identified a wide range of structural and process explanations for climate policy. To a large extent, the policy windows theory requires testing and support through qualitative, longitudinal case studies, which will be the focus of Chapters 6 through 9. In this chapter, the structural theories will be tested and partially supported, in two steps. First, I will assess four major political-structural barriers to climate policy making for the U.S. national case and then for California and New York State. The purpose is to test the extent to which the weaknesses of U.S. climate policies can be attributed to these barriers, and whether their absence helps to account for the relatively strong policies of those two states. Are the barriers present to a greater degree in federal policy making than in the leading states, and to a greater degree in California than in New York? Since the answer is largely "no," the remainder of the chapter assesses a number of other structural factors in the two state cases, most of which do promote climate policy making there.

STRUCTURAL BARRIERS TO FEDERAL CLIMATE POLICY

The literature on U.S. national climate policy has advanced four main structural arguments for its relative weakness compared with other industrialized democratic countries (see Chapter 3). Each of these has a large degree of validity for explaining U.S. federal policy making, but closer examination shows that the operation of each of them also has depended on the emergence of contingent political processes that heightened their effects.

First, the separation of powers fragments governmental authority and increases the number of veto points, which has allowed opponents of climate policy to block it. The United States has a presidential system with exceptionally fragmented authority, including an unusually strong upper chamber of the national legislature that has strong rights for the minority party through rules on filibusters and private holds on legislation. The independence of Congress from the executive branch and the strong powers of the minority party in the Senate have led to the weakening or blockage of climate legislation, in part because they magnify the power of fossil fuel energy producers. In addition, congressional power is fragmented because Congress is elected through single-member districts and represents a large country that is heterogeneous with regard to income, local economic composition, and climate. This makes it more difficult to assemble winning coalitions than in small, relatively homogeneous countries, such as the Netherlands or the Scandinavian countries. The fragmentation of authority in the United States helped to ensure that Congress whittled down President Carter's ambitious energy conservation plans in 1978, defeated President Clinton's energy tax proposal (the "BTU tax") in 1993, blocked the Kyoto Protocol in 1997, and failed to pass the many cap-and-trade bills proposed since 2003.

However, the effects of the separation of powers on climate policy have not been fixed and have also depended on political processes. The effects have increased as the Republican Party moved to the right on environmental issues, first under Reagan in 1980, then in 1994 through Gingrich's leadership of the congressional Republicans, and then more sharply with the rise of the Tea Party faction within the Republican Party beginning in 2009.[1] Republicans moved to the right largely in response to campaigns by conservative think tanks and media outlets, which denied the existence of anthropogenic climate change and attacked climate policy as too economically costly.[2] In addition, during these three decades, Democrats have moved to the left on climate change and other environmental issues; their shift was influenced by the growing scientific consensus on climate change, the mobilization of environmental organizations, and the decline of their party's conservative Southern wing. Elite polarization can be seen in environmental policy voting in Congress, as measured by the League of Conservation Voters, which shows that the Democrats moved fairly steadily to the left on these issues from 1971 to 2011 while Republicans moved to the right after 1990.[3] Congressional Republicans were instrumental in blocking the efforts to adopt climate policies described earlier, and the George W. Bush administration also blocked greenhouse gas regulation by the EPA and by state governments seeking to adopt the Pavley standards. Public opinion, responding to elite cues, also has become more polarized on climate change since the late 1990s.[4]

At the same time, another process has been reducing the impact of the separation of powers on climate policy. The scope for unilateral executive action has risen strongly in the postwar period, and in climate policy it was strengthened by the Supreme Court's *Massachusetts v. EPA* decision (2007) authorizing that agency to regulate greenhouse gases under the Clean Air Act.[5] As a result of Congress's blocking role, executive action has become more important for climate policy, especially in recent greenhouse gas standards for motor vehicles and power plants, although the executive is also limited in what it can accomplish by itself. For example, an emissions trading program or a national renewable portfolio standard would require legislation, and the courts can block executive action if it strays too far beyond statutory limits and is opposed by state governments or powerful economic interests.

In addition, the separation of powers is strongest as a barrier in periods of divided government, and hence its role has been affected by election results. When the Democrats won control of both Congress and the White House in 1993–95 and again in 2009–11, this opened brief opportunities for climate policy. When Republicans won control of Congress (in 1994 and 2010) or the presidency (in 2000), opportunities narrowed sharply and for longer periods.

Two additional structural barriers to strong national climate policy operate together: the highly pluralist interest group system and a strong domestic fossil fuel industry. Business groups have ready access to the political parties and elected officials in Congress, which is buttressed by a strong role for private campaign finance and expensive, professional lobbying in the U.S. political system.[6] Moreover, the United States has large domestic fossil fuel industries (oil, natural gas, and coal). They spent about $135 million on campaign contributions and $700 million on lobbying in 2010–13. These amounts dwarfed those spent by the alternative energy industry and environmental organizations combined, which totaled $25 million on campaigns and $190 million on lobbying in that period.[7] Fossil fuel and manufacturing industries mobilized to weaken or block climate policy during several windows of opportunity analyzed in Chapter 9, including the 1970s energy policy reforms, the proposed BTU tax in 1993, and President Bush's reversal on greenhouse gas regulation in 2001. Recently, the National Association of Manufacturers and the Chamber of Commerce have opposed emissions trading bills and greenhouse gas regulation by the EPA.[8]

Moreover, fossil fuel interests, the car industry, other manufacturers, and electric utilities supported public campaigns denying climate change and opposing climate policy, beginning with the formation of the Global Climate Coalition in 1989.[9] They funded and publicized the work of a handful of scientists who deny the existence of anthropogenic climate change (*skeptics* or *deniers*). Skeptics were able to strongly influence media and public opinion in the 1990s and early 2000s, and they have had an enduring impact on public opinion.[10] They argue that global warming does not exist, is largely due to natural causes, or would benefit the United States, that climate scientists are divided, and that climate policy would severely harm the economy.[11] During 1988–2003, about 60% of the articles in U.S. newspapers gave equal space and credibility to the views of skeptics and of mainstream scientists, or else reported only the skeptics' viewpoint, although coverage shifted back to reflect the scientific consensus in over 90% of articles beginning in 2004.[12] The skeptics have found a ready audience in congressional Republicans, led by Senator James Inhofe of Oklahoma, who declared scientific claims of climate change to be a hoax.[13] President George W. Bush amplified the skeptics' message by asserting that the findings of climate science were too uncertain to justify vigorous governmental action.[14]

However, the pluralist interest group system also makes possible the activity and influence of a large number of environmental organizations, which increasingly have focused on climate change and have promoted climate policy. By 2010, more than 200 organizations organized in 21 coalitions participated in congressional hearings or United Nations meetings on climate change. Fifteen organizations were responsible for half of the congressional testimony on climate change and two-thirds of the mentions of environmental organizations in *New York Times* articles on this issue during 1980–2010. Those most likely to appear before Congress were the Natural Resources Defense Council (NRDC), Environmental Defense Fund (EDF), Resources for the Future, the Alliance to Save Energy, and the Pew Charitable Trust; those most often mentioned in the press also included Greenpeace, the Sierra Club, the World Resources Institute, and the World Wildlife Federation.[15] Through their campaigns and testimony, environmental organizations have been key actors in influencing elected officials (mostly Democrats) and the public debate on climate change and climate policy.[16]

Another structural factor to consider is the size of domestic fossil fuel production, which makes the U.S. national dependence on fossil fuel imports lower than in most European countries and than in many U.S. states. As a result, the national interest in reducing fossil fuel consumption has been relatively limited.[17] Net fossil fuel imports in the United States (driven by oil imports) are about 25% of total energy consumption, compared with about 60% for the European Union, 55% for California, and 70% to 80% for New York and most other Northeastern states.[18] Nonetheless, import dependence was a salient feature of national energy policy debates in the 1970s and early 1990s, because of oil-supply disruptions and price increases, and again in the late 2000s, due to price increases. The role of the Organization of Petroleum Exporting Countries (OPEC) in influencing international oil prices, the importance of gasoline and fuel oil prices for U.S. consumers and voters, and the military role of the United States in the Middle East combined to give climate policy proponents arguments for using energy efficiency, renewable energy, and carbon markets or regulation to reduce dependence on foreign oil in these periods.[19]

However, fossil fuel import dependence has been a variable, not a fixed feature of U.S. energy politics, and changes in dependence have not correlated with the ups and downs of climate policy making. Due to increasing consumption and the depletion of conventional oil and natural gas reserves, oil imports rose sharply from 36% of consumption in 1985, to 73% in 2006, before declining to 51% of production in 2014.[20] Hence, dependence on oil imports was higher in the mid-2000s, when little climate policy was adopted, than in the late 1970s, in the early 1990s, or during the Obama administration, periods in which major policies were adopted.

Although high import dependence led to a renewed focus on energy security in the 2000s,[21] this did not benefit climate policy at the time, because President Bush and Republicans in Congress favored using the issue to promote domestic fossil fuel development. Hence, the energy policies of the Bush administration, along with rising prices, helped lead to an increase in natural gas and oil production in the United States, including through hydraulic fracturing, especially in Texas, Oklahoma, Pennsylvania, and North Dakota. Together with reduced car driving and consumption, this has led to declining oil-import dependence since 2006, which, ironically, is when a surge in climate policy making began.[22*]

A fourth structural barrier is a political culture or dominant ideology that is skeptical of government regulation and prefers private sector, market-based solutions more than in most other industrialized democracies. Although this structural feature is supposed to act as a brake on environmental regulation, it did not prevent major regulatory laws from being enacted in the late 1960s and early 1970s, when the United States became an international leader in environmental policy.[23] However, the business backlash against regulation that began in the late 1970s[24] arguably put the center of gravity in U.S. politics more clearly on the market-liberal side of the spectrum. This shift helped to stymie the enforcement of air pollution standards, the tightening of fuel economy standards, and later the adoption of new measures such as a carbon tax or cap-and-trade system. As the Republican Party moved farther to the right in 1994 and 2009, adopting uncompromising positions against taxes and most kinds of federal government activity, congressional Republicans began to intensely oppose even market-based climate policy instruments.[25*] Rather than operating as an autonomous structural force that determines the course of policy, the anti-statist political culture of the United States made possible the strong anti-regulatory mobilization by business and the corresponding shifts by the Republican Party.

Similar to the argument about political culture, it has been argued that the nature of public opinion on climate change in the United States helps to explain the weakness of U.S. policies compared with European countries.[26] However, by the mid-2000s, the U.S.-European differences were not very large in surveys that provide comparable data. For example, in the 2005–09 World Values Survey, 57% of respondents in six West European countries with relatively strong climate policies said that climate change is a very serious problem, compared with 47% in the United States.[27] Similarly, the 2006 BBC World Service/Globe Scan survey found that 60% to 75% of respondents in five Western European countries with relatively strong climate policies considered climate change to be a "very serious" problem, but nearly 50% in the U.S. also said this. On the question of whether "human activity is a significant cause of climate change," the West European countries registered 80% to 90% agreement and the United States 70%.[28]

Moreover, public opinion in the United States is broadly supportive of the scientific consensus, with majority support on most questions during the last 10–15 years.[29*] Public opinion on the issue has been fairly stable since the late 1990s, except for a moderate decline in support during 2008–11 and a partial rebound to previous levels since then.[30] Over a number of years, an average of 68% say that climate change is happening already or that it will in their lifetimes;[31*] however, 60% to 65% disagree that global warming "will pose a serious threat to you or your way of life in your lifetime."[32] An average of 56% of those surveyed say that observed warming is more due to human than natural causes,[33] and since 2007, 50% say global warming is a proven fact and mostly caused by cars and industry.[34] However, when respondents were asked about warming and its causes separately, only 41% said there is solid evidence for warming *and* that it is mainly anthropogenic.[35] An average of 71% say that climate change is a very or somewhat serious problem,[36] while 59% say they worry a great deal or a fair amount about it.[37] However, respondents do not give climate change a high priority compared with other issues; for example, only 28% to 35% thought global warming should be a top priority for the Obama administration, lower than 21 other issues that they were asked about.[38]

Furthermore, there is strong support for climate policy in general.[39] Even in 2010, near the nadir of public support, 71% of respondents (and even 55% of Republicans) thought that the government should regulate greenhouse gas emissions.[40] Respondents also were strongly in favor of state action in 2008–09. Three-quarters favored states being allowed to adopt stricter greenhouse gas emission standards than the federal government, and 60% to 66% believed that their state should adopt climate policies even if neighboring states or the federal government failed to act.[41]

Public opinion also very strongly supports a variety of specific climate policies, if they involve regulation or tax credits rather than tax increases. The support often cuts across partisan divisions and demographic categories. This probably reflects the co-benefits for energy savings, job creation, and air pollution reduction associated with these policies; the co-benefits evidently are more popular than action purely to prevent global warming. In 2013–14, between 60% and 70% thought the federal government should regulate greenhouse gas emissions from power plants.[42] In transportation, an average of 72% favored government increasing emission standards for automobiles, and 79% favored the same for industry.[43] Increased spending on subway, bus, and rail systems is supported by 66% of respondents.[44] An average of 81% strongly or somewhat supported states increasing fuel efficiency standards for cars, even if these would increase vehicle costs, and 76% strongly or somewhat supported state RPSs, including 67% of Republicans.[45] Hybrid vehicle tax credits were favored by 75%.[46] In 2012, 73% of respondents, including 60% of Republicans, supported increasing automobile fuel economy to the same extent that the Obama administration had announced in 2011. Support dropped only slightly, to 63%, if a 5% increase in new car prices was included in the question.[47]

Indeed, many studies show that people are willing to pay something for climate policy. However, surveys have found that linking renewable energy policies or a cap-and-trade program to specific costs for individuals sharply reduces support for those measures; support also declined with the onset of the 2008–09 recession.[48] Nonetheless, in a review of 27 studies, the median finding was that the average household is willing to pay $135 a year for climate policy.[49]

In energy policy, the public favors the development of renewable energy much more than fossil fuels or nuclear power. In 2011-13, respondents preferred the development of alternative energy sources such as wind and solar rather than oil, coal, and natural gas, by an average margin of 61% to 30%.[50] Since 2001, an average of 77% favored increased government spending more for wind and solar energy.[51] More specifically, majorities in 2013 thought that solar (76%), wind (71%), and natural gas (65%) should get more emphasis, much higher than the shares who favored this for oil (46%), nuclear power (37%), and coal (31%).[52] An average of 77%, including 58% of Republicans, supported a national RPS in 2012.[53] Energy conservation also has been popular in recent years, with an average margin of 53% to 36% favoring it instead of more energy production.[54] There was also extremely high support for tax deductions for residential energy efficiency (88%) and for companies investing in renewable energy (88%), in 2009 polls.[55]

Cap and trade was less popular than these measures, with only 55% supporting it strongly or somewhat in 2008, and support dropped to 39% in 2009; however, a different set of polls found support at about 51% in 2009–10, "even if it may mean higher energy prices."[56] Finally, tax increases are very unpopular, with 61% opposing them for fossil fuels and 75% for gasoline in 2012. A narrower margin of 50% to 44% opposed a carbon tax that would cost $15 per month per individual in 2009.[57]

In sum, public opinion has not been a strong, structural barrier to U.S. climate policy making. Public support for climate science and climate policy, when framed in general terms, has been strong enough to give political leaders substantial scope for climate policy making. At the same time, as will be shown in the public agenda section of Chapter 9, the generally low importance that the public places on global warming means that public opinion has not been a reliable, independent source of political support, either. Finally, although it is fairly stable, public opinion on climate change and climate policies also has responded to political and economic events, making it an intervening variable that will be included in the case studies.

The nature of public opinion gives policy makers the ability to frame climate policies in politically acceptable terms. That is, politicians are more likely to gain public acceptance if they seek regulatory and financial incentive policies rather than new taxes, and if they emphasize co-benefits rather than focusing solely on addressing climate change. Moreover, the studies on the public's willingness to pay suggest that, contrary to conventional political wisdom, politicians can call for climate policies that require financial sacrifices – if the sacrifices are small ones. Households are currently willing to bear some of the costs of climate policy, and if they gain experience with doing so, they might become willing to bear larger costs in the future.

THE STRUCTURAL BARRIERS IN CALIFORNIA AND NEW YORK

While structural barriers go a long way toward explaining the relative weakness of U.S. climate policy, the successes of California and New York State are not due to the absence of those barriers. Indeed, they have operated in California to about the same degree as they have nationally, and to a greater degree in California than in New York, even though the former has substantially stronger climate policies than the latter.

According to the structural theory, when governmental authority is unified and hence fewer veto points are available to opponents, strong climate and other environmental policies are more likely (see Chapter 3). However, neither California nor New York has a political system with concentrated governmental authority. Indeed, California has a governmental system that is even more fragmented than those in most other U.S. states. First, California has a highly plural rather than a unitary executive, currently with ten statewide elected offices, including the governor, lieutenant governor, treasurer, controller, attorney general, secretary of state, insurance commissioner, and members of the Board of Equalization (which oversees tax policy). The officials in these offices are frequently from different parties, often conflict over policy, and sometimes litigate against each other.[58] In New York, by contrast, executive power is more clearly concentrated with the governor, who shares the stage only with the lieutenant governor, attorney general, and state comptroller, a smaller number than in most states.

In several other ways, California's governors have average institutional powers, while New York's are more powerful than average. Governors are limited to two four-year terms in California, but there is no term limit in New York. The governor's budgetary powers are co-equal with the legislature in California, but New York's governor has a line-item veto. As a result, New York ranked first among all states in the formal power of the governor in the late 1990s, scoring 4.3 points on a 1 to 5 scale; only five other states scored at least 4 points. California scored 3.2 points, slightly below average, in this index.[59]

Moreover, as in all other states except Nebraska, the executives in California and New York share power with a bicameral legislature. The potential for conflict and deadlock between governor and legislature is, if anything, even greater in these states than in the average U.S. state, since the California and New York legislatures have unusually high capacities. Along with only eight other states, they both have full-time legislatures. In the early 2000s, they ranked in the top three states in legislative salaries and the size of permanent legislative staff, and in Squire's legislative professionalism index, they ranked first and second in the country, respectively.[60]

In addition, a strong role for direct democratic procedures in California limits the executive, by making the governor subject to recall, and also diffuses the legislative power. Ballot initiatives (since 1911) and referenda are important methods of policy making and policy blockage in California, with about a dozen measures on the ballot in an average year since 1990. New York is much more restrictive with direct democratic procedures, allowing only legislatively referred constitutional amendments and averaging only two ballot measures per year since 1996.[61] In short, if fragmented government is a recipe for inaction on climate and other environmental policies, then California should be a laggard rather than a leader in these areas. Similarly, New York, though showing some signs of concentrated gubernatorial authority, still has a system where the governor must contend with other statewide executive elected officials and a highly professional legislature.

The pluralist-neocorporatist continuum is another institutional feature that is prominent in structural accounts. The U.S. national political system ranks quite low in neocorporatism, whether in the context of economic or environmental policy making.[62] California and New York State show no signs of departing from the U.S. model. Their interest groups are multiple and competitive, engage heavily in electoral politics through campaign finance, and lobby the legislature. In California, interest groups write more than 60% of the bills introduced in the legislature and are often in adversarial relationships with government, including litigation.[63] In New York, interest groups are very numerous and spend much on lobbying; 3,600 registered groups spent $120 million in 2003. However, they are mostly defensive in their lobbying and initiate only an estimated 20% of bills.[64] Accordingly, Thomas and Hrebenar classified California as a system with stronger power for interest groups than New York in the late 1990s.[65]

The rampant pluralism of California politics is evident in the climate policy arena, where public utilities and the oil and gas industries were the third and fourth largest contributors to electoral campaigns in 2008, respectively, totaling $60 million in spending.[66] Oil companies, car dealers, and conservative talk shows fought fiercely against the Pavley bill.[67] Interest groups are also very active in direct democratic politics. The oil and gas industry spent $100 million in 2006 to defeat Proposition 87, which would have taxed oil profits to support alternative energy projects. Pacific Gas and Electric spent $46 million in 2010 in a losing effort to pass Proposition 16, which would have required a two-thirds vote of ratepayers before a local government could create an agency to buy renewable-source electricity.[68] The Global Warming Solutions Act was targeted by the oil industry, which spent over $10 million in support of Proposition 23, an attempt to overturn that climate law in 2010.[69] In addition, a variety of businesses were active in supporting the Global Warming Solutions Act and other climate legislation, as detailed in Chapter 7. With New York State's climate policy pursued mainly through executive rather than legislative action, the role of interest groups has been less visible, but still important, in both supporting and opposing gubernatorial initiatives (as described in Chapter 8).

Another institutional barrier to consider is an electoral system that marginalizes green parties. The cross-national literature finds that green parties are important drivers of climate and other environmental policies, and hence that their shares of the vote and of parliamentary seats are consistent correlates of such policies. But this factor is essentially absent in the United States. Nationally, the green party vote share has averaged 0.8% in presidential elections since 1996 and about 0.5% in congressional elections, which could help to explain the weakness of U.S. climate policy. But California and New York have equally weak green parties, which is not surprising, given their single-member-district, plurality electoral systems. In California, the Greens averaged only 2.6% of the vote in the last five governor's races and the party has elected only one member of the state legislature since its founding in 1991. In New York, the Green party has been on the gubernatorial ballot since 1998, but it has averaged only 1.1% of the vote since then.[70]

The energy sector is another structural feature that has been related to climate policy. Indeed, the earliest and strongest state-level climate policies mostly have been in states that are not major producers of fossil fuels, in the Northeast and on the West Coast.[71] New York fits this pattern well. While the state has a small amount of natural gas production, which increased after 2000, it ranked only 23rd in the country in that activity in 2012, and it has no coal mining and virtually no oil production.[72] Hence, New York State produced only 1.0 million British Thermal Units (BTUs) of fossil fuel energy per person in 2000, compared with 174 million BTUs/person produced in the United States as a whole. New York ranked 33rd among the states, but even this overstates its fossil fuel production, which was negligible.

However, California is an important exception to the generally negative relationship between fossil fuel production and climate policy, since it is a major fossil fuel producer, with large oil producing and refining industries.[73*] The state had 16% of U.S. crude oil reserves, with 3,389 million barrels estimated, in 2006. In both 1981 and 2005, California had the third largest oil production among U.S. states, behind Alaska and Texas. It has produced about 12% of U.S. oil over the last three decades, basically the same as its share of the U.S. population. Refineries are also major economic actors in California; 14% of the nation's refinery capacity was located there in 1982, a figure that fell only slightly to 11% in 2011. The state's total fossil fuel production was 58 million BTUs/person in 2000, ranking it 22nd in the nation.[74] California's fossil fuel industries are large enough to form formidable political actors, as noted in the previous discussion of pluralist interest group politics in the state. Therefore, the strength of California climate policy is evidence against the argument that large fossil fuel industries can block vigorous climate policy through their lobbying strength or structural power.[75]

However, despite its role as a producer, California is a net importer of fossil fuel energy. It consumes more petroleum than it produces, and its net imports of fossil fuels equaled 49% of its total energy consumption in 1995–2000 and 56% in 2010. A state's reliance on fossil fuel imports may promote climate policy by making investments in conservation or in domestic renewable energy sources more attractive. But other states with relatively strong climate policies, such as New York (78%) and Massachusetts (83%), have been much more dependent on such imports than California. Some degree of dependence on fossil fuel imports may be a necessary condition for strong climate policies, but this is a condition that is present in many states: 32 of them were at least 50% dependent on imports in 1995–2000. [76] Hence, it cannot explain why California's policies are stronger than those of every other state.

The final structural barrier to consider is the United States' political culture, which generally favors market forces over government action. Indeed, this barrier is lower in California and New York, whose citizens are more "liberal" (i.e., more supportive of government intervention) than in the average U.S. state. The Berry index of citizen ideology, which is derived from election results for congressional candidates, shows that California's voters were more liberal than the U.S. average in 1967, 1974, and 2005, dates chosen because they marked the adoption of major air pollution, energy, and climate policy in that state. However, California is not extraordinarily liberal when compared with the United States as a whole, it is less liberal than New York, and its relative liberalness has been declining. California ranked only between 16th and 21st out of 50 states during the above-mentioned years, with scores about 25% higher than the national average in 1967 but only about 10% higher in 2005. New York is actually more liberal than California. It ranked between 5th and 11th among all states during those years, although its scores also declined relative to the national average, moving from 61% above average in 1967 to 27% above average in 2005.[77]

Similarly, California and New York rank relatively high in environmental public opinion and in congressional voting records on environmental issues. California ranked 12th among U.S. states in environmental public opinion during 1973–96, and 14th in terms of the environmental voting records of its congressional delegation, as scored by the League of Conservation Voters during 1970–90. But here too, New York, as for the other ideology measures, ranked much higher than California: 3rd in public opinion and 8th in congressional voting on environmental issues.[78]

Moreover, greater liberalism in general does not always directly translate into greater public support for, or interest in, climate policy. In 1998–2001, just before California's first burst of explicit climate policy adoption, its residents were actually less supportive of environmental policy than a national sample. Moreover, they were only about as likely as a national sample to assess global warming as a serious threat or to favor federal regulation of greenhouse gas emissions.[79*] In California, public interest in environmental policy, and especially in climate policy, was very low in 1999–2002, with less than 1% naming global warming the most important problem in the state. Public interest rose only after the adoption of strong climate policies in the next few years.[80] Public opinion in California was not an autonomous force driving climate policy making; rather, it initially responded to policy actors, though it later influenced them (see Chapter 7).

STRUCTURAL FACTORS PROMOTING CLIMATE POLICY IN CALIFORNIA AND NEW YORK

It appears that the climate policy successes of California and New York State are not due to the absence of the main structural barriers found at the federal level. Moreover, differences between the two states in those four structural features would predict stronger policy in New York than in California, and hence cannot explain California's pre-eminent role in climate policy making. However, several other structural factors, which tend to promote climate policy, could help to explain their relative successes.

According to the literatures reviewed in Chapter 3, larger administrative capacities make strong climate policy more likely. What is most relevant to climate policies is government personnel rather than revenue, since most climate policies involve regulation, taxation, or tax incentives rather than heavy spending. In terms of state employees per thousand population, California and New York had lower administrative capacity (10.4 and 13.2, respectively) than the national average (14.5) in 2000.[81]

However, California and New York rank 1st and 3rd in the country in the total number of government workers, with 355,000 and 251,000 full-time-equivalent employees that year, respectively. There are several ways in which absolute administrative capacity could affect climate policy. A state with a large absolute number of personnel may be able to more easily, quickly, and reliably set up new functions than smaller states can. A large state may already possess agencies that are able to undertake the complex studies, regulation writing, and administrative oversight involved in climate policies. Moreover, such a state may also be in a good position to create new agencies to implement and develop air pollution, energy efficiency, renewable energy, and explicit climate policies.[82] Furthermore, past experiences with agency creation or expansion may give legislators and executive branch officials the confidence to create new policies that require new or expanded agencies.

Historically, state governments have regulated energy and areas related to it, such as economic development, land-use planning, and disaster planning.[83] Hence, in the 1990s, many states, including California and New York, already possessed relatively strong administrative capacities to regulate energy production and activity in other climate policy areas. In California, CARB, the CPUC, and the CEC have been well-staffed, independent, science-based agencies that have been very important in developing and implementing climate policies (see Chapters 6 and 7). Similarly, in New York, the State Energy Office (until 1996), the Public Service Commission, and the New York State Energy Research Development Agency have been large agencies involved in energy and climate policy (see Chapter 8). In the case study chapters, I will argue that these agencies are not simply structural features that promoted the climate policies of the 2000s, but also are the products of earlier political processes that created the first rounds of climate policy.

The literatures also posit that high incomes and high education levels promote environmental policies by providing greater private and public capacities for bearing the costs of those policies and by creating greater awareness of environmental problems. California and New York are indeed above average in personal income per capita, although California is not near the top of the state ranking and its advantage has been shrinking. In 1970 and 2000, it ranked 7th and 8th, with incomes 18% and 9% above the national averages, respectively. New York ranked slightly higher, at 4th or 5th place, about 18% above the national average. The picture for education is similar. In 2000, California's higher education index, which combines bachelor and advanced degrees (see Chapter 5), placed it in 11th place, while New York was in 8th place among all states.[84]

Another possible explanation is that their economic compositions predispose California and New York to climate policy. Some have cited a relative lack of heavy industry, including the weakness of the car industry, as a factor making California's climate policies possible.[85] However, in terms of manufacturing, the differences between California and the rest of the country are not very large. Manufacturing industries make up the same share of California's economy as in the United States as a whole: 9% of the workforce and 12% of the gross state product in 2011.[86] The state's share of heavy industry (15%) was only slightly below the U.S. average (17%) in 2000. In 2006, the second and third largest manufacturing industries in California were traditional, high-pollution industries: petroleum refining ($33 billion) and chemicals ($19 billion).[87]

In addition, California has substantial energy-intensive industries, including chemicals, glass, and forest products.[88] Greenhouse gas emissions from industry made up essentially the same share of California's total emissions (20%) as they did in the United States as a whole (19%) in 2005.[89] However, California's share of highly polluting industries (3.8% of GDP) is well below the U.S. average (6.3% of GDP).[90] As with several other factors, while California is fairly close to the national average in economic composition, New York has features that should make strong climate policy more likely there: a manufacturing share of employment 28% below the national average; and heavy industry and polluting industry shares of GDP about half the national average in 2000.[91]

Another possibility is that a large technical potential for renewable energy is important for climate policy. States with the largest renewable energy potentials may also have the sources that are most economically viable, and there is evidence that wind power development depends in part on wind potential.[92] Hence, the relative size of a state's renewable potential could affect its willingness to adopt vigorous renewable energy and climate policies.[93] However, in wind power, California and New York are not exceptional, ranking 19th and 21st in the U.S., respectively; in fact, 12 states have wind potentials that are more than ten times as large as California's. However, in rooftop solar potential, California is ranked 1st in absolute capacity, and in urban utility-scale solar potential, it is 2nd (behind Texas). Even when calculated as a share of retail electricity sales in 2010, California's ranks for these solar potentials are

TABLE 4.1 *Comparison of the California and New York Cases Concerning the Structural Hypotheses*

Factor	California	New York State
Political institutions and structures		
Governmental authority	Highly fragmented	Fragmented
Interest group system	Highly pluralist	Pluralist
Electoral system and green parties	Majoritarian; very weak green parties	Majoritarian; very weak green parties
Political culture	Moderately liberal (i.e., favoring government intervention)*	Highly liberal**
Administrative capacities	Small relative size Large absolute size** Large specialized capacities**	Medium relative size Large absolute size** Large specialized capacities**
Socioeconomic structures		
Fossil fuel industry size	Large	Small**
Fossil fuel import dependence	Moderate*	High**
Income per capita	High**	High**
Education	High**	High**
Manufacturing industry share	Medium	Low**
Heavy industry share	Medium	Low**
Polluting industry share	Low**	Low**
Renewable energy potential	Medium* to High**	Medium*
Total number of factors that are:		
Strongly supportive	5 to 6	10
Weakly supportive	2 to 3	1
Not supportive	7	4

Note: ** = provides strong support; * = provides weak support.

still 1st and 6th, respectively.[94] Therefore, physical potential could help to explain California's exceptional solar energy policies, but not its exceptional policies on wind power. In 2012, the state generated more than three times as much power from wind as from solar energy.[95]

Table 4.1 summarizes this discussion of institutional and socioeconomic-structural factors in California and New York. Several patterns are evident. First, basic political institutions do poorly in explaining the strength of climate policies. Governmental authority is fragmented, interest group systems are pluralist, and green parties are insignificant in both states. On these factors, California and (to a lesser degree) New York seem to share with the federal government the profile of a likely climate policy laggard. However, two political-structural variables do point toward strong climate policy in both states: their relatively liberal political cultures and their large absolute administrative capacities.

Second, several socioeconomic factors are also plausible explanations of their strong climate policies, with income and education high, and polluting industries low, in both states. New York also has advantages in a weak fossil fuel industry, high fossil fuel import dependence, and a low manufacturing industry share of GDP. Third, New York appears to be more disposed to climate policy making than California is, on both sets of variables. For New York, 10 of the 15 factors point clearly toward climate policy, while in California, only 5 or 6 of them do so. Since California's climate policies are more rigorous than New York's, this suggests some limitations of the structural theories.

In sum, the strongest support for the structural theory comes from the factors that are favorable for climate policy making and are clearly present in both California and New York: high absolute administrative capacities, liberal citizen ideology, high incomes and education, and the relative absence of polluting industries. Also, another favorable factor is present to some degree in both states, fossil fuel import dependence. However, even these factors do not distinguish California from many other states in the Northeast and on the West Coast, many of which have relatively strong climate policies but are far from California in the scope and ambitiousness of their interventions. While California is above average on these dimensions, many other states have even higher incomes or education levels, more liberal citizens, and more dependence on fossil fuel imports. Hence, these may be necessary conditions, but they are not sufficient for strong climate policies. These, and the other factors considered in this chapter, will be explored further in a statistical analysis of climate policies in all fifty states, in the next chapter.

5

Causes of Climate Policies in the Fifty States

In this chapter, I present the results of a multiple regression analysis of climate policies in the fifty U.S. states. States vary greatly in the extent to which they have developed climate policies, and the theoretical literatures examined in Chapter 3 suggest many hypotheses to explain these differences. A multiple regression analysis can help to test them as well as to assess more generally the relative contributions of structural and process factors, through a systematic comparison of all the U.S. states.

While this approach has advantages, there are also several limitations to the analysis presented here. First, it cannot be used to test many of the factors in the process theories, such as focusing events or framing, because comparable, meaningful data on those process variables are not available for all the states. In addition, because the climate policy measure used as the dependent variable in this analysis is available for only one limited period (2008–14), the analysis must be cross-sectional and hence cannot take into account the timing of policy adoption in different states. A pooled cross-sectional time-series approach is not viable because there is too little information available about how the dependent variable has changed over time.[1*] A longitudinal approach to climate policy adoption will be provided in Chapters 6 through 9.

Second, the index of climate policies is based on a simple count of whether policies have been adopted and does not distinguish between policies of different stringency, ambitiousness, or degrees of implementation. This misses important differences between states, for example, in the ambitiousness of their renewable portfolio standards and the size of their public benefits funds and utility programs.[2] In particular, the unique nature of California's emissions reduction targets – which are legally binding and cover 85% of the economy – is not reflected in the index, which assesses California's climate policies as being roughly as strong as those in Massachusetts, Connecticut, New York, and several other states. Hence, as a tool for understanding differences between state policies, it is necessary to supplement the index with more detailed, qualitative comparisons such as those in Chapter 2.

THE DEPENDENT VARIABLE: THE CCES CLIMATE POLICY INDEX

To measure the strength of state-level climate policies, I use an index published by the Center for Climate and Energy Solutions (CCES) and updated in February 2014,[3*] with some modifications. The CCES index is a count of the number of policies that are in operation in each state.[4*] While it is not perfect, the index is quite adequate for this analysis because it captures the main climate policies and the main differences between states.[5*] The CCES index includes cross-sectoral policies such as emissions reduction goals, electricity policies such as renewable portfolio standards, transportation policies such as low-carbon fuel standards, and buildings policies such as commercial and residential construction codes.[6*] The policies included are, in almost all instances, important state-level policies that can reasonably be expected to tend to reduce greenhouse gas emissions, as discussed in Chapter 2. Moreover, all the policies, with minor exceptions noted in the following, are of roughly equal importance, so the index can be interpreted as an interval-level variable. This means, for example, that two states which are rated as having 10 and 15 policies, respectively, are about the same distance apart in climate policy strength as two states that have 15 and 20 policies, respectively. Although it is based on the simple presence or absence of a policy in each state, the inclusion of 25 different policies means that the index differentiates reasonably well between strong, middling, and weak climate policy states. The index values for particular states agree well with the few available comparative studies of states' climate policies.[7]

The CCES index largely reflects climate policy making since 2000. The vast majority of state-level climate policies in the index were adopted between 2001 and 2014, with a midpoint of about 2006.[8*] Of the 25 policies in the modified CCES index, 22 are explicit climate policies that states began to adopt after the mid-1990s and which only a handful had adopted before 2000. Even for the three policy areas[9*] that predate the 1990s, the CCES index counts only the adoption of recent policies.

I have modified the CCES index in several ways. First, I excluded two policies that are not related to greenhouse gas emissions reductions (climate change adaptation plans and fracking chemical disclosure regulations) and averaged the scores for five different carbon capture-and-storage items. By reducing that nascent policy area to one item, I gave it an importance equal to the other policies in the index. Second, in the unmodified CCES index, each policy is weighted equally, which among other things gives disproportionate weight to the electricity sector. Because greenhouse gas emissions in the United States are distributed roughly equally across the electricity, transportation, and buildings sectors,[10*] and because states have been about equally active in each of them, I weighted those three sectors equally. Since cross-sectoral policies such as targets or emissions trading are important in reinforcing and guiding other climate policies, I gave cross-sectoral policies a weight equal to that of each of the three main sectors.

That is, I weighted the index to give equal importance to its four main categories (cross-sectoral, electricity, transportation, and buildings), while keeping the policies within each sector weighted equally in relation to each other.[11*] Actually, this weighting process made almost no difference in the relative scores of the states. The 2014 weighted index correlates extremely highly with the unweighted index (R = 0.998), and regression analysis with the unweighted index as the dependent variable produced nearly identical results to those reported in the main analyses in this chapter; the results with the unweighted index are in Table 5.4, in the Appendix to this chapter. Weighted scores for all the states are given in Figure 2.1, in Chapter 2; the index has an approximately normal distribution.[12*] As that figure also shows, the states with high levels of climate policy are concentrated in the Northeast and on the West Coast, while those with the lowest scores are mainly in the South and Great Plains.

THE INDEPENDENT VARIABLES

When defining independent variables, I aimed to find the most appropriate measure to represent each of the main theoretical factors in Chapter 3. I was able to locate or construct reasonably valid measures for all the major institutional and social structural factors and also for several processes advanced by the literature. Variables were measured at times that accorded with theoretical expectations, subject to data availability.[13] I assumed that structural factors would produce political effects with a lag of at least several years,[14*] so I measured the structural variables before the main period of climate policy adoption. By contrast, I assumed that process variables operate with little or no lag, and hence measured them at a time point that was during policy adoption. Since the midpoint for the adoption of the state-level climate policies in the CCES index was about 2006 (see note 8), the structural variables were measured for the year 2000 or as close to it as possible, while the process variables were measured as close to 2005 as possible. Where feasible, I used period rather than point measurement in order to smooth out annual fluctuations.

Political Institutions and Structures

There are five variables representing the main political institutions and other political structures described in Chapter 3; all variables are summarized in Table 5.3, in the Appendix to this chapter. The first variable, the governor's institutional powers, addresses two views about the effects of a concentration of governmental authority. One view holds that concentrated authority makes strong government intervention possible by reducing the number of veto points and increasing coordination among governmental actors. The contrary view is that the separation of powers provides multiple pathways for possible reform and may lead to a competitive bidding up, or a locking in, of environmental standards. I will test hypotheses reflecting both these perspectives by using two-tailed significance tests. The independent variable is an index of the institutional powers of governors in the late 1990s. This is an average of five sub-indices based on the number of elected executive officials, the governor's potential number of terms in office, and his or her appointment, budget, and veto powers in relation to state legislatures.[15] It varies from 1 (low) to 5 (high institutional power).[16*]

The second institution to be considered is the nature of the interest group system. Neocorporatism is held to promote climate policy by reducing the number of potential veto points and the risk of backlash against environmental policies. By contrast, pluralist systems make it more likely that interests which would be harmed by climate policy will be able to block its adoption or implementation. The U.S. states have interest group systems that are much closer to the pluralist than to the neocorporatist end of the continuum, but there are important differences across states in the degree to which interest groups influence or dominate policy making. The variable used to measure the impact of pluralist interest group systems is the Thomas-Hrebenar five-point scale, which measures the power of interest groups relative to other political actors. It is based on surveys of political practitioners and observers in each state; the version used here was measured in 1998.[17] On this scale, interest groups may be dominant (as they were in 5 states), dominant/complementary (25 states), complementary (16 states), complementary/subordinate (4 states), or subordinate (no states). I translated these categories into a 1 to 5 scale, with dominant interest group states scoring "5."[18*]

The third variable concerns the openness of the electoral system, since this would favor green parties or other third-party challenges to the two major parties in the U.S. states. Green electoral strength promotes environmental policy by getting politicians with environmental commitments into office and by influencing other parties to adopt environmental positions in response to electoral competition. While green parties have negligible statewide electoral strength in almost all U.S. states, the potential for third parties to gain substantial vote shares might still make the two major parties more open to environmental interests. That potential depends largely on an underlying structural feature, the openness of the electoral system. Plurality elections in single-member districts are the norm in state-level as well as national elections in the United States. However, states vary greatly in the costs of ballot access for third parties, such as the number of signatures required to get onto the ballot and the size of filing fees. As a proxy for these complex institutional differences in the openness of the electoral system, I have used a measure of the average number of third-party candidates on the ballot in gubernatorial elections during 1980–2005.[19] The unit of measurement is candidates per election.[20*]

Fourth, large governmental capacities (or "state capacities") make vigorous intervention in the environment more likely, since government personnel and funds are needed to develop and implement policies. Here, governmental capacities are measured in two ways. As an indicator of *administrative* capacity, I use the absolute number of state government employees (full-time equivalent) in 2000, with data from the U.S. Census. The measure is employees rather than state spending because climate policies are mainly regulatory and often highly technical, so large specialized staffs are likely to be more important than large budgets. Theoretically, a state with a large number of personnel has a necessary critical mass or excess capacity that can be used for new kinds of policy initiatives, as discussed in the last section of Chapter 4.[21*]

A second measure of governmental capacity concerns state legislatures, whose specialized committees and staffs may be important sources of climate policy making capacity. *Legislative* capacities vary greatly across the states; for example, only ten states currently have full-time legislatures.[22] To capture these differences, I use Squire's index of legislative professionalism, which gives equal weight to legislative salaries, days in session per year, and staff, expressed as a ratio to the corresponding numbers for the U.S. Congress. I have taken the average of the Squire indices for 1996 and 2003.

Fifth, the literature holds that climate policies are more likely if the dominant ideology or political culture favors government intervention, in contrast with free-market ideology. The degree to which the public supports government intervention is indicated by the index of state-level citizen ideology developed and updated by Berry and colleagues.[23] This rather complex index is based on election results and congressional roll call votes scored by two liberal groups, Americans for Democratic Action and the AFL-CIO's Committee on Political Education. Those scores are used to assess the ideology of winning and losing candidates for congressional elections in each district in the House of Representatives. Citizen ideology is assessed by weighting the popular votes for those candidates by the candidates' scores, or, in the case of challengers, by the average score for congressional representatives of that party in that state.[24]

Hence, this index measures ideology on a conservative to liberal (0 to 100) scale for the active electorate, where liberal means supporting government intervention to increase socioeconomic equality and to pursue other goals in a wide range of policy areas, including environmental protection. Unlike government ideology (described later in this chapter, under political processes), citizen ideology is a structural variable, which measures largely stable differences among states rather than change over time. Although there are year-to-year changes in the index, the rank orderings of the states change relatively little, and if the averages of five-year periods are compared, the index appears quite stable. Pearson's R correlations among the index in adjacent five-year periods from 1990 to 2009 are all above 0.93, and the correlation across the 1990–94 and 2005–09 periods is 0.90. Granted, this index shows enough variation over time to be used in longitudinal studies to explain the timing of policy adoption in particular states.[25] However, the citizen ideology index is also stable enough to be used in this cross-sectional study as a measure of political culture as a structural variable; the strong correlations noted here show that it varies across states in a manner that is relatively fixed over time. I have used the average values for citizen ideology during 1995–2000, since they measure political culture just before the period in which most state-level climate policy adoption occurred.

Socioeconomic Structures

I also include four variables that represent the main socioeconomic structures advanced in the theoretical literature. First, countries with high incomes are seen as having strong citizen demand for environmental policies and sufficient economic capacity to adjust to the costs of domestic climate policies. For this variable, I used personal income per capita for each state from U.S. Census data for 2000, in nominal dollars per person. Second, high education levels increase public understanding of environmental issues, as well as scientific capacities for addressing problems. To measure education levels, I created an index of higher education attainment based on U.S. Census data. In the United States as a whole in 2000, 24% of the population aged 25 and over had at least a bachelor's degree, and 9% had an advanced degree. In theoretical terms, education should make strong climate policy more likely in two ways: by making the voting public more supportive; and by increasing the economic capacity (which includes a highly educated workforce) needed to make a transition to a low-carbon economy. In terms of the census data, the relevant voting public is best represented by the percentage with at least a bachelor's degree, while economic capacity is better represented by the share with an advanced degree. Therefore, I created a higher education index by adding those two measures.[26*]

Third, economies based in manufacturing or other heavy industry are held to hinder climate and other environmental policies because those sectors have relatively high costs of adjustment to such policies and hence tend to oppose them. The differences among states on this dimension are captured by a heavy industry variable, which includes mining, utilities, and manufacturing. The variable is the total GDP of those sectors as a percentage of the GDP in all economic sectors, as reported by the Bureau of Economic Analysis and averaged over 1997–2000.

Fourth, according to the structural theory, a large fossil fuel sector increases the strength of the fossil fuel lobby and its structural power to block consideration of climate policy; it also reduces fossil fuel import dependence. Conversely, if a jurisdiction's fossil fuel production is low, its import dependence is high, and hence it has a collective interest in reducing fossil fuel consumption through climate policies in order to reduce the costs of imports and the risks of supply disruption. To measure this interest relative to the size of a state's economy, I use a fossil fuel production variable, which is the amount of coal, oil, and natural gas produced per unit of GDP, averaged over 1997–2000. It is measured in thousands of British Thermal Units (BTUs) per dollar of GDP.[27]

Fifth, the severity of environmental problems can be an important motivator of citizen and elite action for policies to ameliorate them. The problem most closely linked to climate policy is air pollution, since about 80% of greenhouse gas emissions come from fossil fuel burning – which also produces sulfur dioxide, nitrogen oxides, particulate matter, and other pollutants that have been known to harm human health and economic activity since at least the 1950s.[28*] To measure air pollution problems, I used an indicator developed and provided by Daniel Matisoff. This is based on county-level EPA data concerning non-compliance with federal air quality standards for six major "criteria" pollutants: NO_x, SO_2, carbon monoxide, lead, one-hour ground-level ozone, and particulate matter (PM-10).[29] The variable is the percentage of a state's population living in non-attainment areas, averaged over 1990–2002.[30*]

Political and Economic Processes

Although obtaining reliable cross-state data on political process variables is more difficult than for structural variables, I was able to include variables representing several factors of interest in the policy windows and advocacy coalition theories. First, the theoretical literature on state-level climate policy emphasizes leadership by officials with environmental interests or commitments. In this study, I use a government ideology variable to measure the extent of environmentally committed leadership among elected officials. This variable taps into the potential role of governors, their appointees, and legislators as leaders in initiating climate policy, as well as the importance of election outcomes in placing individuals with environmental commitments or interests in office. The literature also posits that Democratic control of government increases the chances of policy adoption. However, the important roles of Republicans in recent state-level climate policy making, including those of governors in California and New York, makes government ideology a theoretically more plausible factor than party control[31*] – especially since most policies in the CCES index were adopted during the 2000s, when relatively moderate state-level Republican officials were not yet a rarity.

The Berry et al. measure of state government ideology captures the results of elections for governor and for both state legislative chambers, while also taking into account ideological differences across states within the Republican and Democratic parties.[32] Similar to the citizen ideology variable, this index is a 0-to-100 conservative-liberal scale based on scores of roll-call voting by congressional representatives; I use the version that is based on a complete list of roll-call votes rather than interest group scores.[33*] The index assumes that state elected officials have the same ideology as the congressional representatives from their respective parties. It gives a one-half weight to the governor and one-quarter weight to each legislative chamber (in the 49 states with bicameral systems). The legislative chambers are scored by scoring the majority and minority parties and then weighting them according to their seat shares, but with a minimum 60% weight for the majority party.[34]

For the government ideology variable used in this analysis, I took the average of the 2001–10 values for the Berry et al. index, which reflects the approximate period in which most state climate policies were adopted. Since this is a process variable, it should be measured at the time of policy making rather than clearly before it, as was done for the structural variables. Unlike the citizen ideology measure, government ideology varies significantly over time, in part reflecting election results, although it also displays a degree of stability. The correlations across the adjacent five-year periods between 1990 and 2009 are 0.59 to 0.74, while across the 1990–94 and 2005–09 periods the correlation is only 0.34; these are much lower than the comparable values for the citizen ideology variable, described in the earlier section on political institutions.

Interest group processes are also prominent in theoretical explanations of climate and other environmental policy. Interest group mobilization concerning climate policy is indicated by two variables. While the fossil fuel industry measure discussed in the previous section indicates the *potential* political strength of climate policy opponents, the former is a structural variable, that is, one that changes only very slowly over time. By contrast, here I use two process variables that help to test the advocacy coalition theory. Environmental organization membership is a relatively direct indicator of political mobilization. It is constructed from data provided by Robert Lowry, which are the most comprehensive available.[35] I have totaled the membership in seven national environmental organizations that had a major focus on climate change throughout 2001–13. The organizations included were the Environmental Defense Fund, Friends of the Earth, Natural Resources Defense Council, National Wildlife Federation, Sierra Club, Nature Conservancy, and Wilderness Society; excluded were the National Audubon Society (active on the issue only since 2007) and the National Parks and Conservation Association (active on the issue since 2009).[36*] The variable is defined as members per thousand population, measured during 1990–93; unfortunately, similarly comprehensive data for a later period were not available.

The other indicator of interest group mobilization, clean energy jobs, indicates the strength of the clean energy or clean technology sector of each state's economy, which may be a lobbying force for climate policy.[37*] It is also an indicator of the first-mover advantages that a state may possess relative to other states. If a state has a clean energy sector that already has reached a critical mass, this may give more weight to arguments for building up that sector through climate policy than in states with small clean energy sectors. The variable used here is based on data in a study of the clean energy economy by the Pew Charitable Trusts, which counted 770,000 jobs nationally in energy efficiency, renewable energy, environmentally friendly production, conservation and pollution mitigation, and training and support in 2007.[38] The variable is defined as clean energy jobs per thousand jobs; the 1998 and 2007 values were averaged.

Finally, I include two process variables concerning economic conditions, economic growth and unemployment.[39*] Unemployment has potentially ambivalent effects on climate policy making. High unemployment permits justifying renewable energy and energy efficiency policies in terms of co-benefits for local economic development and job creation. However, high unemployment may retard climate policy by shifting the public agenda away from environmental issues toward economic ones. Hence, I will test this variable for both positive and negative impacts on climate policies, using two-tailed significance tests. Economic growth is more straightforward in its expected effects. Rapid economic growth provides slack resources that could be used for investments in clean energy and it also makes bearing additional costs more politically viable, while slow or negative growth tends to crowd environmental issues off the agenda. For the unemployment rate, I use the average unemployment rate from 2001 to 2012;[40] for economic growth, I use the percentage change in real GDP from 1997 to 2012.[41]

ANALYSIS AND METHOD

I organized the analysis around four sets of theoretical questions:

- Do structural factors affect climate policy adoption? Which structural factors do so, and which do not? I addressed these questions by developing a structural model.
- Do process factors affect climate policy adoption? For this question, I created a model using only process variables.
- Do process factors matter even if structural factors are taken into account? To address this question, I tested the process variables while controlling for all the structural variables, in a combined structural-process model.
- To what extent are the relevant processes driven by structural factors? To address this question, I performed a path analysis, in which the process variables that were strong predictors were treated as intervening variables and the theoretically plausible structural variables in the dataset were used in an effort to explain them.

Before entering variables into regression equations, I checked the independent variables for linear relationships with the climate policy index. If a variable showed a non-linear relationship, of diminishing returns at extreme values, I transformed it with natural logarithms to reduce skew; this was done for three variables.[42*] Using SPSS, I ran Ordinary Least Squares regressions.[43*] Significance tests were generally one-tailed, which is appropriate because the literature provided clear expectations about the nature of the relationships between the independent and dependent variables.[44*] Exceptions were governor's institutional powers and unemployment, where the theoretical literature is ambivalent about the direction of the independent variables' effects, and hence I used two-tailed tests. I used the .05 level as the cutoff for significance, which means that there is less than a 5% chance that a result for a significant variable was due to chance.

I used a relatively large number of independent variables (16), because of the broad theoretical questions that motivate this study, including those concerning the relative importance of institutional, socioeconomic-structural, and process explanations. While such a large number generally would be a methodological concern, the degrees of freedom are sufficient to produce significant results, and collinearity overall was not a major problem in this dataset.[45*] To detect potential problems due to multicollinearity, I examined bivariate correlations, observed changes in coefficients and standard errors when variables were first entered into models, examined multicollinearity diagnostics (VIF values), and regressed each suspect independent variable on all others in the model.[46]

Results from the Structural, Process, and Combined Models

The results of the structural model are shown in the first column of Table 5.1. Two structural variables – income and legislative professionalism – are not included here because each of them showed severe multicollinearity with one other variable and each of them was statistically insignificant when entered in the absence of the other variable. Comprehensive testing revealed that income did not add any information to the structural model or the combined structural-process model, beyond what is provided by education, and that legislative professionalism added no information beyond what is provided by administrative capacity. See the Appendix to this chapter for details.

In the structural model, citizen ideology, education, and air pollution are significant at the .01 level, while governor's institutional powers and fossil fuel production are significant at the .05 level. The coefficient on governor's institutional powers is negative, which supports the theory that a separation of powers leads to a competitive raising of environmental standards. The model is highly predictive, with an adjusted R-squared value of 77%. Among the socioeconomic structures, the income and heavy industry variables were insignificant,[47*] as were three of the political-institutional variables (interest group power, electoral system openness, and administrative capacity).

The process model is shown in the second column of Table 5.1. Three of the five process variables are significant, the first two of them at the .01 level: government ideology, environmental organization membership, and the unemployment rate, the latter with a positive sign. The adjusted R-squared is 58%. Clean energy jobs and economic growth had coefficients with positive signs, as expected, but both were far from significant.

Finally, the combined structural-process model is shown in the last column of Table 5.1. Four of the significant variables from the structural model remain significant at the .01 or .05 level, but governor's institutional powers loses

TABLE 5.1 *Regression Results: Structural, Process, and Combined Models*

	(1) Structural model	(2) Process model	(3) Combined model
Governor's institutional powers	−2.120*		−1.572
	(.961)		(.924)
Interest group power	.820		.629
	(.625)		(.651)
Electoral system openness	.143		−.063
	(.333)		(.337)
Administrative capacity (absolute)	.008		.017**
	(.006)		(.007)
Citizen ideology (liberal)	.188**		.103*
	(.036)		(.044)
Higher education index	.369**		.189*
	(.088)		(.098)
Heavy industry share of economy	.094		.037
	(.081)		(.091)
Fossil fuel production (per GDP)	−.665*		−.741**
	(.303)		(.293)
Air pollution	2.158**		1.817*
	(.689)		(.809)
Government ideology (liberal)		.086**	.061*
		(.030)	(.026)
Environmental organization membership (per capita)		6.406**	2.820*
		(1.364)	(1.231)
Clean energy job share of economy		.235	.069
		(.498)	(.399)
Unemployment rate		1.156*	−.174
		(.567)	(.496)
Economic growth rate		.020	.000
		(.046)	(.034)
Constant	−5.937	−12.356**	−4.678
	(5.584)	(4.437)	(6.196)
N of cases	50	50	50
F of the model	19.276**	14.673**	15.973**
R-squared	.813	.625	.865
Adjusted R-squared	.770	.583	.811

Entries are unstandardized coefficients, with standard errors (OLS) in parentheses.
* $p < .05$; ** $p < .01$; one-tailed tests except two-tailed for governor's institutional powers and unemployment rate.

significance (p = .098, two-tailed). All of those variables except fossil fuel production have reduced coefficients when compared to the structural model; the coefficients for citizen ideology and education drop by about half. On the other hand, administrative capacity becomes a much stronger predictor when the process variables are added; its coefficient doubles and becomes significant at the .01 level.

In the combined model, two of the three process variables – government ideology and environmental organization membership – remain significant at the .05 level, although the coefficient on the latter is halved. The coefficient on the unemployment rate is reduced to nearly zero and is now highly insignificant (p = .73, two-tailed). Altogether, seven variables are significant at least at the .05 level in the combined model: administrative capacity, citizen ideology, education, fossil fuel production, air pollution, government ideology, and environmental organization membership. The combined set of variables is highly predictive, with an adjusted R-squared of 81%.

Marginal Effects of the Independent Variables

The combined structural-process model estimates the impact of each variable on state-level climate policy, in terms of the unstandardized coefficients, which refer to the units in which the variables were measured in the dataset.[48*] Table 5.2 shows that each of the following "produces"[49*] one additional climate policy in a state when compared to other states, when all other variables are held constant: a state government payroll that has 59,000 more full-time employees; a political culture that is 10 percentage points more liberal; or a higher-education index that is 5.3 percentage points greater.[50*] Similarly, one additional climate policy would be produced by a 74% reduction in fossil fuel production per GDP or a 73% increase in the share of the population living in high air pollution areas. Turning to the process variables, a government ideology that is 16 percentage points more liberal or 43% more environmental organization members per thousand population would also produce one additional climate policy.

The last three columns of the table show the difference in the effect of each significant variable in a typical state (i.e., one ranked near the middle of the fifty states) and in a state with a high value for the variable.[51*] For example, New York State's more liberal citizen ideology, compared with a typical state such as Florida, is responsible for 2.1 additional climate policies in New York. This part of the table also shows that all seven significant variables have effects of very roughly similar size on the climate policy index, when comparing typical states with states that have a high value for the variable in question. While the effect of any one variable is limited, the effects are additive, which makes it possible for the combined model to do well at explaining the differences in climate policy from state to state.

TABLE 5.2 *Marginal Effects of the Explanatory Variables*

Independent variable	Unstandardized coefficient	Difference in the independent variable that explains a one-policy difference in the climate policy index	Typical state (one ranked between 24th and 27th)
Administrative capacity (thousands of employees)	.017	59,000 more full-time employees	Colorado
Citizen ideology (% liberal)	.103	10 points more liberal population	Florida
Higher education index (% with degrees)	.189	5.3% higher index (5.3% more with bachelor's degrees or 2.6% more with advanced degrees)	Texas
Fossil fuel production (ln of thousands BTU/$ GDP)	–.665	74% decrease in fossil fuel production per GDP	Michigan
Air pollution (ln of % non-attainment areas)	1.817	73% increase in the percentage of the population living in non-attainment areas	Ohio
Government ideology (% liberal)	.061	16 percentage points more liberal elected officials	Virginia
Environmental organizations (ln of members per thousand population)	2.820	43% more members per thousand population	Illinois

Path Analysis

Understanding the relative contributions of structural and process theories also requires considering the possibility that apparently independent processes are actually strongly influenced by structural features. To examine this possibility, I regressed the two significant process variables (government ideology and environmental organization membership) on the structural variables that are theoretically plausible for explaining them.[52*] Only two relationships were highly significant, both at the .01 level: citizen ideology helps to explain government ideology, and education helps to explain environmental organization membership. In addition, the citizen ideology and air pollution variables were significant at the .10 level in predicting environmental organization membership. The adjusted R-squared values for these models were 36% and 49% respectively, showing that government ideology and environmental organization membership were conditioned to some extent by the structural variables employed here, but were far from determined by them. This supports the interpretation that these process variables were intervening variables between structural variables and climate policy index, while also having their own, independent influences on climate policy making.

Robustness Tests

I tested the robustness of the combined structural-process model in several ways; full results are in Table 5.4 in this chapter's Appendix. First, I identified cases that were potentially highly influential on the results, by examining Cook's Distance, standardized and studentized residuals, and dfit values.[53*] This led me to exclude Alaska, Oklahoma, and Utah; each of these failed 3 of the 4 tests, and they were clearly the most outlying cases for the combined model.

With those 3 cases excluded, the independent variables from the combined model predicted the climate policy index even more strongly than before; the adjusted R-squared value rose from 81% to 89%. Moreover, all seven variables that were significant in the combined model were significant in this slightly smaller dataset. Coefficients for most variables were similar to those in the 50-state combined model, although they rose markedly for environmental organization membership and dropped substantially for administrative capacity and air pollution. In short, with the 3 most influential cases excluded, the model became even stronger. The results reported earlier, in Table 5.1, were not driven by the outlying cases, but rather were weakened by them.

In a second check on the results in the combined structural-process model, I used the independent variables from that model to predict two alternative dependent variables (now including all 50 states). The first of these dependent variables was the unweighted CCES index for 2014. It produced results nearly identical to those for the weighted CCES index. All seven variables that were significant in the combined model had similar coefficients with this dependent variable, and six of them were significant at the .05 level; education was barely insignificant ($p = .051$).

The second alternative dependent variable that I used was the ACEEE's index of energy efficiency policies in 2013,[54] which is described in note 5. Despite the conceptual differences in the CCES and ACEEE indices, the regression results were quite similar. Five of the seven variables in the combined model that were significant predictors of the weighted CCES index were also significant predictors of the ACEEE index; only administrative capacity ($p = .07$) and environmental organization membership ($p = .053$) were not. The unemployment and economic growth rate variables also became significant, both with positive signs.

Finally, because the results of the Breusch-Pagan test for heteroskedasticity in the combined model were not *highly* insignificant ($p = .12$), I checked the robustness of the model by using heteroskedasticity-consistent standard errors. Since there were several cases with high leverage, I used the HC4 method.[55] The results were very similar to the Ordinary Least Squares results reported earlier in Table 5.1; all seven variables were significant, though only at the .05 level, and all other variables were insignificant.

DISCUSSION AND CONCLUSIONS

Institutions, Structures, and Processes

In this book, I am arguing that climate policy making is affected both by certain structures – though not all the structures that have been advanced in the literature – and by the processes that are indicated by the Kingdon and Sabatier theories. The analysis of California and New York in the previous chapter shows that most political-institutional factors are not related to state-level climate policy. That conclusion is also supported by this chapter's statistical analysis, in which most of the political-institutional variables examined were not statistically significant, namely the concentration of governmental authority, interest group power, electoral system openness, and legislative professionalism. Given the overall framework of a separation of powers and a two-party system in the U.S. states, the remaining formal institutional differences across the states apparently do not matter much for climate policy. The two political-structural variables that were significant variables in this analysis (and which also helped to explain California and New York in the previous chapter) did not concern the separation of power or the size of the party system. Administrative capacity is closely related to population size, while citizen ideology is rooted in economic, demographic, and historical factors; the latter helps to explain, for example, the low values for liberalism in the states of the former Confederacy.

By contrast, the socioeconomic-structural explanations fared much better, as three of the five variables tested were significant in both the structural and the combined structural-process model (education, fossil fuel production, and air pollution). Economic composition, measured by the heavy industry share in the economy,[56*] was not significant; nor was income per capita. In short, an educated citizenry and workforce, and problem pressure from urban air pollution, are important drivers of climate policy, while fossil fuel production is an important brake.

The results of the structural analysis present a mixed picture for the future of state-level climate policy making in the United States. Two facilitative factors have been growing gradually. State administrative capacity increased at a rate of 5% per decade during 2000–12, despite the effects of the recent recession on state government finances, while the education index grew at 8% per decade during 2000–13, even faster than in the previous decade.[57] Air pollution presents a more mixed picture. All common pollutants declined from 1980 to 2012, and marked declines have continued since 2000, except for ground-level ozone. This important pollutant, an ingredient of smog that endangers public health, has declined only 6% per decade since 1990 and only 8% per decade since 2000.[58] Since ground-level ozone pollution has been roughly stable nationally and in all major regions of the country, it may continue to act as a spur to a variety of policies that reduce common air pollution while also reducing greenhouse gas emissions.

Two other structural factors do not show recent trends that indicate an upcoming change in their effects on climate policy making. Berry et al.'s citizen ideology index shows no clear trend over the last decade, although it has been unusually volatile in recent years. A sharp liberal shift in state politics, which was connected to the success of the Democratic Party and Barack Obama in the 2006 and 2008 elections, was followed by rapid conservative gains in the following elections. The unweighted average of states' citizen ideology scores rose from 43% in 2000 to 62% in 2008, but fell back to 47% in 2010.[59*] Finally, fossil fuel production recently has grown overall, but in a mixed way with unclear implications for climate policy. Coal production has been fairly stable, falling at a rate of 6% per decade. But oil production has risen at a rate of 21% per decade, and natural gas production has grown at a rate of 16% per decade.[60] Hence, while fossil fuel production as a whole is rising, its composition is shifting from more to less carbon intensive fuels. The latter may somewhat reduce the fossil fuel industry's tendency to hinder climate policy, by increasing the divergence of the natural gas industry's interests from those of the oil and coal industry.

This book also argues that certain political processes affect climate policy making, and that their effects are substantially independent of structural factors. There is scope for political actors to affect policies, within structural constraints. The statistical analysis supports these claims. Government ideology and environmental organization membership were significant in the combined model, which shows that those processes affect climate policy even when controlling for all the structural factors. Both structures and processes matter, independently of each other. On the other hand, economic processes seem not to be important for climate policy. Economic growth and the unemployment rate had very small coefficients and were very far from significant when controlling for the structural conditions. Similarly, the path analysis shows that the two significant process variables are to some extent intervening variables between certain structural factors and climate policy making. A relatively liberal citizen ideology helps to produce a relatively liberal government ideology, and education increases environmental organization membership. At the same time, the modest R-squared values for those regressions suggest that government ideology and environmental organization membership are substantially independent of structural causes.

Linking the Significant Variables to the Case Studies

I will explore the significant variables from this analysis further in the case chapters.[61*] How the relationship between the governor and legislature affects climate policy making will be examined through the contrasting experiences of

California (competition between the branches) and New York (executive dominance). The role of administrative capacity will be explored in terms of the creation of specialized agencies for energy and air pollution policy in California and New York, both of which have large administrative personnel.

The statistical significance of the air pollution index, which measures the percentage of the population living in areas with poor air quality, points toward the importance of the objective severity of air pollution, its politicization, and the creation of early policies to address it. Those will form major parts of the California and New York case chapters. Fossil fuel production per capita indicates the importance of the fossil fuel industry as a political force opposed to state climate policies. California is partly an exception to this generalization, and how climate policy advocates overcame the attempts of its large fossil fuel industry to weaken or prevent climate policy in that state will be a subject of Chapters 6 and 7. The statistical significance of air pollution and a weak fossil fuel industry indicate the importance of the co-benefits of climate policies for air pollution control and energy security, and these, too, will be explored in the case studies.

The process variables found significant here will also reappear in the case chapters, and frequently. Government ideology points toward the importance of election results and the ideological commitments or positions of elected officials. Finally, environmental organization membership indicates the key role of those groups in advocacy coalitions, and the ways that air pollution and earlier environmental conflicts contributed to organizational creation and expansion.

More Detailed Findings

This statistical study of climate policies also points to more specific theoretical conclusions about the significant variables and several insignificant ones. The variable measuring the concentration of governmental power had a positive coefficient that was significant at the .05 level in the structural model but only at the .10 level in the combined model. These results indicate weak support for the theory that a separation of powers promotes higher environmental standards by giving advocates multiple paths for advancing reforms or by encouraging competition between the governor and legislature. This contrasts with the prominence of the opposing theory, that a concentration of power promotes climate policy, in cross-national studies.[62] The weakness of the statistical result could mean that the differences in this variable across U.S. states are not large enough to affect climate policy making. It could also mean that there are alternative paths toward relatively strong climate policy in different institutional contexts. Some states, such as California, may develop extensive climate policy through competition between a governor and a legislature that have relatively equal powers, while others, such as New York, may develop them through strong governors who dominate or evade their legislatures.

Administrative capacity is somewhat neglected among the political-institutional variables in the literature. State-level energy and climate policy studies emphasize the role of specialized administrative agencies, such as state energy offices,[63] and they are an important topic in Chapters 6 through 8. In this chapter's statistical analyses, the absolute size of a state's administrative capacity was a significant predictor, though not the number of employees as a share of population. This may be because states with larger absolute administrative capacities can more easily set up new regulatory functions than smaller states can, and hence policy makers have more confidence in creating new policies that require expanded specialized capacities (see the last section of Chapter 4). Alternatively, the administrative capacity variable, which correlates extremely highly ($R = 0.98$) with population, may be picking up some other advantage that large states have. For example, a state with a large economy may have influence over its neighbors (e.g., leading them into a regional climate policy compact) or may have little to fear from its neighbors in terms of economic competition if it adopts policies that raise the costs for its in-state businesses.

Citizen ideology was a strong predictor in the structural and combined models, a finding that supports a number of other climate policy studies.[64] However, its coefficient dropped by one-third when the government ideology variable was added to the model. This suggests that much of the apparent effect of political culture among citizens is really due to liberal government officials, and that one needs to control for the latter in order to assess the real effects of citizen ideology. However, the path analysis also shows that citizen ideology has an indirect effect on climate policy, by influencing government ideology.

Education was a strong predictor of climate policies, with an additional indirect effect through its influence on environmental organization membership. By contrast, per capita income was not significant in the structural or combined models, although it had a strong bivariate correlation with the CCES index, and several climate policy studies found it to be significant.[65] Nor did an income-education index perform better than the education variable by itself (see the Appendix to this chapter for details). As an explanatory variable, income does not add any information not already provided by education. And yet few studies of climate or environmental policy include education as an explanatory variable.[66] Education and income are closely related to each other, and both are theoretically related to postmaterialism. But the results here suggest that education may be a more important indicator of postmaterialist values than income is. More research on the potential impacts of education, including possible effects of the "creative class," would be helpful.[67]

The air pollution variable was significant in the structural and combined models. This confirms the results of Matisoff's study of renewable portfolio standards,[68] and it supports the observation from case studies that air pollution policy making is an important forerunner and adjunct to other kinds of climate policy making. However, the coefficient of this variable dropped by one-quarter when environmental organization membership was added, and the path analysis also suggests that air pollution affects climate policy partly by encouraging the growth of environmental organizations. Severe air pollution problems also have broader effects, affecting economic elites, the public, and a wide range of elected officials, as shown in Chapters 6 and 8.

The significance of fossil fuel production in the structural and combined models confirms a result found in other climate policy studies.[69] There are several mechanisms by which this variable could affect climate policy. High levels of production would increase the structural power of the fossil fuel industry and its lobbying power, providing a brake on climate policy development. Alternatively, low levels of fossil fuel production would increase dependence on fossil fuel imports and associated import expenditures, and hence would create incentives to increase energy efficiency and renewable energy production.

Two process variables were statistically significant: state government ideology and environmental organization membership. These findings confirm the results of a number of studies which conclude that elected officials committed to environmental policy making and advocacy by environmental organizations are important causes of climate policy making.[70] These are process variables, which change over relatively short periods and are potentially influenced by political actors. State government ideology varies over time according to election results and the strategic decisions of elected officials; the statistical significance of this variable suggests that the leadership of elected officials matters. Environmental organization membership was measured in this analysis at one point in time, the early 1990s. However, this is an indicator of the power resources of environmental organizations, which generally vary over time. For example, the total memberships of eleven major environmental organizations doubled from 1979 to 1990, and the annual lobbying expenses by the Sierra Club and the Environmental Defense Fund more than tripled from 2006 to 2010.[71] The case study chapters will examine the mobilization of environmental organizations in advocacy coalitions.

Appendix to Chapter 5

Independent Variables Not Included in the Analysis

There are several potential independent variables that I did not include, for reasons worth clarifying. Population is not included as a control variable because it correlates extremely strongly with the administrative capacity variable; theoretically, if large-population states have an advantage, it is because they have large government administrative capacities (although I discuss some alternative interpretations in this chapter's conclusions). Also, any independent variable that could have been strongly affected by population size, such as fossil fuel production or clean energy jobs, has been adjusted in per-capita or per-GDP terms.

Environmental public opinion is not included because of difficulty in obtaining a good state-level measure; in addition, the effects of this factor are partially included in the citizen ideology and environmental organization membership variables. Economic losses due to environmental disasters are not included, because the best available data include only hurricanes, floods, and tornadoes;[72] drought and wildfire losses are substantial relative to the former categories, but state-level data on them are not publicly available. Finally, energy intensity or CO_2 intensity of the economy is not included although it is strongly, negatively associated with climate policy, because it is an intervening variable rather than an independent variable. It is driven by economic structures, such as fossil fuel production and heavy industry share, and is a product of energy efficiency and energy pricing policies, which are part of climate policy (see the structural theories section of Chapter 3).

Variables Excluded Due to Multicollinearity

Income and education were found to have a serious multicollinearity problem,[73*] so using both variables together would have produced unreliable coefficient estimates and significance tests. Therefore, I created separate structural models, each including only one of these variables along with all the other structural variables. While education remained significant when income was excluded, income was not a significant variable in the absence of education; the same held when the procedure was repeated in the combined model. Finally, to try to capture the effects of both variables, I constructed an index of socioeconomic development by summing the standardized values of the education and income variables. However, this index performed less well than education in the structural model, and it was not significant in the combined model.[74*] I concluded that, as an explanatory variable, income does not add any information not already provided by education, and hence I dropped income from further analysis.

I took a similar approach to address the serious multicollinearity problem[75*] between legislative professionalism and administrative capacity when both variables were entered in the same model. When entered one at a time in the structural model, neither was significant. However, in the combined model, administrative capacity was highly significant ($p < .01$) and legislative professionalism was not ($p = .16$). I also tested an index of overall governmental capacity, which was the sum of the standardized values of the administrative capacity and legislative professionalism variables. It performed less well than the administrative capacity variable in both the structural and combined models. Similar to the situation with the income variable, legislative professionalism did not add any information to the model that the administrative capacity variable was not already supplying;[76*] hence, I dropped legislative professionalism from further analysis.

TABLE 5.3 *Descriptive Statistics for the Variables in the Analysis*

	Variable	Minimum	Maximum	Mean	Standard deviation
Dependent variable					
Climate policy index	Modified CCES index, February 2014	3.65	24.22	14.21	5.63
Political institutions					
Concentration of governmental authority	Governor's institutional powers, late 1990s (1–5 scale)	2.40	4.30	3.40	0.48
Pluralist interest group systems	Interest group power, 1998 (1–5 scale)	2.00	5.00	3.62	0.78
Electoral system openness	Third-party candidates, 1980–2005 average (number per election)	0.17	6.50	1.90	1.34
Governmental capacity	State government employees, absolute number, 2000 (in thousands)	11	355	82	69
	Legislative professionalism index, average of 1996 and 2003 (ratio to the score for the U.S. Congress)	0.03	0.60	0.18	0.12
Political culture	Citizen ideology, 1995–2000 average (% liberal)	19.52	79.32	46.53	14.06
Socioeconomic structures					
Income	Income per capita, 2000 (nominal $)	21,017	41,392	28,261	4,409
Education	Higher education index, 2000 (sum of % with bachelor's and % with advanced degree)	20.77	46.85	32.16	6.18
Heavy industry	Mining, utilities, and manufacturing sectors, 1997–2000 average (% of GDP)	4.16	31.75	18.56	6.07
Fossil fuel industry	Fossil fuel production, 1997–2000 average (thousands BTUs/$GDP) (ln)	0.00	6.10	1.30	1.61

Environmental problem severity	Air pollution non-compliance areas, 1990–2002 (% of population living in) (ln)	0.00	2.60	0.91	0.68
Political and economic processes					
Environmentally committed leadership among elected officials	Government ideology, 2001–10 (% liberal)	11.46	76.45	49.42	19.03
Interest group mobilization in favor of climate policies	Environmental organization membership, 1990–93 (per thousand residents) (ln)	0.90	3.37	2.16	0.50
	Clean energy jobs, average of 1998 and 2007 (per thousand jobs)	2.37	9.13	5.05	1.41
Economic conditions	Unemployment rate, 2001–12 average (% of workforce)	3.39	8.36	5.99	1.16
	Economic growth rate (% change 1997–2012)	−1.62	85.14	19.65	14.09

TABLE 5.4 *Regression Results: Robustness Tests of the Combined Structural-Process Model*

	(1) Combined model (from Table 5.1, for reference)	(2) Excluding three influential cases	(3) Unweighted CCES climate policy index as dependent variable	(4) ACEEE energy efficiency policy index as dependent variable	(5) Combined model, with heteroskedasticity-consistent standard errors (HC4)
Governor's institutional powers	−1.572	−.882	−1.521	−1.649	−1.572
	(.924)	(.736)	(.915)	(1.682)	(1.003)
Interest group power	.629	.914	.531	.926	.629
	(.651)	(.502)	(.645)	(1.186)	(.620)
Electoral system openness	− .063	.131	−.047	.656	−.063
	(.337)	(.289)	(.334)	(.614)	(.486)
Administrative capacity (absolute)	.017**	.013*	.017**	.018	.017*
	(.007)	(.005)	(.006)	(.012)	(.009)
Citizen ideology (liberal)	.103*	.115**	.105*	.154*	.103*
	(.044)	(.036)	(.044)	(.081)	(.053)
Higher education index	.189*	.183*	.164	.473**	.189*
	(.098)	(.076)	(.097)	(.179)	(.083)
Heavy industry share of economy	.037	.066	.049	.237	.037
	(.091)	(.072)	(.090)	(.166)	(.085)
Fossil fuel production (per GDP)	− .741**	−.680**	−.676*	−1.394**	−.741*
	(.293)	(.243)	(.290)	(.533)	(.399)
Air pollution	1.817*	1.319*	1.832*	−3.436*	1.817*
	(.809)	(.641)	(.801)	(1.473)	(.770)

Government ideology (liberal)	.061*	.048*	.057*	.107*	.061*
	(.026)	(.021)	(.026)	(.048)	(.028)
Environmental organization	2.820*	4.163**	2.936*	3.719	2.820*
membership (per capita)	(1.231)	(.976)	(1.219)	(2.241)	(1.315)
Clean energy job share of	.069	.147	.010	−.545	.069
economy	(.399)	(.305)	(.395)	(.726)	(.381)
Unemployment rate	−.174	.293	−.223	1.566*	−.174
	(.496)	(.400)	(.491)	(.902)	(.685)
Economic growth rate	.000	−.003	.005	.161**	.000
	(.034)	(.026)	(.034)	(.062)	(.058)
Constant	−4.678	−14.087**	−3.548	−32.434**	−4.678
	(6.196)	(5.099)	(6.136)	(11.278)	(8.781)
N of cases	50	47	50	50	50
F of the model	15.973**	27.463**	15.027**	14.621**	17.511**
R-squared	.865	.923	.857	.854	.865
Adjusted R-squared	.811	.890	.800	.796	.811

Entries are unstandardized coefficients, with standard errors in parentheses; the latter are from OLS regressions, except column 5.
* $p < .05$, ** $p < .01$; one-tailed tests except two-tailed for governor's institutional powers and unemployment rate.

6

California: Air Pollution and Energy Policy Making (the 1940s to the 1980s)

This and the next chapter will explain why California has become a national leader in climate policy, adopting stronger policies than all other states and often influencing them and the federal government. After a description of the relevant policy-making episodes in the state, this chapter will focus on the adoption of air pollution legislation from the late 1940s to the late 1960s and on energy efficiency and renewable energy policies in the 1970s and 1980s.

To understand the development of climate policy in California ultimately requires examining ten rather diverse episodes of policy making that occurred over a sixty-year period. This chapter will analyze seven of them – concerning air pollution control (1947, 1960, 1967–68), the California Environmental Quality Act (1970), the Warren-Alquist Act (1974), the energy policies of the Brown administration (1976–82), and the reversal of renewable energy policies under his Republican successor, George Deukmejian (1983–85). Chapter 7 then will analyze the three most recent episodes: the explicit climate policies adopted in 2002 and in 2005–08, and the attempt to block the implementation of the Global Warming Solutions Act through an initiative and referendum in 2010. Four of the ten episodes involved the adoption of major policies, four involved the creation of administrative and legal capacities upon which later climate policies depended, one was a policy reversal, and one was a failed attempt at policy reversal. Why these are all important to the California case study requires some explanation.

In the first place, major California climate policies – those which had impacts on greenhouse gas emissions – were adopted, reversed, or targeted by reversal attempts in six periods. First, in 1974, the Warren-Alquist Act created the California Energy Commission (CEC) and directed it to develop energy efficiency standards and engage in energy planning. Second, under the Brown administration in 1976–82, the state adopted strong energy efficiency and renewable energy policies. These included the first energy efficiency standards for appliances (1976) and buildings (1978), and tax credits and targets for renewable energy (1978, 1980). They also included the uncoupling of utility profits from sales and the enforcement of utility non-compliance with conservation and alternative energy requirements (1978–82), and generous feed-in tariff rates for renewable energy (1982). (See Chapter 2 for details on the policies.) Third, renewable energy policies were reversed in 1983–85 under the Deukmejian administration.

The fourth episode of major policy making was in 2002, when the state legislature passed both the Pavley bill, which directed CARB to develop standards to reduce vehicle emissions of greenhouse gases, and an ambitious renewable portfolio standard with a target of 20% by 2017. Although statewide emissions targets had not yet been adopted, the Pavley bill was a major explicit climate policy because it directly regulated greenhouse gas emissions as such, albeit with a planned seven-year delay before implementation would begin.

Fifth, during the Schwarzenegger administration, the 2005–08 period saw the adoption of a series of major measures, most adopted by the legislature. These included statewide greenhouse gas emissions targets aiming to return emissions to 1990 levels by 2020 (by executive order in 2005), the conversion of the targets into legally binding measures (in the Global Warming Solutions Act in 2006), and the adoption of a low-carbon electricity standard (2006). These measures were followed quickly by the acceleration of the renewable portfolio standard to 20% by 2010 (2006), a low-carbon vehicle fuel standard (by executive order in 2007), and a law orienting regional transportation planning to greenhouse gas reduction targets set by CARB (2008). Sixth, an attempt was made to block the implementation of the Global Warming Solutions Act via Proposition 23 (a 2010 referendum), but it failed by a decisive margin.

However, to understand how the major climate policies outlined here could be adopted and implemented, it is also necessary to explain the creation of California's administrative capacities in the air pollution and energy areas as well as its nationally unique legal capacities to adopt motor-vehicle pollution standards stricter than those of the federal government. The capacities of the main agencies involved in California climate policy have proved crucial for monitoring greenhouse gas emissions, assessing their impacts on the state and its economy, proposing policy solutions, and implementing them. The three key agencies are quite large: CARB (1,200 employees and a $740 million budget in 2007), the CEC (560 employees and a $420 million budget), and the California Public Utilities Commission (CPUC), with 1,000 employees and a $1.3 billion budget.[1]

In air pollution policy, CARB is an aggressive, independent, science-based agency with a wide jurisdiction. It has the lead role in making rules on greenhouse gas emissions, which it does while drawing on the expertise of the CEC and CPUC.[2] The CEC has authority to develop appliance and building efficiency standards, promote renewable energy, regulate municipal utilities, and oversee power plant siting. The CPUC has regulatory authority over the electricity, natural gas, telecommunications, water, and public transportation sectors, including the state's large investor-owned electricity utilities.[3] After resisting change for a long period beginning in the 1970s,[4] it has worked closely with the CEC on energy efficiency standards, utility programs, and energy planning.

California also has a legal capacity to regulate air pollutants that is unique in the United States. The state gained an exemption from the federal preemption of motor vehicle air pollution standards in the federal Air Quality Act of 1967 and again in the Clean Air Act of 1970. These federal laws allow California authorities to set higher standards than the federal government, and the 1977 amendments to the Clean Air Act allow other states to adopt California's standards, giving it a potential role as a leader among the states.[5]

More precisely, section 209 of the Clean Air Act gives California the right to apply for a waiver from federal requirements in order to implement its own, stricter air pollution standards, if they are needed to address the state's "compelling and extraordinary conditions." The burden of proof is on the EPA to prove otherwise, and hence it has granted over one hundred waivers.[6*] Other states often have adopted the California standards, which is permitted by section 177. For example, other states or the federal government copied California's NO_x standard in 1971, its catalytic converter requirement and unleaded gas standard in 1975 and 1976, its low-emission vehicle standard in 1994, and ultimately a version of its Pavley standards in 2009.[7]

California's legal right to set higher standards. along with the rights of other states to follow it, have had far-reaching effects on state and national climate policy. These rights made the Pavley bill a viable approach to initiating greenhouse gas regulation in the United States. The Pavley law had reasonably good chances of being upheld by higher authorities, as it was by the Supreme Court in 2007 and the EPA two years later. In the context of the early 2000s, the Pavley law also had good chances of being copied by other states, and 17 states had announced their intention to do so by 2007; this gave the Pavley law a large potential impact on the U.S. car industry and probably influenced the Supreme Court decision. And, if it could succeed in California and other states, the Pavley law had good chances eventually of being adopted by the federal government, because of the car industry's general preference for uniform standards.

In terms of the policy-making episodes to be explained in California, the growth of the administrative capacities of the CEC and CPUC fall under the two 1970s subcases, because the CEC was created in 1974, and the CPUC, though created in 1945, was reoriented toward energy conservation during the Brown administration in the late 1970s. However, CARB is a different matter. It was created in 1967, in the Mulford-Carrell Act, and the next year, the Pure Air Act specified the air emissions standards for motor vehicles that it was to implement. In passing those laws, the state legislature in part was trying to preserve an independent role for California state government in the face of possible preemption by pending federal air pollution legislation and regulation.

At this time, California policy makers and congressional representatives mobilized in Congress and succeeded in gaining an exemption for California from the emerging federal air pollution standards.

Hence, I will also analyze an earlier California case, of air pollution policy making in 1967–68, because administrative and legal capacities that proved crucial for later climate policy making in the state were created at that time. In turn, explaining the late 1960s legislation requires understanding two earlier episodes of air pollution policy making, in 1947 and 1960, which helped to show the need for a powerful state agency to tackle the state's smog problem. Similarly, the 1974 Warren-Alquist Act was made possible by the passage of the California Environmental Quality Act in 1970, which created barriers to power plant siting and hence showed the importance of energy conservation. Therefore, in addition to six policy-making episodes that directly concerned climate policy, I will analyze these four cases of precursor legislation.

THE PUBLIC AGENDA

The policy reforms in these episodes were driven in part by events in the problem stream, which increased public attention on problems for which the new policies were seen as solutions. Here I will describe the public agenda only for air pollution and energy problems, with the global warming issue treated in Chapter 7. The emergence and growth of smog since the 1940s put air pollution on the public agenda in California, as seen in the coverage of the *Los Angeles Times*. Public attention on the air pollution problem rose to peaks in 1947, 1954–55, 1959, and 1967–70, as shown in Figure 6.1.[8*] Each of the peaks in attention occurred during or just after periods of intense smog events, which are described in the 1947 policy-making episode, in the next section. During or soon after three of these periods, the issue was politicized and major reforms were adopted by state laws that: permitted counties to create air pollution control districts (1947); created a state agency with authority to regulate motor vehicle emissions (1960); and established CARB and tightened air pollution standards (1967–68).

This figure also shows that public attention on energy problems hit large peaks in California, as it did nationwide, in 1973 and 1979, due to energy supply crises and energy price increases.[9*] These periods of intense concern helped lead to the Warren-Alquist Act in 1974 and to the energy efficiency and renewable energy policies of the late 1970s.

The patterns in these figures largely confirm the theoretical expectations of Chapter 3.[10*] Attention on air pollution and energy problems tends to cluster in response to problem events (smog episodes and energy shortages). Further, the peaks of public attention to these issues were also those with the largest policy changes, with the exception of the 1954–55 peak in attention to the smog problem, which was not accompanied by major reform.[11*]

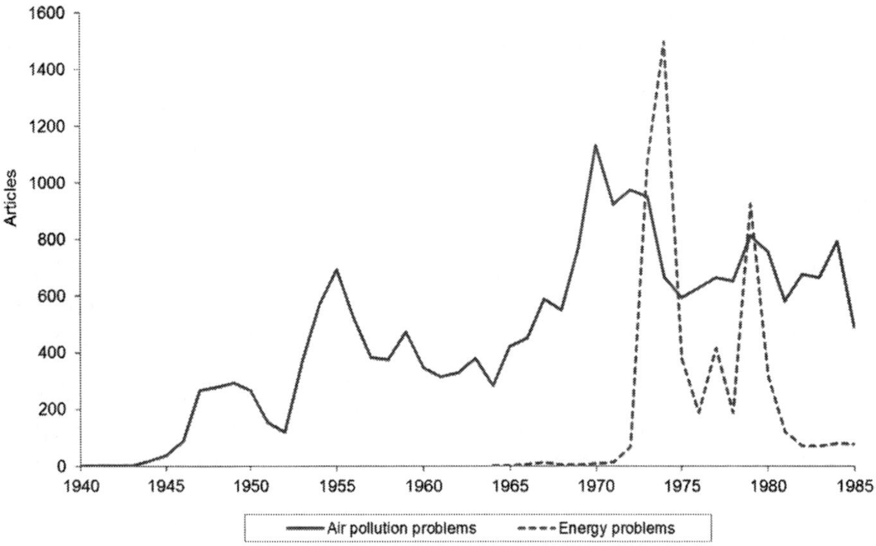

FIGURE 6.1 Public Agenda in California: Air Pollution and Energy Problems in the *Los Angeles Times*, 1940–1985
Source: see notes 8–9.

SMOG EVENTS AND MOBILIZATION: AIR POLLUTION POLICY FROM THE 1940S TO THE 1960S

Several episodes of air pollution policy making that began in the late 1940s ultimately led, in the late 1960s, to the establishment of CARB and to the federal exemption for California's air pollution standards. During two decades of grappling with smog problems, policy makers created county air pollution boards and a state motor vehicle pollution control board. But the perceived failures of those agencies convinced them that California needed a powerful state agency to set air pollution standards and ward off federal pre-emption.

County Air Pollution Control Districts (1947)

The main driver of these changes was a set of new problems. Rapid population growth, industrial growth, and increased car travel in the Los Angeles area, and to a lesser extent in other urban areas of California, produced sudden, intense episodes of photochemical smog, the causes of which initially were not understood. The Los Angeles Basin is unusually susceptible to smog formation, because it is bounded by mountains on the north and east, and the prevailing light westerly winds often prevent polluted air from dissipating to the west.[12] Beginning in 1943 and continuing into the late 1960s,[13*] dense smog events occurred in Los Angeles in the summer and fall, producing eye irritation, breathing difficulties, coughing, and significant economic harm. Smog threatened public health and California's image as a desirable, healthy destination for visitors and new residents. The most severe smog events, lasting for weeks at a stretch, occurred in 1943–46, 1949, 1954–55, 1958–59, 1965–67, and 1970.[14]

During severe smog events, hospital admissions for respiratory complaints rose, airplanes had difficulty landing, and outdoor shooting by the film industry was hampered. Long-term economic harm was also apparent or threatened, as agricultural crops in Southern California were damaged and many residents moved out of Los Angeles, citing smog as the reason. In addition, the deaths of 20 residents in Donora, Pennsylvania due to industrial smog in 1948 raised fears of a similar catastrophe in Southern California. One expert testified that 6,000 people could die there.[15]

The smog problem produced a relentless series of focusing events that raised public awareness and triggered political responses from citizens and elites, which in turn provided a recurring source of pressure on policy makers. The initial policy responses that this pressure evoked were largely ineffective, which was evident in the poor air quality during much of the year as well as the intense, headline-grabbing smog events that occurred at irregular intervals. Therefore, the smog problem combined with political mobilization to produce what were, at the time, significant policy reforms in three periods: 1947, 1960, and 1967–68.

In each of these policy-making episodes, mass participation occurred alongside elite mobilization, and both were driven by the sudden grievances[16*] presented by the reappearance of intense smog. The first citizens group opposed to smog formed in 1944. Protesters circulated petitions against industrial polluters, hundreds of citizens marched on the city halls in Los Angeles and Pasadena in 1946, and angry residents disrupted a meeting of the county board of supervisors the next year.[17] In 1949, after 21 days of intense smog, residents threatened to recall the supervisors. Protests spread outside Southern California as the problem did. In San Francisco in 1949, unusual weather conditions and a traffic jam due to a football game in Berkeley combined to create a large smog event. Citizens became alarmed, taped their windows shut, and called elected officials and police.[18]

The smog problem was severe enough to rapidly produce political commitment from elites in Southern California. The Los Angeles Chamber of Commerce called for uniform, county-wide regulation of air pollution in 1944, and a Pasadena hotel owner, Stephen Royce, took the lead in organizing business support for state legislation that was needed to authorize it.[19] Furthermore, the Chandler family, which owned the *Los Angeles Times*, initiated an air quality campaign in 1946. The newspaper published a series of articles on the problem and brought in an air pollution expert (Raymond Tucker), who had been successful in a similar newspaper-led campaign against air pollution in St. Louis. The *Times* published Tucker's findings in a front-page article in

January 1947, which mostly blamed industrial and population growth and called for the creation of a county-level air pollution control district.[20] A state law to this effect was needed to overcome the objections of municipalities within Los Angeles County.

Supporters of this legislation included the Los Angeles and Pasadena chambers of commerce and the Automobile Club of Southern California. The coalition for reform overcame what was initially strong opposition from oil companies, railroad companies, and lumber companies (which burned waste sawdust). Those interests held up the 1947 bill until the *Times* printed two highly critical articles about the oil industry. Then the oil and lumber companies immediately dropped their opposition, which led to unanimous passage of the bill in the State Senate and its signing by the Republican governor, Earl Warren.[21]

State Regulation of Motor Vehicles (1960)

However, these initial policies proved to be too modest to reduce smog, because of limited powers for the new agency, weak enforcement, resistance by homeowners, and continued population growth. Pursuant to the new state law, Los Angeles County created an Air Pollution Control District in 1947 with powers to set rules and enforce them, including issuing permits to polluters, but only for stationary sources; motor vehicles were exempted. The new agency had slow and uneven effects on the smog problem. It was lax about industrial sources until new leadership was more willing to confront polluters after 1954. When the agency targeted backyard incineration of trash by residents, strong public resistance delayed a ban on the practice until 1957.[22]

Moreover, when experiments in 1950 by Professor Haagen-Smit, funded by the pollution control district, showed that car exhaust and sunlight combined to form smog, resistance by the car industry and the Western Oil and Gas Association slowed acceptance of the finding. The latter's experts at the Stanford Research Institute initially disputed Haagen-Smit's results. The controversy was not officially settled until the industry-backed Air Pollution Foundation accepted Haagen-Smit's results in 1958. In addition, Los Angeles County and its air pollution control district were not legally empowered to regulate these mobile sources. Despite reductions in pollution from refineries, steel mills, other industrial sources, and trash burning, smog remained a serious problem that had not been reduced much, if at all. If anything, the problem seemed increasingly serious as local doctors became convinced that smog was causing respiratory illnesses and perhaps cancer, and the medical profession began to voice its concerns publicly.[23]

Given this problem setting, citizen protests and elite mobilization came together again to produce major state legislation in 1960. Citizens wanted a quick solution to the smog problem and did not trust authorities to be strict enough with polluters; they wanted officials to target industrial sources and (later) the major carmakers in Detroit. The depth of citizen concern was very clear in 1953–54. Norris Poulson (a Republican) used the smog issue to win the Los Angeles mayor's race, housewives paraded with gas masks, and between four and six thousand residents attended a meeting at the Pasadena civic auditorium to demand action. Pressure from the Citizens Anti-Smog Action Committee forced the head of the county's pollution control district to resign. Protests also forced the incumbent Republican governor, Goodwin Knight, to make concessions to anti-pollution sentiment in his successful bid for re-election.[24]

In 1958, citizen protests reached a new level of visibility with the formation of the Stamp Out Smog group, which was founded and led by Beverly Hills housewives and quickly dominated the anti-smog movement. The new organization created a network of 20 groups and eventually, at its peak in 1970, claimed 5,000 members and ties to 600 citizen groups with hundreds of thousands of members. Its leaders were wealthy and well-connected to public officials. They also gained helpful expertise and used an effective strategy that combined grassroots campaigns with lobbying. The group attended and chanted at hearings, engaged in letter writing and phone calls, lobbied intensively in Los Angeles and the state capital, and got the ear of Governor Pat Brown, a Democrat. Favorable publicity from the news media magnified the group's influence.[25]

On the elite side, the Los Angeles Air Pollution Control District and other local officials led the charge for the reduction of motor vehicle pollution, which would require state legislation. In this, they were supported by the business-backed Air Pollution Foundation, which had concluded that automobiles were largely responsible for smog and then focused its attention on motor vehicle pollution control. As Haagen-Smit's findings became accepted, the district first tried to get the major carmakers to produce pollution control devices. Since the latter proceeded very slowly, forming a consortium that critics said prevented any of them from developing technology too quickly, the district formally requested state legislation in 1958. Other local officials also were active on the issue. The Los Angeles city council in 1958 called for state government to ban the sale of cars by 1961 if they were not fitted with control devices, and Mayor Poulson called for vigilante committees to pressure the car industry to adopt smog-control devices.[26]

In addition, during the 1950s, the car industry's rationale for opposing regulation began to weaken. Motivated by citizen action that they saw as too zealous and concerned about their industry's public image, the carmakers responded to a five-day smog event in 1953 by joining with leading Los Angeles businesses to create the Air Pollution Foundation. After even *its* studies showed a large role for automobile exhaust in smog creation, the car industry finally, in 1958, accepted that cars were largely responsible. However, it still denied that smog caused health problems, could occur anywhere but Southern California, or could be remedied by affordable automotive technology.[27]

Citizen and elite mobilization, which was reinforced by severe smog events in 1959, led to the passage of a 1960 state law authorizing pollution control requirements for cars, to be implemented by a new state agency. Once the new Motor Vehicle Pollution Control Board had certified at least two devices as effective in controlling air pollution, new cars sold in the state would be required to be equipped with one such device. The same would apply to used cars, but only in counties that did not opt out of that provision of the law.[28]

CARB and Exemption from Federal Standards (1967–68)

However, as had happened with the 1947 law, weaknesses in the 1960 law allowed the smog problem to continue, and this ultimately led to the mobilization of new political commitment for a stronger law. During the 1960s, the smog problem did not improve, and it even worsened beginning in 1965.[29] The new state Motor Vehicle Pollution Control Board seemed inadequate for the task of reducing smog. Lobbying by the carmakers had prevented the new agency from getting strong enforcement powers, such as through vehicle inspections. Moreover, the 1960 act specified that a lay board – dominated by representatives of major interest groups rather than air pollution experts – would run the agency. Representatives of automobile clubs, carmakers, and the trucking industry sat on the board, among others, and the board depended on the major carmakers for technical advice.[30] While the board was dependent on cooperation with the carmakers, the latter dragged their feet on developing emissions control technology.

In this context, the Los Angeles Air Pollution Control District and its head, Louis Fuller, led elite mobilization in favor of a new, more powerful state-level agency. The district publicly criticized the board concerning the effectiveness of the first control device that the board had certified, the speed at which devices were certified, and a delay in adopting new, tighter emissions standards announced by the state health department. The district claimed that the board was too conciliatory to the car industry, which, together with the smog problem, helped to create a negative public impression of the board. A revolving door between the board and industry, along with trips to Europe paid by carmakers, reinforced this impression. The controversy boiled over in 1966, when Fuller denounced the board's policy as a total failure.[31]

In addition, pollution problems and anti-pollution movements in other cities created a general public perception that air pollution was a serious problem requiring urgent action. With rising car ownership, air pollution problems became evident in other U.S. cities in the 1960s. Knowledge of health effects grew and doctors became involved in the debate, framing air pollution as a health problem. Strong reform movements developed in a number of cities with major air pollution problems, including New York and Pittsburgh.[32]

The widespread local mobilization against air pollution spurred the federal policy-making process. The issue skyrocketed in national news coverage and as a priority issue in public opinion during the 1960s.[33] By 1965, Pennsylvania and New York State were considering bills, the latter with standards tougher than California's. As a result, the carmakers stopped opposing federal legislation, since they preferred uniform national standards. The federal Motor Vehicle Air Pollution Control Act was passed that year, and it threatened to preempt state authority over air pollution from cars. After two decades as the national leader in air pollution control, California now was at risk of seeing its standards replaced by new, weaker federal standards and of permanently losing the ability to set its own. In response, both the California Motor Vehicle Pollution Control Board and the press called for the creation of a broader air pollution agency in California in order to preserve a role for the state in air pollution policy.[34]

The result was the 1967 Mulford-Carrell Act, which created CARB and gave it authority to develop air quality standards in different air basins that it would define. The new agency also had authority to regulate motor vehicles and stationary sources, although the latter only if local government failed to meet CARB's standards.[35] From 1968 to 1973, CARB was headed by Haagen-Smit, who by this time had been a pioneer not only in smog research but also in advocacy and regulation in Southern California. CARB initially was headed by a board of five agency heads and nine gubernatorial appointees, who were required to have some expertise in air pollution, unlike the old motor vehicle board, and it had a technical advisory committee as well.

CARB's role and independence were soon strengthened by additional legislation. The 1968 Pure Air Act set specific standards for motor vehicle emissions of hydrocarbons, nitrogen oxides, and carbon monoxide, while increasing CARB's powers to set and enforce stricter standards. Although the car industry lobbied against the bill, mobilization by radio journalist Al Wiman, the Stamp Out Smog group, and the press helped to prevent significant weakening of it. Later, in 1971, the legislature reformed the board to consist of five full-time members. The initial board had been criticized as ineffective, since many of its members were part-time, uncompensated, and alleged to be politically motivated appointees of Governor Ronald Reagan.[36] These legislative efforts were successful, as the agency became a highly independent regulator of air quality in the state and grew to more than a thousand employees by the time the Global Warming Solutions Act was passed in 2006.

The smog problem was severe enough to create a degree of cross-party consensus, so these bills had bipartisan support. A Democrat and a Republican sponsored the Mulford-Carrell Act, and the Pure Air Act had Democratic sponsorship in the legislature; both were signed by a Republican governor, Reagan. Although he was a conservative Republican and had gained a reputation as an anti-environmentalist for his campaign statements concerning redwood forest preservation ("A tree is a tree, how many more do you need to look at?"), Reagan began moving toward popular environmental positions soon after taking office.[37] He paid attention to the smog issue throughout his eight years in office, supported state regulatory standards, and in 1967 had proposed an even broader agency that would control air, water, and solid waste. However, after CARB was formed, Reagan deferred to the car industry, for example by questioning the need for catalytic converters, so the legislature often took the lead on smog control.[38]

The timing of the Mulford-Carrell Act helped to preserve an independent role for California in air pollution regulation. The 1967 federal Air Quality Act gave the state the right to apply for an exemption from federal standards – putting the burden of proof on the federal Department of Health, Education, and Welfare to show that California's stricter standards were not needed to meet its "compelling and extraordinary conditions." However, this provision of the federal law was far from inevitable. It was adopted only after an intense struggle and with the help of fortunate timing. The latter took the form of a coincidence, in Fall 1967, of severe smog in Southern California and congressional debate on the bill in Washington.

Los Angeles officials testified in Congress in favor of minimum federal air pollution standards that states could make stricter as needed. As a result, the U.S. Senate adopted the Murphy amendment, which allowed California to set stricter standards. However, the amendment was strongly opposed by the Automobile Manufacturers Association, which preferred uniform standards. The manufacturers wrote an amendment that gutted the exemption for California; Representative Dingell of Michigan introduced this amendment in the House of Representatives. Outrage in California united Democrats, Republicans, and even the feuding state Motor Vehicle Board and Los Angeles Air Pollution Control District in a vigorous lobbying campaign in Washington. Moreover, the Dingell amendment came to light during an intense, long smog episode from late August to October 1967, and Dingell visited Southern California at that time, further inflaming opinion against the car industry.[39]

The lobbying campaign also received invaluable support from a multi-part radio series, called the "Breath of Death," that was developed by an enterprising journalist, Al Wiman, at the station KLAC.[40] The series focused on the health effects of smog (which it called "Dingell's dust"), evidence of the car industry's ability to reduce emissions, and the industry's efforts to block legislation. The series prompted constituents to write more than 300,000 letters, which were dumped on the steps of the Capitol. Inside Congress, the air pollution control officials used states' rights arguments and won over several Southern states as well as New York. To the surprise of its advocates, the California exemption was restored and strengthened in the House version of the bill and then survived the conference committee process.[41] California's right to an exemption from federal standards was also included in the Clean Air Act of 1970, and the 1977 Clean Air Act Amendments permitted other states to adopt California standards.

The outcome of this crucial conflict in 1967 was that administrative capacity was created in California and the legal capacity to adopt stricter standards there was preserved, institutionalized, and later legally extended to other states. These outcomes were made possible by a combination of intense problems and political mobilization.

THE ENERGY CRISIS AND ELECTIONS: ENERGY POLICIES IN THE 1970S

During the 1970s, problem events and political action combined to create three windows of opportunity that led to three episodes of policy reform, concerning: the 1970 California Environmental Quality Act, the 1974 Warren-Alquist Act, and the energy efficiency and renewable energy policies of the Brown administration in 1976–82. The last two of these episodes created the CEC and marked the beginning of California's implicit climate policies, while the first episode is included here because it directly contributed to the Warren-Alquist Act.

The California Environmental Quality Act (1970)

In the late 1960s, intense public attention on environmental issues and several political developments produced the Environmental Quality Act. In California, as in the rest of the nation, growing environmental problems and scientific knowledge triggered an increase in awareness of environmental issues, made them appear urgent, and led to the growth of environmental organizations.[42] The passage of the National Environmental Policy Act in 1969 and the staging of Earth Day in April 1970 had nationwide impacts. In California, conflicts over nuclear power plant siting and redwood logging since the early 1960s led to the growth of environmental organizations, which also were inspired by the activism of the civil rights and anti-Vietnam war movements.[43] Moreover, the 1969 Santa Barbara oil spill focused attention on the value and fragility of the state's coastal areas, which the Planning and Conservation League moved to protect through state-level coastal planning legislation.[44] At this time in California, longstanding environmental organizations saw a surge in membership, with the Sierra Club growing from 15,000 members in 1960 to over 130,000 in 1972.[45] Also, longstanding environmental issues such as smog gained the attention of new organizations, such as the Group Against Smog Pollution and student groups. They used somewhat more disruptive tactics and made more radical demands, like calling for a boycott of new car purchases, than did the large environmental organizations.[46]

Several political actors took the opportunity presented by this upsurge in attention to press for a strong environmental review law in the state. During the late 1960s, the Sierra Club became increasingly hostile to business interests as a new generation of activists joined it. Led by the organization's new president, David Brower, the newcomers came mostly from universities, nonprofits, and government employment rather than from business, which had long been the club's main recruiting ground. Under the impetus of the new activists, the Sierra Club broadened its agenda beyond its traditional focus on wilderness preservation to other issues, including the regulation of corporate polluters, energy, and toxic chemicals.[47] The Sierra Club worked with Democratic

Assemblyman John Knox on the environmental quality bill, which was modeled on the National Environmental Policy Act and passed with little legislative opposition.[48]

The amount of political commitment required to pass the legislation was small, but it required much more commitment to implement once its implications became clear. The act required state agencies to conduct detailed reviews of the environmental impacts of new projects and to adopt measures to reduce the impacts. Despite attempts by Governor Reagan's administration to limit its scope through broad exemptions, the law quickly proved to be a powerful tool to slow and call into question both private and public construction projects. Attorney General Evelle Younger implemented the law vigorously and defended it against private interest groups and his fellow Republicans. State courts also supported the law. A September 1972 California Supreme Court ruling in the *Friends of Mammoth* case held that the law applied to private developments. In response, opponents tried to obtain a moratorium on the law's implementation and a legislative exemption for private projects, but both efforts failed.[49] The law was left mostly unscathed, and with it, environmental groups had gained leverage to influence large construction projects, including new power plants.

The Warren-Alquist Act (1974)

The next episode of policy reform, the passage of the Warren-Alquist Act, was due to a convergence of several major problems and several political developments. The first problem was that population and economic growth led the state's three large investor-owned utilities to project enormous long-term increases in electricity demand (6% per year) and hence a need for massive new power plant construction.[50] A Rand Corporation report estimated that the state would need between 80 and 120 new nuclear power plants by 2000, one every seven miles along the Pacific Coast.[51]

The second problem was that the Environmental Quality Act required the utilities to submit proposals for new plants to detailed environmental review, which opponents could use to slow or block the permits that were required from a variety of local and state agencies. Seeking to avoid these difficulties of compliance, the utilities sought a reformed, centralized siting law from the state legislature, arguing that it was necessary to avoid electricity blackouts and brownouts.[52] This created a sense of urgency that was used by Democratic Assemblyman Charles Warren to pass the first edition of his bill in 1973. It would have given the utilities more centralized siting procedures, while also initiating an energy efficiency policy.[53]

Third, just a few weeks after Governor Reagan vetoed that bill in October 1973, the five-month-long Arab oil embargo began. This created a sense of crisis due to the quadrupling of oil prices, shortages, and long lines at gas stations. Although the West Coast was less affected by supply disruptions than the Northeast, California was forced to adopt odd-even gas rationing in February 1974.[54] The crisis made comprehensive energy planning suddenly seem very desirable.[55]

These problem events coincided with several political developments that increased the political commitment for reform, thus creating a window of opportunity. First, opposition to nuclear power increased in California. The state had seen strong opposition to proposed coastal nuclear plants at Bodega Bay, Malibu, and San Clemente beginning in 1962.[56] In the late 1960s, the Sierra Club became internally divided on the issue, and ultimately adopted an anti-nuclear position in 1974.[57]

Second, nuclear opponents found influential allies in Charles Warren, a Democrat who became the chair of the Assembly's natural resources and energy committee, and in the Rand Corporation.[58] When the California State Senate adopted a utility-sponsored centralized siting bill in 1971, the Assembly's land use committee commissioned a study of energy demand and planning from Rand.[59] Several Sierra Club members helped to produce the Rand report, which criticized the utilities for relying too heavily on nuclear power, given emerging safety concerns. Crucially, the report estimated that increased energy efficiency and alternative energy sources could obviate the need for most of the new power plant capacity that the utilities anticipated they would require.[60]

Warren, who later became chair of President Carter's Council on Environmental Quality, used the hearing process and concessions on the content of the bill to build support and overcome opposition.[61] The basic architecture of his bill reflected an attempted compromise between opposing interests. It combined centralized site planning, which the utilities favored, with a new regulatory agency to promote conservation and alternative energy sources, which the environmental organizations wanted. Warren considered the creation of a new agency to be essential, given that the CPUC was lax in regulating the utilities. At an early stage, Warren got the support of Al Alquist, a pro-nuclear, Democratic state senator, in order to overcome the hurdles in the Senate, which was less apt to pass environmentalist legislation than the Assembly.

Warren held extensive hearings on the bill and made concessions to meet objections expressed by a range of interest groups.[62] In response to the Spring 1973 hearings on the Rand report, Warren downplayed the anti-nuclear implications of the bill in order to defuse opposition, while still promoting conservation and alternative energy.[63] Vigorous lobbying by General Electric and the Association of Home Appliance Manufacturers led Warren to modify the bill by limiting appliance standards to those shown to be technically feasible.[64] Warren also courted the news media and won endorsements from the *Los Angeles Times*, *Sacramento Bee*, and other newspapers.[65] He also gained support for the bill from environmental organizations, labor unions, major elected officials, local government associations, the publicly owned utilities, and eventually the private utilities.[66] Appliance manufacturers, other manufacturers, and the CPUC were opposed. Warren also worked hard to get the utilities to take a neutral position on the bill, and although he failed in 1973, they did ultimately back the bill in 1974.[67]

Third, after the oil embargo, Governor Reagan switched to support the bill, and his staff worked to get broader support for it. His 1973 veto had come under pressure from industry, utilities, and local government, the latter because they did not want a powerful new state agency.[68] Reagan came under much criticism for this veto, and the Planning and Conservation League threatened to launch an initiative campaign; legislative passage of the bill assured that the measure could be placed on the ballot. Moreover, the oil crisis and its gas lines were creating a sense of panic, and the bill's supporters warned of future electricity blackouts. Within a few months, Reagan's staff contacted Warren, and negotiations began. Reagan agreed that the bill he had just vetoed would be the basis for a new draft of the law, and after detailed negotiations among Warren, Reagan's staff, and the utilities, the bill that emerged was largely unchanged. Reagan supported the final product, despite opposition from his party, and insisted that the utilities do so too. With Reagan's help, the bill passed by one vote in the State Senate, and he signed it in May 1974.[69]

The Warren-Alquist Act created the Energy Resources Conservation and Development Commission (later called the California Energy Commission), with authority to carry out research, forecast energy demand, decide plant siting, regulate appliance energy consumption, set standards for energy use in new buildings, and promote the development of alternative energy technology. The CEC's five commissioners would be appointed by the new governor after the 1974 elections, and hence not by Reagan.

Implementation of appliance and building standards was not smooth, and it required additional mobilization of political commitment. In the face of threatened litigation by trade groups and initial opposition from some CEC commissioners, Charles Warren publicly criticized the agency, and the legislature held hearings to pressure it in March 1976.[70] This led the CEC to adopt the standards according to the schedule specified in the 1974 legislation. The commission issued the first mandatory appliance standards in 1976, for six gas appliances and for refrigerators, freezers, and air conditioners. The former set of standards, which banned pilot lights, was amended by the legislature in 1978.[71] California's appliance standards became models for other states and a spur to the adoption of mandatory appliance standards by the federal government.[72] The CEC also adopted standards for new non-residential and residential buildings in 1976 and 1977, respectively.[73] Despite successful litigation to block the building standards, legislative overrides and amendments allowed most of them to go into effect in July 1978.[74] Local government could not issue a building permit for a new building unless it met the new energy standards.[75]

Energy Efficiency and Renewable Energy Policies (1976–82)

The energy policies of the late 1970s resulted from another convergence of severe problems and political changes that were favorable to reform. The problems were two-fold. First, a series of energy crises occurred during the 1970s. The 1973–74 oil shock had long-term effects, by raising the political salience of energy policy (see Figure 6.1 earlier in this chapter) and calling into question the doctrine of indefinite growth in energy demand.[76] It also made energy efficiency, conservation, and new energy sources, especially alternatives to oil, more attractive. But the shift in attitudes and priorities took almost a decade to produce major changes in private utilities' behavior, even in California, due to institutional resistance by the utilities and the CPUC.[77] The pressure to change was heightened in the first months of 1979, when a second supply shock due to the Iranian Revolution caused oil prices to nearly double and shortages and gas lines to reappear in May.[78] Around the same time, in March, the Three Mile Island nuclear accident made it clear that the moribund nuclear power industry would not be reviving soon. The accident contributed to the sense that there was a crisis in energy policy and that new solutions were needed.[79]

The second problem that emerged in the 1970s heightened the sense of urgency about reforms in the utility sector specifically. In 1979, a combination of rising fossil fuel price increases and skyrocketing interest rates (which reached 15%) helped to bring on a financial crisis in California's three large investor-owned utilities.[80] These already were burdened by the need to pay interest on money borrowed to build new power plants and by construction delays and cost overruns. The latter were caused by environmental reviews and litigation, as well as Pacific Gas and Electric's choice to build the Diablo Canyon nuclear plant on what proved to be an earthquake fault. With energy prices rising and the economy slipping into recession, the utilities' optimistic electricity demand forecasts and hence the financial rationale for building new power plants became untenable.[81]

Around the same time, developments in the political stream were also creating a strong political commitment to alternative energy policy in California. The first was that the Democrat Jerry Brown won the 1974 gubernatorial election, by a 50% to 47% margin over his Republican opponent, Houston Flournoy. Although he was the son of a former governor, Brown's election was something of a fluke, since a centrist Democrat, Robert Moretti, had been widely expected to become the party's nominee. Brown won the Democratic nomination largely because during 1974, the Watergate scandal suddenly elevated the issue of political corruption, which Brown was already known for addressing as California Secretary of State.[82]

Electoral competition led both major-party candidates, as well as Brown's opponents in the Democratic primary, to emphasize their support for state government intervention on energy and environmental issues, which were major campaign issues. However, while Brown and Flournoy both criticized oil companies and called for the CPUC to represent consumer interests, Brown's support for government intervention on the environment and energy was clearly much stronger than Flournoy's; hence, Brown gained the endorsements of the League of Conservation Voters and the Sierra Club.[83] Brown harshly criticized the CPUC as a "private utility commission" and called for it to change utility rate structures in order to reward conservation. Brown also called for the creation of a state-owned oil company to deliver fuel to the electric power companies and a super-agency to oversee energy research and development.[84] In addition, Brown endorsed the development of solar and geothermal power by a state energy corporation, opposed offshore oil drilling, and expressed reservations about nuclear energy.[85]

Jerry Brown generally pursued a strong environmental policy in his first term as governor. He appointed environmental activists and others sympathetic to their cause to key positions, including the heads of the state's Natural Resources Agency, Water Board, Department of Transportation, Air Resources Board, Energy Commission, and Forestry Board.[86] His CARB chief, Tom Quinn, was particularly aggressive in tightening smog rules. On energy policy, Brown created a small Office of Appropriate Technology in 1976, which provided information to promote small-scale wind energy systems.[87]

However, Brown initially did not place much priority on energy policy, and he did not use the levers of the executive branch to promote energy alternatives or efficiency during most of his first term. His first five appointments to the CEC were a mix of those who favored energy development, including nuclear power, and advocates of conservation.[88] In a 1976 ballot measure to restrict nuclear power, Brown remained neutral even as his father, former Governor Pat Brown, took a leading role in the pro-nuclear coalition.[89] Although Jerry Brown had appointed a majority of the CPUC's commissioners by 1977, the agency delayed taking action against the utilities for their foot-dragging on efficiency and renewables until 1979, after Brown had won re-election against the pro-nuclear Evelle Younger.[90]

However, the anti-nuclear movement helped to spur Brown to greater commitment to an alternative energy policy. In 1976, the anti-nuclear ballot measure led to compromise legislation that made permits for nuclear power plants dependent on adequate nuclear waste disposal. Brown was influenced by that conflict, by environmentalists' criticism of his CEC appointees, and by growing public criticism of electricity rate increases, which the CPUC routinely approved. Between Summer 1976 and the end of 1977, he shifted toward a clearly anti-nuclear position. Brown appointed nuclear opponents to the CEC, and in February 1978, he publicly opposed the Sundesert nuclear plant. The plant was killed two months later when Kern County voters rejected the plant in a referendum and an Assembly committee rejected exempting the plant from the 1976 nuclear safeguards laws. Brown had threatened to veto an exemption for the plant.[91]

Given the air quality problems associated with coal burning and the federal ban at the time on burning natural gas for electricity, the end of the Sundesert plans put pressure on the state to promote energy efficiency, renewables, and other energy sources. During 1975–90, California's in-state electricity generation moved sharply away from oil and toward natural gas (since its use for electricity generation was permitted again), nuclear (as two major plants finally came online in the 1980s), and, to a lesser degree, geothermal, biomass, and wind power, while coal consumption was essentially stable.[92*]

Brown's shift on nuclear power led him to support energy efficiency. In his search for justifications to block the Sundesert plant, Brown spoke with Arthur Rosenfeld, who explained that his research for the CEC on refrigerators showed that adopting an efficiency standard for that one appliance alone could save the capacity equivalent of one and a half Sundesert plants. Rosenfeld implies that Brown's support helped to get the appliance standards approved, despite opposition by manufacturers, who had threatened litigation in October 1976.[93]

Brown also embraced renewable energy more fully at this time. He was inspired by "soft-energy" advocates Wilson Clark, who became a Brown advisor, and Amory Lovins, who spent 1977 lecturing and debating in California and who briefed the governor on energy policy. Brown liked the potential for renewable energy to help obviate the need for nuclear power, reduce dependence on oil, and keep down energy prices.[94] In 1977, Brown argued against increased reliance on coal-generated electricity, due to air pollution concerns, and supported a bill to expand the state income tax credit for homeowners who installed solar hot water or space heating devices. The legislature passed a bill raising the credit to 55%, with a maximum of $3,000 per household.[95] Around the time of the nationally declared Sun Day, which was held on May 3, 1978, the Brown administration convened an expert conference on solar energy at the University of California at Los Angeles.[96] At the urging of the Office of Appropriate Technology, the state's solar tax credit law included a 25% tax credit for wind power installations.[97] Moreover, in 1980, Brown called for the creation of an alternative energy financing authority; he reached an agreement with the legislature to issue bonds worth $500 million, and voters approved a corresponding ballot measure.[98] Finally, Brown held an important conference for wind power investors in 1980, which helped lead to the first large wind installations in the state the next year.[99]

The second political factor that helped create a window of opportunity was the advocacy work of the Environmental Defense Fund, Sierra Club, and California Tomorrow, a conservation and planning think tank. These organizations supported alternative energy policy in the 1970s and criticized the utilities' plans to expand electricity generation from nuclear, oil, and coal sources in order to serve the state's growing population.[100] The Environmental Defense Fund also intervened in rate increase proceedings at the CPUC from 1977 onward. It argued that if utilities would invest in alternatives such as cogeneration, attic insulation, more efficient air conditioners, geothermal steam, solar collectors, and wind turbines, they could make more money and keep electricity rates for consumers lower than if they made investments in large new power plants. After Pacific Gas and Electric mounted a counter-offensive advertising campaign that belittled renewable energy in 1978, a coalition of environmental organizations successfully gained equal time to air their own television advertisements for conservation and energy alternatives in northern California. Television stations also ran documentaries on the Environmental Defense Fund's proposals.[101]

Third, without waiting for the governor, the Democratic-controlled legislature also took the initiative in pressing for renewables and efficiency. Legislators proposed and passed a series of laws promoting wind and solar energy in 1977–80, in addition to those just mentioned. These included a requirement for solar energy use in new state buildings, laws to ease solar financing, and ambitious targets for renewable-source electricity (1% by 1987, 10% by 2000). In 1978, the legislature approved funding for the CEC to conduct research on wind resources in California, which proved crucial in attracting investors to California, since it was the only state able to provide this essential technical information.[102] At a time when the CPUC was still unwilling to act, Democratic Assembly Speaker Leo McCarthy pressed the agency and the governor to give priority to an internal investigation of the merits of utility investments in efficiency and energy alternatives. Brown, who at first had tried to keep the utilities away from such investments in order to create opportunities for decentralized producers, finally accepted the idea of utility-financed investments in alternative energy that would be produced by other firms, as part of a $1.8 billion energy program.[103]

Pressure from the governor, legislature, and the Environmental Defense Fund finally converged on the CPUC. In 1979, after years of quiet defiance by the large private utilities, the CPUC fined Pacific Gas and Electric $14.4 million over two years for failure to develop cogeneration, other alternative sources, and energy efficiency.[104] The fine – though tiny compared with the $400 and $900 million rate increases that the utility was winning in those years – was a shock for the company and its counterparts in other parts of California. The utilities immediately became more receptive to alternatives to new plants and began to implement them, a shift facilitated by their financial problems and the comparatively low cost of the alternatives.[105]

Therefore, utility spending on conservation programs increased in subsequent years.[106] Under pressure from the CPUC – which itself was under pressure from the CEC – Pacific Gas and Electric and Southern California Edison withdrew from a large coal plant project in Utah in 1981.[107] Within two years, Pacific Gas and Electric announced plans to use efficiency measures, cogeneration, and other alternative sources to either save or create 5,000 Megawatts in capacity. At this point, its plans did not include any new power plants, in contrast to its 1975 plans to build ten large coal and nuclear plants.[108]

The conservation measures saved not only fuel, but also money for ratepayers, since they cost only 2 to 6 cents per KWh to implement, compared with 6 to 15 cents for new electricity generation.[109]

The CPUC also decoupled utilities' profits from sales volume in 1978 (for gas) and 1982 (for electricity). This removed the utilities' disincentives to pursue efficiency programs and set the stage for long-term curbs on energy demand.[110] The Natural Resources Defense Council and the CEC initiated the concept and developed the rules with the CPUC.[111] Decoupling meant that a utility which saved electricity, for example through education or fluorescent light bulb subsidies, would get revenue from a public goods charge to compensate for the lost electricity sales. As a result, the large investor-owned utilities became advocates of energy efficiency and no longer resisted proposals for new state efficiency policies.[112]

Finally, the CPUC, in implementing a 1978 federal law, also set high prices for renewable electricity in contracts that it developed in negotiation with the utilities. Its 1979 fines had gotten the utilities interested in cooperating with wind power development.[113] The CPUC began to issue standard contracts beginning in 1982, at the generous rate of seven cents per KWh on average, guaranteed for ten years. This feed-in tariff, combined with generous tax credits and CEC research on wind potential, led to a boom in wind power, so that in 1985, 87% of the world's wind power capacity was located in California.[114]

Policy Reversal on Renewable Energy (1983–85)

However, in the early 1980s, both the problem stream and the political stream abruptly reversed in California as well as nationally, creating a window of opportunity for rolling back renewable energy policy. After peaking in 1980, real crude oil prices fell sharply during the 1980s, as did coal and natural gas prices.[115] The low prices sharply reduced public and political interest in alternative energy and made aggressive renewable energy promotion policies difficult to sustain.[116]

At the same time, California saw a political reversal with the election of the conservative Republican George Deukmejian as governor in 1982. Deukmejian sought to end the state's renewable energy tax credits, and although the legislature initially resisted him,[117] the combination of changes in the problem stream and the political stream was irresistible. California's utilities stopped signing preferential contracts for wind power in 1985 and the CPUC gave operators with existing contracts only five years to get their projects online; the state also allowed its tax credit for wind projects to expire in 1988.[118] As a result, wind power capacity stagnated,[119*] and some wind and solar companies went bankrupt. Many skilled and experienced people left the industries, and wind and solar energy became saddled with a stigma of ineffectiveness and financial failure.[120]

CONCLUSIONS

The major reforms adopted in this period – the 1967 Mulford-Carrell Act, the 1970 Environmental Quality Act, and the 1974 Warren-Alquist Act – were made possible by policy windows that opened because of the convergence of problems and political commitment to address them; these are summarized in Table 10.1, in Chapter 10. From the 1940s to the 1960s, the policy reforms were the result of a convergence in the smog problem and a broad coalition of elites and protest groups who wanted regulation of air pollution. In the first two episodes in this period, in 1947 and 1960, the problem largely called forth the political response, because the problem had intense, short-term, highly visible effects on everyday life and the economy. The passage of the Mulford-Carrell Act marked a transition, where the problems that policy makers had to deal with widened to include institutional issues presented by policy-making processes in other states and at the federal level. In this case, political events were largely independent of the problem stream, and the outcome depended partly on contingent events, notably the coincidence of the congressional deliberations in Washington with a severe smog event in California and a carefully orchestrated grassroots and elite lobbying campaign.

However, the policies adopted had broad support in both parties. Hence, election outcomes did not play much of a role, because officials in both parties were aware that public concern about air pollution could cost them at the polls. The policies had the support of leading Republicans in the executive branch (Poulson, Knight, and Reagan) as well as Democrats in the legislature (Don Mulford and Jim Foran for the 1967–68 policies).

In the 1970s, the results of the energy crisis, Brown's election, and the growing environmental movement placed California in the forefront of energy conservation and renewable energy development. A problem stream and a political stream that were highly favorable to implicit climate policy making converged at this time in California, as they did to a lesser extent, and with more legislative resistance, nationally (see Chapter 9). In California, policy entrepreneurs used the windows of opportunity that appeared to create energy policies that, four decades later, are still major components of the state's climate policy. As with the period from the 1940s to the 1960s, intense problems were sometimes able to call forth political commitment. The 1974 Warren-Alquist act did not require a change in electoral politics, since during the oil crisis even Reagan was forced to act, despite the opposition of utility companies, appliance manufacturers, and the building industry. However, Warren's political commitment, which predated the oil embargo, was also crucial to the bill's passage.

In the late 1970s, political events were more independent of the problems. The far-reaching changes in policy in that period, which were carried out by the CEC and CPUC, depended to a large degree on Jerry Brown's election and the commitment of his administration to issues of environmental protection and alternative energy. But Brown's election was far from inevitable. He first benefited from the Watergate scandal, which elevated him above the favorite, Robert Moretti, in the Democratic primary election. Then Brown ran a weak campaign (e.g., he canceled all television advertising for a week) and was hampered by rumors that he was gay. He held on to win the general election by only three percentage points, despite the political advantages that Democrats had in the immediate aftermath of Watergate.[121]

In California, earlier reforms made later reforms more likely, in a form of path dependence. The 1947 state law that authorized the first county-level air pollution control district, in Los Angeles County, created an agency that later pressed for an expansion of state authority through the creation of the motor vehicle pollution control board in 1960 and CARB in 1967. In turn, the creation of CARB made possible California's exemption from federal preemption in federal air pollution laws. Much later, CARB's administrative capacity and unique legal authority allowed it to be a crucial actor in developing and implementing explicit climate policies in the 2000s.

Path dependence also operated in other ways during the 1970s. The environmental quality act drove the public utilities to seek a siting law, which became the Warren-Alquist Act. The reviews under the 1970 act also helped to stall the construction of large new power plants, which made the alternatives of energy efficiency and renewables more viable in the late 1970s. The Warren-Alquist Act created the CEC in 1974 and mandated it to create energy forecasts and advise the CPUC. This helped to push the CPUC and the utilities finally to abandon new power plants in favor of increased efficiency, renewable energy, and cogeneration in the late 1970s and early 1980s.

Although I have emphasized the role of processes and policy windows, most of the structural conditions identified in the statistical analysis shaped the political processes during these early phases of the California case study. In Los Angeles, the local geography and wind patterns, combined with rapid population and economic growth, created the worst air quality of any major U.S. city in the first postwar decades. At the same time, economic growth and in-migration produced a large middle class, attached to its automobiles but also expecting good living conditions. The sprawling growth pattern in Southern California dispersed the air pollution for a time, but when it worsened, it was intense over a wide area that transcended county boundaries and seemed to require a state-level solution.[122] Therefore, these structural conditions made it possible for strong political mobilization against smog to emerge in the late 1940s and to recur for decades.

California's highly educated population helped lead to the growth of the environmental organizations that were important in the conflicts over smog, nuclear power, and the energy efficiency alternatives. Education levels were relatively high in California in the 1960s and early 1970s; in 1960, 9.8% had college degrees in the state, compared with 7.7% nationally.[123] During the late 1960s and the late 1970s, 80% of Sierra Club members were drawn from professional occupations that require high levels of education, with the bulk of them working in the education sector or nonprofit organizations.[124]

California's citizen ideology was also more liberal than the U.S. average in this period. The state had a rating of between 50% and 55% liberal in 1967 and 1974, about 10 percentage points higher than the national average, although it ranked only 16th to 19th among U.S. states. A relatively liberal citizenry helped to produce a relatively liberal set of state government officials, at least in the legislature,[125*] and the combination of liberal citizens and a liberal legislature helped to lead the Republican governor Reagan to a bipartisan accommodation on air pollution control policy. By contrast, fossil fuel production was not a crucial factor in the California case. The large size of California's oil production and refining industries was not an effective barrier to the adoption of cutting-edge air pollution regulations and regulatory institutions in the 1960s and 1970s. The same holds in the present period, as the next chapter helps to show.

7

California: Climate Policy Making (2000s to the Present)

This chapter focuses on three episodes of explicit climate policy making in California: the adoption of vehicle greenhouse gas emissions standards and a renewable portfolio standard in 2002; the setting of binding emissions targets and the adoption of supportive sectoral policies in electricity, transportation, and renewable energy in 2005–08; and the failed attempt to block the Global Warming Solutions Act by a ballot measure in 2010 (see Chapter 2 for details on the policies). In this period, California began the explicit regulation of greenhouse gas emissions and rapidly became a national and international leader in climate policy. It began to serve as a model for other states, a spur to action for the federal government, and a partner or potential partner for other western states and Canadian provinces, RGGI, the United Kingdom, Mexico, and the European Union. Compared with those in the previous chapter, the policy windows that opened in the 2000s were influenced to a greater extent by political developments that operated autonomously from problem events and that helped to shape perceptions of the problem.

THE PUBLIC AGENDA AND THE POLITICAL AGENDA

Public attention on the climate change issue in California rose during several periods to small peaks, in 1989–90, 1992, 1997, and 2001–02, and to a very large peak in 2006–08, as shown in Figure 7.1.[1*] In 2001, attention to energy problems also spiked in California, dwarfing the climate change issue that year and even exceeding the 1970s peaks in attention to energy.[2*] Specific problem events drove those peaks: the emerging scientific consensus about global warming (late 1980s), the Rio conference (1992) and Kyoto Protocol (1997), the California electricity crisis and Bush's rejection of the Kyoto Protocol (2000–01), and a confluence of national events that highlighted climate change in the mid-2000s, which are analyzed in the 2005–08 episode in this chapter.

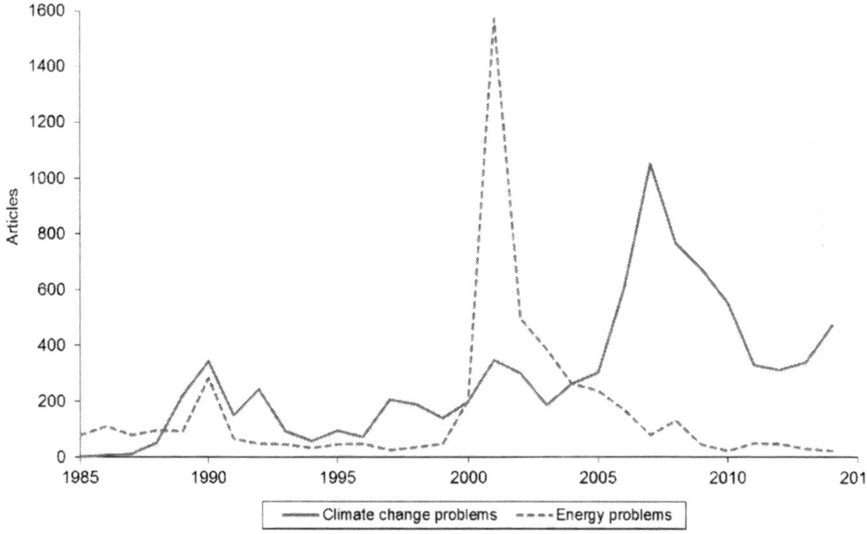

FIGURE 7.1 Public Agenda in California: Climate Change and Energy Problems in the *Los Angeles Times*, 1985–2014
Source: see note 1.

There is a fairly close fit between these periods of rising attention and climate policy making in California. The increase in attention to climate change in 1989–90 helped lead to the first, tentative climate policies, which consisted of legislation directing the CEC to study the problem.[3*] However, the 1992 and 1997 peaks had no effects on climate policy in California. Then, the large rise in attention to both climate change and energy problems in 2001 helped lead to the major reforms of 2002. The last period of rising attention to climate change, in 2006–08, occurred just as the legislature adopted the Global Warming Solutions Act (Assembly Bill 32, or AB 32) and supportive sectoral policies.

The climate change issue rose on the political agendas of legislators and governors in California largely at the same time as public attention did, though with more inertia. The legislative agenda, measured in terms of bills introduced, shows a small but sharp increase in interest in the 2001–02 session and a major increase beginning in 2005–06 and continuing for four more years (see Figure 7.2). A more detailed analysis of the 2005–06 session shows that 31 bills were introduced in the first two months of 2005, three times as many as in the previous two years. This sharp increase occurred several months before Schwarzenegger's announcement of his executive order setting an emissions reduction target. As the 2005–08 policy making episode will show, sometimes the legislature led on the issue, and sometimes the governor did. In any case, the periods of legislative interest correspond to the bursts of policy reform in 2002 and 2005–08.

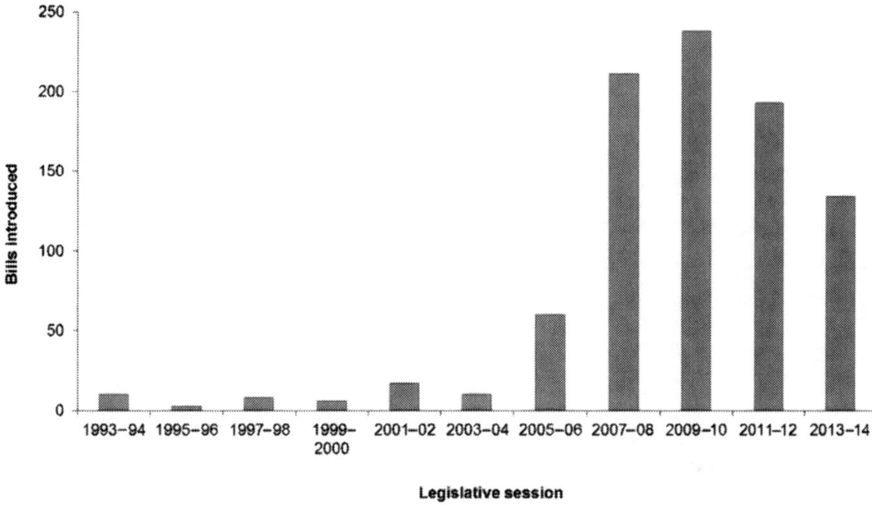

FIGURE 7.2 Legislative Agenda in California: Climate Change Bills, 1993–2014
Source: www.leginfo.ca.gov; keywords were "global warming," "climate change," and "greenhouse gas."

The political agendas of California's governors have corresponded somewhat more loosely to the public agenda on energy and climate problems. Annual State of the State messages are well suited for measuring the governor's agenda, since they are major speeches in which the governor addresses the legislature as well as the public. In State of the State speeches, the electricity crisis led Governor Gray Davis to give energy problems enormous attention in 2001, which presaged the 2002 renewable portfolio standard. However, climate change received significant attention in these speeches only in 2007 (from Schwarzenegger) and in 2012–14 (from Jerry Brown), as shown in Figure 7.3.[4*] Executive orders on climate change mitigation policy are another way of gauging governors' engagement, although to an extent this is also a measure of policy adoption. The figure shows that executive orders on climate change began in 2004, with orders on energy efficiency in state buildings and on infrastructure for hydrogen fuels in transportation. They then continued at a fairly constant pace, averaging one to two per year, through 2012.[5*]

In sum, these three figures generally support the hypotheses offered in Chapter 3. First, attention on energy and climate problems tends to cluster in response to problem events, such as the electricity crisis and President Bush's rejection of the Kyoto Protocol. The latter increased the pressure on, and opportunities for, the states to act and hence is part of the problem context for California. However, problem perception became somewhat uncoupled from specific problem events by the mid-2000s, as political shifts by a variety of actors helped to shape perceptions of the climate change problem. This is

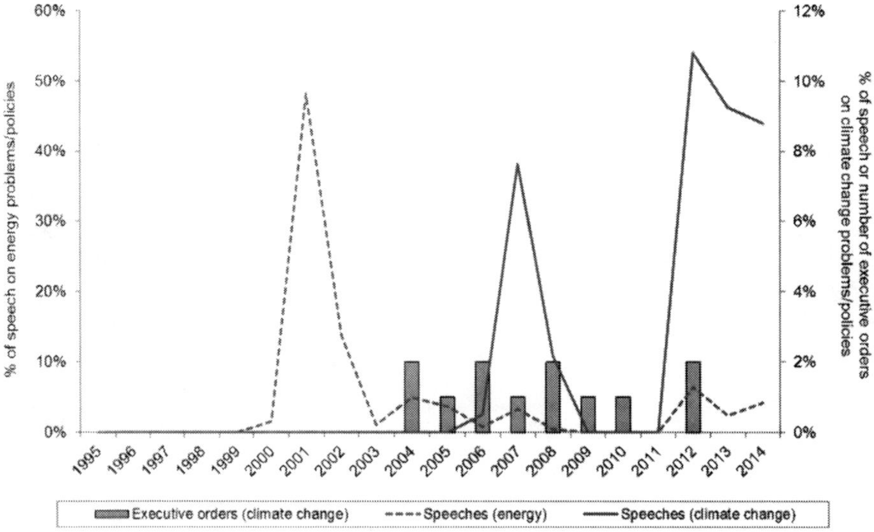

FIGURE 7.3 Governor's Agenda in California: Climate Change and Energy Issues in State of the State Speeches and Executive Orders, 1995–2014
Source: see notes 4–5.

described in the 2005-08 episode in this chapter and, in more detail, in the section on the post-2009 period of federal policy making in Chapter 9.

Second, the peaks of public attention to the global warming issue preceded or occurred simultaneously with the issues' peaks on the legislature's and governor's decision-making agendas, because political commitment was, or quickly became, high at these times. Hence, legislators responded to the electricity crisis by attempting major reforms, and both Schwarzenegger and legislative Democrats responded to the national rise of the global warming issue with major climate policy initiatives. Third, increases in attention by policy makers preceded the largest policy changes, in 2002 and 2005–08, although climate change became even more prominent on their decision-making agendas after 2008 and remained a high priority through 2014. Policy makers' attention to climate change displayed more inertia than public attention did.

POLICIES ON VEHICLE GREENHOUSE GAS EMISSIONS AND RENEWABLE ENERGY (2002)

In the early 2000s, the problem stream and the political stream came together and created a policy window, during which the legislature passed major legislation. Assembly Bill 1493, sponsored by Fran Pavley and usually called the "Pavley bill," created an innovative policy to regulate carbon dioxide emissions from motor vehicles, while Senate Bills 1078 and 1038, sponsored by Byron

Sher, created an ambitious renewable portfolio standard; all three bills were passed in 2002. At this time, there were three major events that heightened attention on energy and climate change problems and helped to put them high on policy makers' agendas.

Problem Awareness: The Electricity Crisis, Bush, and Climate Change Impacts

First, California experienced a major electricity crisis in 2000–01, with large price increases, electricity emergencies, and rolling blackouts. After the state's major utilities were required to sell their generating facilities and purchase electricity on the spot market, the wholesale cost of electricity to California rose from $7 billion in 1999 to $27 billion in each of the next two years, partly due to market manipulations by energy traders.[6] Natural gas prices also spiked at ten times previous levels in December 2000, due to high demand for the fuel in electricity generation. Since the CPUC did not grant corresponding rate increases to the investor-owned utilities, these were unable to make their payments to power producers, who responded by taking more than 10,000 Megawatts in generating capacity offline after October 2000. Pacific Gas and Electric went bankrupt, Southern California Edison nearly did, and the state government had to step in to buy electricity for their customers. For twelve months, the public was confronted with recurring emergency announcements that called on them to curtail consumption. Potential blackouts were a major public concern, and three rolling blackouts occurred in January, March, and May 2001.[7]

The electricity crisis had major short- and long-term effects on many aspects of California energy policy, which extended forward to the 2005–08 period of climate policy making. The crisis led to a spike in public attention on energy issues in 2001 that was even larger than the 1974 peak during the Arab oil embargo. The electricity crisis increased elite and public interest in energy security and diversification, paving the way for a renewable portfolio standard to be adopted with those justifications.[8] The crisis also demonstrated the potential and the importance of energy conservation as a way to address supply constraints and high prices. In addition, the crisis shook confidence in market solutions to energy problems in California, increasing the credibility of regulatory approaches. A *Los Angeles Times* poll in April 2001 found that California residents blamed the crisis mostly on deregulation (35%), state government (26%), and local power companies (22%), rather than on the federal government or out-of-state energy suppliers.[9*]

The effects of the electricity crisis on the RPS legislation are apparent in the text of the two bills and in the arguments made by advocates. Senate Bill 1078 referred to "stable electricity prices" as the first of several goals, while Senate Bill 1038 pursued "environmentally sound, safe, reliable, and affordable energy." The environmental organizations which advocated the legislation argued that the RPS would not only put California in the forefront of renewable energy development and cut air pollution, but would also buffer the state from natural gas price spikes.[10] A *Los Angeles Times* editorial in favor of the legislation argued that it was needed in order to diversify energy supply in light of the state's over-reliance on natural gas, which it saw as a major cause of the 2000–01 energy crisis.[11]

Second, international and national events combined around this time to raise the perceived importance of climate change and to motivate action at the state level in California, as in other states. The adoption of the Kyoto Protocol in 1997 spurred the interest of California's elites in climate policy, since they saw the state as part of the global community.[12] This helped lead to the passage of a 2000 act directing the CEC to conduct regular inventories of greenhouse gas emissions and establishing the California Climate Action Registry. The latter was a voluntary program through which companies could claim credit for emissions reductions made since 1990, which were potentially valuable in the event of a later adoption of an emissions trading system.[13*]

More directly relevant to the 2002 reforms was President Bush's retreat from mandatory climate policy measures and his rejection of the Kyoto Protocol, which he did through two major announcements in March and June 2001 (see Chapter 9). Bush's actions were widely criticized, which led to an increase in attention on climate change in California, as in the rest of the country. The number of *Los Angeles Times* articles on the topic more than doubled from 1999 to 2001, as shown in Figure 7.1, in the previous section. In response to Bush's policy reversal, officials in many state governments became more active on the issue, including George Pataki in New York (see Chapter 8). The news media and environmental organizations focused attention on the states, instead of the federal government, as venues for action on climate change.[14]

President Bush's announcements also led to criticism by political actors in California. Jerry Brown (at this point a former governor and the mayor of Oakland) argued that "global warming is not going away" and criticized Bush for not focusing more on alternative energy and conservation in his energy plan.[15] Environmental organizations such as the Environmental Defense Fund reported that membership was surging and that activists had greatly increased their interest in global warming in response to Bush.[16] Attorney General Bill Lockyer joined a lawsuit by environmental organizations that opposed the Bush administration's blocking of new refrigerator efficiency standards in June 2001.[17] In California as in the United States as a whole, public opinion ran against Bush's environmental policies, with a 56% to 34% majority opposing his reversal on climate policy.[18]

Bush's rejection of the Kyoto Protocol set the stage for the Pavley bill, by putting attention on global warming and by creating a policy void that the states could fill for at least the next eight years. After Bush's controversial election (by the electoral college) in December 2000 – and even before his reversal on climate policy – Russell Long, the executive director of the Bluewater Network, decided to focus on a state-level strategy for climate change mitigation policy. It would seek state regulation of greenhouse gas emissions from motor vehicles by making use of California's exemption from the Clean Air Act. By February 2001, the Bluewater Network had written the bill that became the Pavley law and had circulated it in the state legislature.[19]

After the Assembly's transportation committee approved this bill, national environmental organizations, including the NRDC and the Sierra Club, mobilized on it, as did the car industry. This nationalized the conflict over the bill, in what was a prelude to the 2003 struggle in Congress over the Climate Stewardship Act proposed by Senators John McCain (a Republican) and Joe Lieberman (a Democrat).[20] Advocates in California, arguing for the Pavley bill, linked it to Bush's policies. The NRDC said "the Bush administration is missing in action, and California is uniquely positioned to provide leadership and set a standard for the rest of the country."[21] Similarly, the *Los Angeles Times* and *San Francisco Chronicle* endorsed the Pavley bill, citing the threats posed by global warming, the fact that Bush was unlikely to act, and the U.S. Senate's rejection in early 2002 of vehicle fuel efficiency increases.[22]

The third major problem event was the publication of a report on the likely effects of global warming on California's ecosystems, which drew elites' attention to the specific threats in the state. The report was initiated and coordinated by the Union of Concerned Scientists (UCS) and the Ecological Society of America, and was written by eleven leading California scientists in 1999. It was widely distributed to state agencies and led the Natural Resources Agency to hold internal discussions. According to an energy commissioner, the report played a large role in educating policy makers about climate change, and one scholarly study concludes that it probably built support for climate policy in the legislature, too.[23] Although the report focused on ecosystem impacts, the executive summary highlighted threats to the state's economy and its residents due to impacts on water supplies, agriculture, fishing, flooding, and landslides.[24] Reports on the same topic had been issued by the CEC in 1989 and 1991. But the political context at that time, including Republicans Deukmejian and Pete Wilson in the governor's office and the expectation of federal government leadership under the George H. W. Bush administration, was not as favorable as the situation in the early 2000s.[25]

The 1999 UCS report clearly influenced the Pavley bill, the text of which provided a detailed justification largely based on the expected impacts of global warming in the state, including water supply problems, agricultural losses, wildfire risks, and threats to coastlines posed by storms and sea level rise. The public debate on the bill also emphasized global warming impacts in California, especially threats to the Sierra snowpack and hence to the state's water supply.[26] Pavley argued that California should lead the nation partly because "we are going to be more impacted by global warming than other states," an argument implicit in the UCS report.[27]

Political Commitment: Environmental Organizations and Fran Pavley

Attention on energy problems and global warming was high in 2001 and 2002, but to pass legislation would also require political commitment. However, this was not forthcoming from Governor Gray Davis (1999–2003), who took a very cautious approach to environmental policy. Davis was a Democrat and former top aide to Jerry Brown, but he also took major contributions from the investor-owned utilities and other businesses. Davis had no apparent commitment to climate policy, and in the first years of his administration, he pursued energy policies that tended to increase carbon dioxide emissions. When Davis took office, he retained the CEC chair who had been appointed by his Republican predecessor, the former oil industry lobbyist William Keese. Moreover, the long-term contracts, worth $43 billion, that Davis negotiated during the electricity crisis ignored renewable energy (creating a "green blackout") and gave long-term support to fossil fuel plants.[28] Davis lacked the background, interest, and credibility to resist President Bush on climate policy, and he took no public position on the Pavley bill until after it had passed the legislature.

Rather, the necessary political commitment came from environmental organizations and legislative leaders. They became engaged in the issue for somewhat idiosyncratic reasons, which suggest that it was not inevitable that sufficient political leadership would emerge at this time. The Bluewater Network was a small national organization, a project that had spun off from the Earth Island Institute, with a staff of about ten people based in California. Before it authored and promoted the Pavley bill, it had worked mainly to limit pollution from marine vessels and the use of jet skis and snowmobiles in national parks. The Bluewater Network's shift to state-level regulation of greenhouse gas emissions from vehicles initially was not supported by other major environmental organizations, although they were very sympathetic with the bill's aims and had themselves been campaigning against sport utility vehicles since about 1999.[29] The major environmental organizations viewed the stakes to be too high and the approach too risky given Bluewater's lack of influential support within the legislature, so they asked the organization to withdraw the bill. However, Bluewater persisted and within two months had gained the formal endorsements of Greenpeace, Friends of the Earth, the Sierra Club California, and other major state-level environmental organizations. After the Assembly's transportation committee approved the bill in April, the major environmental organizations mobilized strongly in support of it, as they did for the renewable energy bills.[30]

That the Bluewater Network found capable, committed legislative allies was also somewhat accidental. When the organization's staff sent the proposed bill to key members of the legislature, they met with little interest. They cast a wider net and found help from a new Assembly member, Fran Pavley, who had been elected in 2000 after her predecessor was forced out due to term limits. The bill's opponents initially underestimated Pavley, a retired middle school teacher and former mayor of the small, affluent city of Agoura Hills.[31] Pavley worked hard on the bill and made effective strategic and tactical decisions, described in the next section; legislative leaders also gave crucial support for final passage in June 2002. Byron Sher, a Democratic state senator who had authored major water and air pollution bills in the 1980s, sponsored the two RPS bills.

A final political factor that influenced passage of the Pavley bill was the prior existence and credibility of CARB as a regulatory agency. The bill was very brief (only eight pages long) and gave CARB the authority to regulate automotive greenhouse gas emissions without specifying standards or detailed procedures for developing them. Pavley considered the brevity of the bill and legislators' confidence in CARB to have been important for its passage.[32]

By 2002, CARB was a large, independent agency with major reservoirs of expertise. It is run by a board of political appointees with professional backgrounds related to air pollution, which balances political accountability and a degree of insulation from partisan politics. The agency is funded by dedicated revenue that comes from the entities it regulates, which has helped it to hire a large, well-paid, professional staff. For example, CARB's implementation of the Global Warming Solutions Act is funded by a small carbon tax on major emitters.[33*] Moreover, by the early 2000s, the agency was popular for its successes in reducing air pollution over the previous three decades. For example, ozone violations in the South Coast air quality district fell from 644 in 1973–80 to just 2 in 2003–11.[34]

Coalitions, Strategies, and Tactics

Finally, the Pavley and RPS bills passed in 2002 because proponents mobilized broader coalitions than opponents did, and because proponents used effective strategies and tactics. Climate policy advocates began that year with a favorable partisan context, with Democratic majorities in the State Assembly (50 to 30) and State Senate (24 to 16) – as they have had for all but two years since 1971 – as well as a Democratic governor. However, moderate, business-oriented Democrats held the swing votes in the legislature, and the Pavley bill passed very narrowly, gaining only 41 affirmative votes in the Assembly and 23 in the Senate; 41 and 21 votes, respectively, were needed to ensure passage.[35] As noted in the previous section, Governor Davis was neutral on the Pavley bill until after passage, although he supported the RPS bills during the legislative process. Therefore, the interest group coalitions and their strategies were crucial for the enactment of the Pavley bill.

The coalition in favor of the Pavley bill included environmental and health organizations, government bodies, and some businesses. By contrast, the opponents drew almost exclusively from business. This pattern was repeated for the RPS bills, AB 32, and the other major climate laws adopted in 2005–08. Table 7.1 presents an overview of the number of interest groups in these categories that formally endorsed or opposed climate policy bills. Although

TABLE 7.1 *Interest Group Mobilization on California Climate Change Legislation, 2002–08*

Bill	Year	Position	Environmental	Health	Business	Labor unions	Government	Misc.	Total
AB 1493 (Pavley bill)	2002	Proponents	6	1	3	0	5	1	16
		Opponents	0	0	12	0	0	1	13
		Bill total	6	1	15	0	5	1	29
SB 1078 (RPS)	2002	Proponents	5	0	6	0	2	0	13
		Opponents	0	0	5	5	2	0	12
		Bill total	5	0	11	5	4	0	25
SB 1038 (RPS funding)	2002	Proponents	1	0	3	0	1	0	5
		Opponents	0	0	1	0	1	0	2
		Bill total	1	0	4	0	2	0	7
AB 32 (Global Warming Solutions Act)	2006	Proponents	55	8	205	5	67	12	352
		Opponents	0	0	117	0	0	0	117
		Bill total	55	8	322	5	67	12	469
SB 107 (RPS)	2006	Proponents	4	0	2	0	1	0	7
		Opponents	0	0	6	0	0	0	6
		Bill total	4	0	8	0	1	0	13

SB 1368 (Power plant standards)	2006	Proponents	11	1	5	0	1	0	18
		Opponents	0	0	5	0	1	0	6
		Bill total	11	1	10	0	2	0	24
SB 375 (Transportation planning)	2008	Proponents	19	5	18	6	21	1	70
		Opponents	1	0	14	0	8	0	23
		Bill total	20	5	32	6	29	1	93
Totals		*Total proponents*	*101*	*15*	*242*	*11*	*98*	*14*	*481*
		Total opponents	*1*	*0*	*160*	*5*	*12*	*1*	*179*
		Grand totals	*102*	*15*	*402*	*16*	*110*	*15*	*660*
		as % of all proponents	*21.0*	*3.1*	*50.3*	*2.3*	*20.4*	*2.9*	*100.0*
		as % of all opponents	*0.6*	*0.0*	*89.4*	*2.8*	*6.7*	*0.6*	*100.0*

Sources: The legislative analyses of each bill, available at leginfo.ca.gov. I used the most recent analysis that contained verified lists of supporters and opponents. The last two rows may not total 100% because of rounding.

these data give equal weight to each organization, which does not account for the superior financial resources and legislative connections that some groups have, the basic patterns evident in the table also apply to the most powerful interest groups that mobilized on these bills.

The first pattern evident in the table is that supporters and opponents mobilized about equally on the Pavley and RPS bills, although the former greatly outnumbered the latter on the 2006–08 legislation. Second, by far the most heavily contested bill was AB 32, followed by the transportation planning bill, the RPS bills, and the Pavley bill. Third, while environmental organizations were key supporters, the climate policy bills also drew significant support from businesses, especially clean technology and alternative energy companies, and from government agencies, especially local governments and water and air control districts. Hence, as the last lines of the table show, the coalition of supporters was quite diverse, with about 50% of them coming from business, 20% each from environmental and government agencies, and 2% to 3% each from health and labor. By contrast, almost 90% of the opponents were drawn from business, with another 7% from government bodies.

The greater diversity of the proponents' coalition gave them an advantage in convincing legislators to vote for the 2002 legislation, because it seemed that "many independent voices were calling for the same end," in contrast to the almost exclusively economic interests that the opponents represented.[36] An array of powerful environmental organizations endorsed the Pavley Bill. These included the NRDC, Sierra Club, and the Coalition for Clean Air, which ultimately were listed as "co-sources" of the bill along with the Bluewater Network; the California League of Conservation Voters and the Planning and Conservation League also supported it.[37] The large environmental organizations' advertising resources, experience, and credibility were important factors in the campaign.[38]

But the coalition for the bill went well beyond environmental organizations. The American Lung Association, the Bay Area Air Quality Management District, two major city governments (Los Angeles and San Jose), and several business groups, including Architects for Social Responsibility and the Clean Power Campaign (a business-environmental hybrid organization) supported it. In addition, Environmental Entrepreneurs, a group of Silicon Valley businesses in partnership with the NRDC, issued press releases and had representatives at events to speak in favor of the Pavley bill. This group was important for convincing moderate legislators about the bill's economic impacts.[39] The main RPS bill, which grew out of a 1995 proposal by the UCS and the American Wind Energy Association, was supported by a coalition of environmental organizations and renewable energy producers.[40]

By contrast, the opposition to the Pavley bill drew heavily on business associations, including the California Chamber of Commerce, the state car dealers association, the national Alliance of Automobile Associations, and the Western States Petroleum Association.[41] Opposition to the RPS bill was somewhat broader than to the Pavley bill, as it included five labor unions and a similar number of energy-related businesses. But opposition to the RPS legislation was greatly weakened when municipal utilities were exempted from its provisions and San Diego Gas and Electric dropped its opposition; the other two large investor-owned utilities supported the bill.[42]

Finally, the proponents ultimately were more effective in their strategies and tactics, including how they influenced the framing of the conflicts. Their success was not a foregone conclusion. In the early months of 2002, it appeared that the Pavley bill had been stopped by a vigorous advertising campaign that emphasized the economic costs of the regulations and appealed to consumers' desire to choose what kind of car they would drive. The car industry's campaign gained support from conservative radio hosts, who broadcast to car drivers during the morning commuting hours, and from a well-known car dealer, Cal Worthington, and his equally famous tiger. The campaign included grassroots activities, including a protest caravan to Sacramento and tens of thousands of voice and email messages to legislators. This mass mobilization undermined legislative support; two Republicans and two Democrats who had voted for the bill in the Assembly announced their opposition, and Pavley was forced to withdraw it in May 2002.[43]

However, the opponents' campaign ultimately was unsuccessful because the climate policy proponents counterattacked against the car industry and succeeded in framing the conflict in terms favorable to the advocacy coalition. The opponents were vulnerable because the car industry played a prominent role in the opposing coalition, and that industry had low credibility because of its involvement in previous conflicts over regulatory policies. In addition, the Enron scandal, which unfolded in October 2001, reduced public confidence in major corporations at this time in California. In regard to the Pavley bill, the Sierra Club and League of Conservation Voters argued that the car industry was trying to deny consumers the choice of less polluting vehicles. More generally, proponents put the bill in the context of other technological improvements that the car industry had vigorously resisted, including smog-control devices and air bags. The proponents' two main messages were that global warming would harm California and the automobile industry was standing in the way of progress, once again.[44] As a result, a dominant theme in the media coverage was that there was a conflict between the interests of California and those of the national car industry.[45]

In other ways, the proponents were strategically effective. They aimed for a broad coalition and made concessions to defuse opposition. When the Pavley bill first ran into opposition from the oil and car industries, Pavley began to revise it to meet the concerns of the moderate Democrats who were the swing votes on the bill. The bill was amended to include a one-year delay between CARB's issuing regulations and their taking effect, in order to give the legislature a chance to override them. The phrase "cost-effective" was added to the requirement that CARB adopt regulations for the "maximum technologically feasible" reduction of emissions. Moreover, CARB was explicitly prevented from banning any type of vehicle (such as sport utility vehicles), requiring reductions in vehicle weights, speed limits, or vehicle miles traveled, or imposing new taxes.[46] Pavley held hearings in order to counter opponents' charges about the bill's economic impacts; she worked hard in efforts to reassure interest groups and to round up legislative support.[47] Similarly, Sher modified the main RPS bill in response to criticism. The original bill called for 20% new renewables by 2010 and included municipal utilities, but in response to objections, the municipal utilities were dropped and the compliance period was lengthened by seven years, to 2017.[48*]

The advocates also minimized economic costs and included popular co-benefits in their arguments for both bills. Pavley limited the economic costs of her bill by requiring CARB to achieve a cost-effective reduction in emissions; that is, increases in new car prices borne by consumers would be paid for with savings in gasoline costs.[49] Also, both bills called for delayed or at least gradual implementation, which postponed most costs to later years. The Pavley rules were to be finalized in 2005 and to go into effect no sooner than 2009, while the 20% target in the RPS was to be gradually achieved during 2003–17 at the rate of 1% per year. Although the Pavley bill was aimed at global warming, the justifications in the bill text and the lobbying campaign for it included major emphases on air pollution and health, which helped to gain legislative support for it. The RPS bills were justified mainly on the basis of energy security and air quality, with global warming only an afterthought that was not even mentioned in the legislation (see Table 7.2 in the next section).[50]

The strategy of emphasizing co-benefits was in tune with public opinion in 2000–02. Although climate change was not a salient issue for most California residents, the public was very concerned about air pollution and electricity supply.[51] And the proponents' campaigns apparently were successful with public opinion. A poll by the Public Policy Institute of California in June 2002 showed massive (80% to 85%) support for both bills when they were pending in the legislature, including over 80% of Republicans for the renewable energy standard and over 75% of sport utility vehicle drivers for the Pavley bill.[52] The release of that survey, as well as lobbying by environmental organizations, a key constituency during Davis's 2002 re-election campaign, prompted him to sign the Pavley bill.[53] At the signing, Davis touted the bill's effects on global warming and clear air, and he mocked the carmakers for their opposition to the bill – given their previous opposition to unleaded gasoline, catalytic converters, seat belts, and air bags.[54]

Finally, proponents used effective surprise tactics to win passage of the Pavley bill. After the bill ran into fierce opposition, Pavley withdrew it in May 2002. John Burton, the Senate majority leader, then unexpectedly revived the bill by using a controversial "gut and amend" procedure to replace the contents of an existing, unrelated bill. The content was also amended to meet some objections; more important is that it got a new bill number. Thus, "AB 1058," which had been the bill number targeted and mentioned specifically in the opponents' fierce advertising and letter-writing campaigns, was replaced by the innocuous-sounding "AB 1493."[55] Legislators had not received a barrage of constituent complaints associated with the new bill number, and it passed both chambers on nearly party-line votes within three days.[56]

POLICIES ON STATEWIDE GREENHOUSE GAS EMISSIONS (2005–08)

Schwarzenegger's executive order in 2005 and the enactment of AB 32 the following year set greenhouse gas emissions reduction targets and provided for implementation by CARB through regulatory measures and a cap-and-trade program. These actions were part of a larger burst of policy reforms that included a law creating an electricity performance standard that barred new coal plants or long-term contracts with the operators of existing plants (in 2006), and laws that accelerated the RPS and created a solar energy initiative (2006). This policy-making episode also included an executive order creating a low-carbon fuel standard (2007) and a transportation planning law (2008). These dramatic policy changes, which suddenly put California in the forefront of state-level climate policy and made it a globally significant leader in this area, were made possible by the convergence of many developments in the problem stream and the political stream.

Problem Awareness: A Conjuncture of Causes

Problem awareness increased dramatically because five aspects of the problem stream came together in the early and mid-2000s. First, the after-effects of the 2000–01 electricity crisis were still being felt in California years later. The crisis had durably revived the interest of policy makers and the public in energy conservation and energy supply diversification, in a reprise of the 1970s. It also led to new leadership and initiatives on energy and climate policy by state agencies.[57] The CEC and CPUC, together with the California Power Authority, jointly issued a California Energy Action Plan in 2003. A direct response to the electricity crisis, it attempted to chart a path that would make energy reliable, affordable, and environmentally sound, which included reducing impacts on climate change.[58] It called for increased energy efficiency, reduced per-capita energy consumption, and an accelerated schedule for introducing renewable energy as the preferred strategies for avoiding future price spikes and power outages.[59]

The same year, the CEC issued two other reports, one on reducing petroleum dependence and the other a detailed energy plan. Both reports emphasized energy security, and the former also linked energy use to global warming and the effects of air pollution on public health.[60] They called for increased motor vehicle fuel efficiency, the development of low-carbon fuels, and the inclusion of greenhouse gas emissions reporting and avoidance in utilities' and state agencies' planning documents.[61]

Second, droughts and wildfires in California in the early 2000s helped to focus elite attention on the economic threats that climate change posed to the state. Severe or extreme droughts hit large parts of the state in 2001 and 2002–03. The Central Valley, the state's key agricultural region, was at least abnormally dry for most of the 2001–04 period, triggering states of emergency in a number of counties.[62*] As a corollary of drought, wildfires became a major problem in the early 2000s, with a major peak in news reporting of the phenomenon in 2002–04. In 2004, for the first time in at least twenty years, wildfires were a year-round problem rather than just a two-month, late summer event.[63*] In 2003, wildfires caused $2.6 billion in damage in Southern California.[64]

These natural disasters in California affected the debate on the Pavley bill. The bill's text and news articles about it included concerns about the effects of climate change on snowpack, water supplies, and wildfire risks.[65] Davis signed the bill just as a second severe dry spell was beginning. Three years of drought also had a lasting effect on political discourse, as global warming became linked to drought risks and threats to water supplies. For example, the AB 32 text and statements by supporters referred prominently to the threats to the state's snowpack.[66]

Third, the publication of major reports on climate change and recommended solutions by expert bodies in 2004 and 2006 increased the perceived importance of the issue by California elites. The UCS published a second report on the impacts of climate change in California, in both scientific and public versions, in 2004.[67] This report was more detailed, with finer-grained quantitative analysis than the 1999 version, and it made grimmer predictions concerning heat waves and water shortages.[68] It predicted that temperatures could increase up to 15 degrees Fahrenheit in summertime by the 2074–99 period, with a doubling of heat waves, an 89% snowpack loss, and hence drastic threats to the state's water supply.[69]

The UCS report was used to brief state agencies and the governor's office, gained media coverage, and shaped the climate debate toward a focus on ongoing and future climate impacts in California. It was used in testimony at the Pavley implementation hearings, where it helped to build support for CARB's draft regulations in 2004, and was used in recommendations by the Western Governors' Association.[70] The UCS report was influential because it was based on peer-reviewed science, represented a unified scientific voice on climate change, was presented in an accessible way, and focused on solutions.[71] In addition, Schwarzenegger's 2005 executive order created the California Climate Center, which ran a scenarios exercise involving seventy scientists. It delivered a high-profile report on climate policy strategy to the legislature and governor in March 2006. This report provided an action plan to achieve the 2020 emissions reduction goals, through a wide range of existing and new measures, including a cap-and-trade program.[72]

These studies were well timed to influence policy making, partly in ways that its authors could not have anticipated. The UCS report was published after several years of unusual drought and wildfires, a coincidence that helped its warnings about the likely effect of global warming on the Sierra snowpack and the state's water supplies to become an accepted part of the discourse on climate policy. The publication of the report also coincided with the debate over the implementation of the Pavley rules, which CARB issued in 2004, and it occurred just a year before Schwarzenegger's executive order put further attention on the issue. For its part, the 2006 scenarios study was deliberately timed to be influential. It was a direct result of the governor's executive order, was conducted on a very compressed timetable, and was released in time to influence the legislative debate on the Global Warming Solutions Act.[73]

Fourth, national and international climate policy decisions continued to spill over into California. Bush's retreat from explicit national climate policy had created political space for other actors to try to initiate or expand climate policies. Moreover, the failure of Congress to enact strong measures in energy efficiency, vehicle efficiency, or renewable energy during Bush's first term meant that the states were the most likely venues for significant action. Moreover, the European Union's implementation of its Emissions Trading System in January 2005 spurred California's policy makers to act.[74] Large corporations began to adjust to the pricing of carbon emissions in Europe and to expect that similar policies were inevitable in the United States. Hence, Schwarzenegger embraced emissions trading as a business-friendly approach to climate policy and emphasized it in his negotiations with the legislature on AB 32. In July 2006, he held a meeting with U.K. Prime Minister Tony Blair and with business leaders from California and the United Kingdom. Schwarzenegger and Blair announced that they had reached an agreement to work toward a possible joint emissions trading system, as well as to cooperate on clean energy research and studies of the economic impacts of climate change.[75]

Fifth, in 2005–07, a number of national developments combined to raise the perceived importance of climate change and support for climate change mitigation policy. The rising salience of the issue can be seen in national news coverage, the congressional agenda, and public opinion, which increasingly saw climate change as an urgent problem requiring government action (see Figures 9.2 and 9.3 in Chapter 9).[76] The developments behind the increased sense of urgency included shifts in attention, position, and activity by many actors – business, news media, environmental organizations and foundations, congressional representatives, California and other state governments, and the European Union. They also included climate events such as the 2005 Atlantic hurricane season and reports by scientists. These developments are described in more detail in Chapter 9.

The events that drove the rise of the climate change issue nationally resonated in California. For example, the May 2006 release of Al Gore's documentary, *An Inconvenient Truth*, had a large influence on the state's elites and the public. California legislators reported that the film made global warming a mainstream issue and helped to move lawmakers to action. Newspaper opinion pages increased their attention to the issue in 2005, and the *San Francisco Chronicle* shifted its position on climate change in 2006.[77]

Thus, the aftermath of the electricity crisis and natural disasters, new expert reports, and international and national developments all focused attention on the climate change issue at this time. The perceived importance of the issue led advocates for legislation to justify the 2005–08 bills largely in terms of climate change impacts in California, along with other rationales. The results of a content analysis[78*] of the justifications used in four major bills and two executive orders in this period are described in Table 7.2. The next-to-last line of the table show that the major justifications offered in 2005–08 were global warming and its impacts (39%), as well as energy security (18%), air pollution and public health (16%), state leadership (14%), and jobs, profits, and economic development (13%).

Political Commitment: Arnold Schwarzenegger and Fran Pavley

Events in the problem stream were creating a golden opportunity for climate policy in California in 2005–06. But if Gray Davis had served out his full second term as governor, which would have ended in January 2007, the opportunity might have passed without enough political commitment congealing to seize it effectively. As in Davis's first term, legislative initiatives would have emerged, and their early proponents probably would have been able to mobilize allies into an advocacy coalition. Ultimately the advocates might have persuaded a reluctant, lame-duck Governor Davis to sign legislation on the eve of the November 2006 election, when a new governor would have been elected. But in this scenario, without the involvement of a Republican governor to press for emissions trading and to reassure and mobilize businesses to support climate policy (or at least to remain neutral on it), businesses probably would have been more unified and vigorous in opposition to the legislature's proposals for the economy-wide regulation of greenhouse gas emissions. Hence, the measures that passed the legislature might have been watered down by a combination of business opposition and the need to get a reluctant governor to sign it. As it was, events unfolded differently. Davis was recalled, Schwarzenegger replaced him and became a champion of climate policy, and the legislature competed with Schwarzenegger to produce more aggressive legislation, upon which they compromised.

In the first place, despite Davis's reelection in November 2002, by a fairly narrow margin of 47% to 42%, Arnold Schwarzenegger replaced him as governor a year later. Schwarzenegger's election was extraordinary in several ways – besides that he was a bodybuilder, businessman, movie star, and Republican who was married to a member of the Kennedy family. The removal of Davis was only the second time in U.S. history that a governor has been

TABLE 7.2 *Justifications Offered in the Texts of California Climate Policy Documents, 2002–08*

Note: Table entries are percentages of the justifications offered in each policy document or set of documents

Bill or executive order	Year	Global warming and its specific impacts	Energy security, reliability, supply	Air pollution, air quality, health, environmental quality	State leadership in the U.S. or globally	Jobs, economic development, profits	Total
AB 1493 (Pavley bill)	2002	78.6	0.0	21.4	0.0	0.0	100
SB 1078 (RPS)	2002	0.0	17.1	60.5	0.0	22.4	100
SB 1038 (RPS funding)	2002	0.0	49.6	50.4	0.0	0.0	100
Executive order on greenhouse gas emissions targets (S-3-05)	2005	40.2	0.0	7.3	12.7	39.9	100
AB 32 (Global Warming Solutions Act)	2006	32.2	6.7	5.9	46.2	9.0	100
SB 107 (RPS)	2006	0.0	100.0	0.0	0.0	0.0	100
SB 1368 (power plant standards)	2006	20.9	43.5	35.6	0.0	0.0	100
Executive order on low-carbon fuel standard (S 01-07)	2007	37.1	51.0	6.6	0.0	5.3	100
SB 375 (transportation planning)	2008	69.1	0.0	30.9	0.0	0.0	100
2002 justifications (all three bills)		*44.5*	*16.1*	*35.7*	*0.0*	*3.7*	*100*
2005–08 justifications (four bills and two executive orders)		*39.4*	*17.8*	*16.0*	*14.1*	*12.7*	*100*
Total justifications, 2002–08		*40.6*	*17.4*	*20.5*	*10.8*	*10.7*	*100*

Sources: Bill texts from leginfo.ca.gov; executive orders from www.gov.ca.gov. Columns may not total 100 because of rounding.

removed in a recall election, and the first time in more than eighty years. Davis's opponents gathered more than 1.3 million valid signatures to force the October 2003 recall election, in which a clear majority of 55% to 45% voted to remove Davis. Schwarzenegger easily won the three-way race to replace him, gaining 49% of the vote versus only 31% for Democrat Cruz Bustamante, who was Lieutenant Governor under both Davis and Schwarzenegger.[79]*

The recall campaign was successful because of a combination of events. These included the highly unpopular electricity crisis, Davis's failure to take early action during it, and, partly as a result, a huge budget deficit that the governor projected at $21 billion in November 2002 and then at $35 billion just a month later.[80] Measures to reduce the deficit, including an $8 billion income tax increase and a tripling of vehicle registration fees, were also unpopular. The governor and legislature were deadlocked on the budget during most of the recall campaign, contributing to a sense of crisis.[81] In addition, the recall campaign created an unusual opportunity for a moderate Republican such as Schwarzenegger to become governor. The recall procedures did not require a Republican primary election, in which lower turnout usually benefits conservative candidates in California, such as Deukmejian and Wilson, the two previous Republican governors of the state.

During the recall campaign, Schwarzenegger became an advocate of climate and other environmental policies. Initially he blundered by saying he would abolish the California Environmental Protection Agency, but he quickly reversed himself. At the urging of his advisor Terry Tamminen, the founder of the environmental organization Santa Monica Baykeeper, Schwarzenegger took strong positions on climate policy during the recall campaign, although he framed them largely in terms of reducing air pollution and enhancing energy security.[82] Symbolically, he said he would have one of his four Hummers retrofitted to run on hydrogen. More substantively, Schwarzenegger endorsed the Pavley rules, saying he would implement and defend them in the courts, and called for the promotion of solar energy and conservation, the construction of a network of hydrogen refueling stations, and an acceleration of the RPS target.[83] By taking these positions, Schwarzenegger set himself apart from the other Republicans in the race and from President Bush, and he challenged Davis for the support of environmentalist voters. Davis belatedly appealed to those voters by announcing a climate policy partnership with the governors of Washington and Oregon and signing a raft of last-minute environmental legislation.[84] The leading environmental organizations endorsed Davis, but Schwarzenegger's campaign statements mollified them.[85]

Although Schwarzenegger did not pursue major climate policy initiatives in his first eighteen months in office, he did lay some groundwork for them. He appointed Tamminen as the head of the California Environmental Protection Agency, although he also tried to appoint a former industry lobbyist, Cindy Tuck, to be head of CARB; the State Senate rejected that appointment.[86] In April 2004, the governor issued an executive order directing state agencies to create a hydrogen fuel network, and in August, he called for a major residential photovoltaic solar program, the Million Solar Roofs Program. When Democratic Assembly Speaker Fabian Nunez sought to add a pro-union provision requiring the payment of prevailing wages, Schwarzenegger tried to evade the legislature by having the CPUC implement the program. Ultimately, first the CPUC and then the legislature, the latter in August 2006, approved the California Solar Initiative; $3.3 billion in funding was authorized, to be paid by a charge on utility customers.[87] Meanwhile, Tamminen began to develop a greenhouse gas emissions target in Summer 2004, and in November he was appointed Cabinet Secretary, which made him the governor's chief policy advisor.[88] Also, in November 2004, the CARB board, almost all of whose members had been appointed or reappointed by Schwarzenegger, approved the Pavley rules to reduce new cars' vehicle emissions 30% by 2016.[89]

Schwarzenegger's political ambitions ultimately drove him to make climate policy a major focus of his administration.[90] With his re-election pending in 2006 in a liberal state, Schwarzenegger's prospects did not look good, given his low approval ratings and the liberal bent of the state. His endorsement of President Bush's 2004 re-election campaign had been unpopular, and he was bogged down in partisan conflict with the Democratic legislature.[91] Schwarzenegger's approval ratings had dropped sharply from 57% in July 2004 to only 34% a year later.[92] Seeking ways to improve his popularity, Schwarzenegger took an aggressive position on global warming, which helped him both to increase his profile and to present himself as a moderate Republican, in contrast to President Bush. Nonetheless, with Republicans in the governor's office and the state Chamber of Commerce resisting stronger climate policy, and his Democratic advisors pressing for it, Schwarzenegger did not decide in favor of greenhouse gas emissions targets until April 2005. He announced the targets two months later, at a ceremony for the United Nation's World Environment Day, on the fiftieth anniversary of the founding of the UN.[93]

Besides establishing the reduction targets, the executive order that Schwarzenegger signed on June 1 directed the head of the California Environmental Protection Agency to report to the legislature every two years on progress in meeting the targets and on the impact of global warming on California.[94] The first report, issued by a reconstituted Climate Action Team in March 2006, recommended 40 different regulatory measures, including a further tightening of the Pavley rules after 2016 and an acceleration of the RPS from 20% to 33% of electricity sold in the state.[95]

Although he initially pursued climate policy through executive actions, Schwarzenegger ultimately adopted a strategy of cooperation with the legislature because of political defeats he suffered late in 2005. During most of that year, Schwarzenegger, like Pataki in New York, clearly preferred to evade the legislature on his climate and other policies. He issued the greenhouse gas reduction targets as an executive order, rather than calling for a law to be passed, and tried to get the CPUC to implement his solar initiative without legislation. As Schwarzenegger battled publicly with Speaker Nunez and other legislative Democrats over the budget and other issues during 2005, he went directly to the voters, asking them to approve four conservative ballot measures on teacher tenure, the use of union dues for political campaigns, state budget limits, and redistricting. But in a special election held in November, voters rejected all four of the measures by large margins. Coming on the heels of his endorsement of President Bush for re-election in 2004, this defeat drove Schwarzenegger's job approval to low levels. With his re-election campaign coming up the next year, Schwarzenegger immediately appointed a Democrat from the Davis administration as his new chief of staff. This began his move toward the political center and toward cooperation with the Democratic-controlled legislature.[96]

Political commitment to climate policy came from the legislature, too. As a result, competition on climate policy between the Republican governor and legislative Democrats developed in April 2006, which produced a stronger climate policy than either actor would have produced alone. In response to Schwarzenegger's move on climate policy in June 2005, and emboldened by the governor's obviously weak position with the electorate after November, Democrats introduced legislation to go beyond his executive order. In April, Assembly Democrats, led again by Fran Pavley, introduced amendments to AB 32, a bill concerning the Climate Action Registry. The amendments would give CARB the authority to implement the emissions reduction targets in Schwarzenegger's executive order. However, the governor initially did not endorse the bill and instead called for a slower approach, warning that introducing emissions regulations too quickly might "scare the business community."[97] He also said that he was unsure if mandatory caps were needed and suggested waiting to introduce them until after 2010.[98]

However, global warming was an increasingly popular issue, and since Schwarzenegger had established the targets, he would have been hard pressed to veto the legislation to implement them. This held especially in a gubernatorial election year, at a time when President Bush's approval ratings were low, and hence when Schwarzenegger needed to burnish, not tarnish, his environmental credentials.

Disagreements between the governor and legislature concerned important details, especially the role of emissions trading. Democrats viewed this market instrument as suspect because of negative experiences with electricity deregulation in the 2000–01 crisis and with an NO_x and SO_x emissions trading system in Southern California in the 1990s. That program led to a backlash by environmental justice activists who charged that it led to high-pollution "hot spots" in poor neighborhoods.[99] The day before the Democrats announced their AB 32 bill, Schwarzenegger had released the final version of his Climate Action Team report, with an executive summary that was revised to emphasize emissions trading rather than the more conventional regulatory measures that made up the bulk of the report.[100]

Besides resisting emissions trading, the Democrats pushed for CARB to be the implementing agency, while Schwarzenegger wanted the California Environmental Protection Agency to play that role. The latter agency would be more directly subject to his authority than CARB, and the Democrats did not trust Schwarzenegger to implement the law aggressively. Schwarzenegger also pressed for the inclusion of a safety valve through which the governor could suspend implementation should the rules prove too economically harmful, and he wanted emissions trading to be required as the primary instrument of the law.[101]

Ultimately, the two sides reached a compromise on August 30, one day before the legislative session expired, after jockeying that showed the high political stakes of the issue. The Democrats pushed for a bill that they thought Schwarzenegger might veto, which would have benefited his Democratic challenger in the governor's race. Schwarzenegger privately told Pavley and Nunez that he would veto the bill, but when they planned a press conference to announce this, he backed down hastily.[102] The bill passed on a nearly party-line vote, with only one Republican assemblyman voting for it.[103]

In the final legislation, the Democrats got CARB as the implementing agency, while Schwarzenegger got authority for the governor to delay implementation indefinitely if there were a "threat of significant economic harm." Emissions trading was included, but only as one option to be considered by CARB.[104] A month after signing the bill, Schwarzenegger tried to undermine CARB's authority over climate policy through an executive order giving the California Environmental Protection Agency a larger role and by interfering with CARB's selection of early action measures to implement the law. However, these efforts generally were not successful. CARB has been the implementing agency for AB 32 and has been able to act with its customary degree of independence from elected officials. Schwarzenegger, for his part, generally showed commitment to implementing climate policy during his second term in office.[105*]

Finally, as with the Pavley bill, an additional political factor in the passage of AB 32 was the expertise and political credibility of CARB. The Global Warming Solutions Act gave the agency extraordinary regulatory capacity with few details about how to exercise it. Indeed, CARB has been crucial in implementing the law. It has been on time with almost all its AB 32 deadlines, including making emissions reporting mandatory, establishing the official 1990 baseline, adopting a scoping plan, implementing early action measures, and initiating the cap-and-trade system.[106]

In addition to the leaders already mentioned – especially Schwarzenegger, Tamminen, Pavley, Nunez, and Nichols – California's attorneys general played key roles in this wave of policy reform and its implementation. Bill Lockyer defended the Pavley law against industry lawsuits, joined the *Massachusetts v. EPA* case, and sued carmakers for public-nuisance damages because of their contribution to climate change. After he became attorney general again in 2007,

Jerry Brown took the side of an environmental organization concerning the consequences for greenhouse gas emissions of the city of Banning's "leapfrog developments" at Black Bench, at the base of the San Bernandino Mountains. The litigation by the Center for Biological Diversity, which was based on the California Environmental Quality Act, led a court to block the development. Brown then sued San Bernandino County over the impacts of its development plans on greenhouse gas emissions. The resulting legal settlement in August 2007 set an important precedent for other local government entities. Its impact was reinforced by workshops that were held to train city and county officials in greener planning decisions, such as shortening commutes and including public transit. It also helped lead to the passage of SB 375 in 2008, under which CARB sets regional greenhouse emissions targets and encourages local planning bodies to meet them.[107]

Public opinion did not drive this wave of policy reform. In the early 2000s, Californians did not place a high priority on addressing global warming, and they were not more likely than national respondents to see global warming as a serious problem or to oppose President Bush's retreat from greenhouse gas regulation.[108*] Rather, the competition between the legislature and Schwarzenegger on climate policy helped to get the public interested in climate change and supportive of California's climate policies.[109] As public opinion shifted, it became an important factor spurring political actors toward a strong policy. By July 2006, energy (at 12%) and global warming (8%) became the second and third "most important environmental issue[s]" named by respondents, up from only 6% and 2%, respectively, in July 2005. Those who saw global warming as a very serious threat to California rose from 39% in 2005, to 49% in 2006, and to 54% in 2007.[110]

Moreover, the targets in Schwarzenegger's 2005 executive order were popular. They were supported by a 69% to 17% margin in 2005, gaining equal support across party lines, and the next year a similar margin of 65% to 27% favored California enacting its own global warming policy separate from the federal government. Once enacted, AB 32 was even more popular; the public supported it by a 78% to 14% margin in 2007. In addition, during this period, the Pavley law remained as popular as it had been just before passage, with public opinion running 78% to 16% in favor of it – including 71% to 24% among Republicans – in 2006.[111]

Framing and Coalitions

While the convergence of problem and political events created a window of opportunity, the framing strategy that was deployed and the broad advocacy coalition that mobilized also helped to ensure the passage of the Global Warming Solutions Act. Proponents of stronger climate policy framed the legislation to minimize economic costs and perceptions of them. They also emphasized the benefits for economic development in California, especially in the clean technology sector, and particularly in the renewable energy and energy efficiency industries. Three studies of the economic costs of AB 32 were presented during the debate on the bill in 2006, by the governor's Climate Action Team, the Washington-based Center for Clean Air Policy, and a research team at the University of California at Berkeley. Their estimates of the economic impacts of the bill ranged from no cost to a gain of several billion dollars.[112*] Hence, similar to the Pavley bill, AB 32 was framed as not requiring sacrifices from Californians.

Proponents argued further that the bill would help California to gain or expand a competitive edge in the clean energy sector, allowing the state's economy to profit from an expected green economic revolution.[113] By adopting the nation's most rigorous climate policy, California would influence other states to adopt policies requiring greenhouse gas emissions reductions, thus enlarging the market for California's low-emissions technologies.[114] Schwarzenegger's executive order emphasized the role of technological development in giving California companies access to worldwide markets, as well as energy cost reductions that would boost profits and consumer spending power across the state's economy. Similarly, the bill text argued that "by exercising a global leadership role, California will also position its economy, technology centers, financial institutions, and businesses to benefit from national and international efforts to reduce emissions of greenhouse gases."[115] In addition, the bill's authors argued that its passage would encourage other states, the federal government, and other countries to act on global warming.

These arguments helped proponents to build a broad coalition for climate policy and to defuse potential opposition to it. Schwarzenegger vigorously attempted to gain business support, or at least acceptance, for AB 32 by emphasizing the benefits of emissions trading and offering reassurances about the costs. As a businessman and investor (in real estate and restaurants), Schwarzenegger enjoyed more credibility with business than most legislative Democrats; in a later interview, he said that "my party affiliation may have been helpful at the beginning of our discussions with certain stakeholders."[116] At an April 2006 global warming summit in San Francisco, Schwarzenegger offered reassurance that "we don't want to go after business and make business leave the state."[117] During 2006, he and his aides often met with businesspeople, even oil company representatives, as well as with Republicans, to try to convince them to accept AB 32.[118] While most of business remained opposed to AB 32, Schwarzenegger's push for emissions trading and his other efforts probably helped to gain some support for climate policy (or neutrality on it) from business interests.

The bulk of business support came from "new economy" firms in an alliance with environmental organizations. Environmental Entrepreneurs, a project of the NRDC, mobilized support from these firms, as did the Silicon Valley Leadership Group. Environmental Entrepreneurs played a crucial role when it argued in the Capitol that the bill would "stimulate innovation, efficiency, and economic benefits."[119] The Environmental Defense Fund was an early and leading advocate of emissions trading, which gained the support of businesses that otherwise were not favorable to climate policy. In 2000, it joined with seven multinationals in the Partnership for Climate Action to promote emissions trading. In regard to AB 32, the oil company BP and the International Climate Change Partnership, a coalition of multinational corporations, pressed for the inclusion of emissions trading in the bill, although they generally did not support state-level carbon regulation. BP also organized the meeting of international and California businesses in July 2006 at which Schwarzenegger and Blair concluded their agreement to cooperate on climate policy.[120]

As a result of the proponents' framing and coalition-building efforts, the advocacy coalition for climate policy had become quite broad by the time AB 32 was debated. High technology industries and venture capitalists, especially in Silicon Valley, were strong supporters because of the bill's implications for renewable energy technology and electric vehicles.[121] Ultimately, the supporters of AB 32 included more than 200 individual companies and associations, representing a wide range of businesses, other than heavy industry. According to a legislative analysis, supporters included 16 companies from the agriculture and food processing sector (mostly restaurants), 58 from clean technology industries, 15 from entertainment, 49 from venture capital and other financial services, and 20 from the technology sector. Support also came from business associations such as San Francisco Chamber of Commerce, the California Ski Industry Association, and state-level wind and solar energy associations.[122]

Of course, the advocacy coalition was much broader than business. Environmental organizations strongly mobilized in support, with 55 groups endorsing AB 32, including all the major California organizations. Major labor unions, including the public workers union AFSCME[123] and the teachers and firefighters unions, did too. The Sierra Club initiated a broad "blue-green alliance" with labor unions in 2006, which called for 1.6 million jobs in renewable energy and engaged in the gubernatorial election that year. The alliance vetted the candidates on climate and other environmental policy, and it ultimately endorsed Schwarzenegger's Democratic opponent.[124] In addition, 49 California cities, both its U.S. senators, and eight of its U.S. representatives also signed onto the legislation, as did a half dozen faith-based organizations, including the California Conference of Catholic Bishops and the state's Lutheran Office of Public Policy. Major newspapers also backed AB 32, including the *San Francisco Chronicle*. When Schwarzenegger and the legislature remained deadlocked with only one week to go in the legislative session, the *Sacramento Bee* and the *Los Angeles Times* argued for passage, which helped lead to the compromise that produced the law.[125]

Nonetheless, the opposition, which came from most of business, was formidable. It included the California Chamber of Commerce, the California Business Roundtable, the state associations of general contractors, of manufacturers, of mining, and of retailers, more than two dozen more specialized trade associations ranging from automobile wholesalers to grocers to poultry to warehouses, and assorted local chambers of commerce. Powerful out-of-state businesses also opposed the bill, including the Alliance of Automobile Manufacturers, the Association of International Automobile Manufacturers, and the Western States Petroleum Association.[126]

As a result of this stiff resistance, Schwarzenegger was very slow to accept AB 32's mandatory controls on emissions, the Republicans in the legislature gave the bill no support, and in the end it passed both chambers on a nearly party-line vote. A broad coalition was needed to overcome business opposition, and this led to important concessions to environmental interests as well as the concessions to Schwarzenegger and business that were noted earlier in this section. Details on emissions trading represented a compromise between Schwarzenegger and environmental justice advocates. While the law specified that CARB must make recommendations regarding emissions trading, it stated that the agency *may* choose to adopt such a program and that it could do so only after analyzing the issues of potential pollution hot spots and increases in toxic emissions.[127] In addition, the law required CARB to take air pollution and energy diversification co-benefits into account and to minimize leakage (i.e., increases in out-of-state emissions as a result of reductions in California).[128]

THE FAILURE OF PROPOSITION 23 (2010)

The passage of AB 32 did not mark the last interest group battle over this law, because oil refineries and other businesses tried to overturn it by referendum. The 2008–09 recession, which helped to scotch efforts at cap-and-trade legislation in Congress, drove California's unemployment rate to a peak of 12% in 2010 and strengthened economic arguments against climate policy in the state. However, the attempt to roll back the Global Warming Solutions Act failed, as Proposition 23 was rejected by a decisive 62% to 38% margin in November 2010. Although the problem stream had turned against climate policy at this time, political developments still favored climate policy in several ways.

First, the implementation of AB 32 had strengthened the advocacy coalition in the four years since its passage.[129] As CARB developed its scoping plan, it conducted inclusive stakeholder processes, with three advisory committees on market instruments, environmental justice, and economic and technology advancement. In a year-long process of drafting the scoping plan, CARB held hundreds of workshops, meetings, and dialogues.[130]

Second, partly as a result, interest groups, especially labor unions, became active in efforts to shape and expand California climate policies after AB 32 was adopted. The Apollo Alliance – a coalition of business, labor, and faith-based organizations formed in 2003 – became active in opposition to Proposition 23 in 2010 and in support of Proposition 39 in 2012. The latter, approved by a 61% to 39% margin, raised taxes on out-of-state corporations by $1.1 billion per year, with half the money dedicated to energy efficiency and clean energy jobs for the next five years.[131]

More generally, labor unions in California embraced AB 32 around 2008, which is also when the national labor federation AFL-CIO reversed its opposition to the Kyoto Protocol. Construction unions tried to shape AB 32 implementation to reduce job loss, compensate workers who lose jobs, and favor jobs that pay family wages and include fringe benefits. The California Labor Federation and the state's construction trades council began to monitor implementation of AB 32 and encouraged unions to contribute to the CARB scoping process. The labor federation also helped to craft the CEC's Alternative and Renewable Fuel and Vehicle Technology Program, which passed in 2007; the union body has a seat on the advisory committee.[132]

Therefore, when climate policy opponents tried to overturn AB 32 via Proposition 23 in 2010, there was a strong base of potential supporters available to defend the law. With California's unemployment rate at 12%, Proposition 23 would have suspended AB 32 until the rate dropped to 5.5% or less for a full year, something that has happened only three times since 1976. Although the proposition was leading or even in the polls through September, its opponents greatly outspent and out-organized its supporters. The "no" campaign spent about $31 million, compared with $11 million by the "yes" campaign.[133]

Those opposing Proposition 23 included venture capitalists, wealthy individuals from a range of companies (including Tom Steyer, Bill Gates, and executives at Google, Intel, and Gap), nonprofits, labor unions, major environmental organizations (which spent about $10 million), and Pacific Gas and Electric.[134] Opponents of the ballot measure also included 130 local ethnic, environmental, health, religious, and clean energy groups organized in Communities United Against the Dirty Energy Proposition, which reportedly fielded 3,000 volunteers.[135] The "no" campaign as a whole said that it sent out 3.4 million pieces of mail and made 2.8 million phone calls to voters, including 900,000 get-out-the vote phone calls and text messages in the final three days.[136]

A third reason the proposition failed is that most of the business groups that opposed AB 32 remained neutral on the ballot measure four years later. The proposition's supporters included the California Manufacturers Association and the Howard Jarvis Taxpayers Association. However, the California Chamber of Commerce and most electric and gas utilities remained neutral, as did the Chevron oil company; the latter claimed recent energy efficiency improvements. As a result, the "yes" campaign was mainly bankrolled by oil companies, including two (Valero and Tesero) that were based in Texas and had major refinery operations in California. Opponents referred to the proponents' funding sources critically during the campaign, calling the proposition an attempt by "Texas oil companies" to interfere with California's policies. Polling showed that this message was effective, as were appeals about the importance of renewable energy jobs.[137]

In sum, although the recession provided a problem context that was favorable to opponents, insufficient political commitment was present in California to overturn the law. Feedback from the law's implementation was one reason why political events did not converge with the problem events of the recession to create a window of opportunity to reverse the policy.

CONCLUSIONS

The major climate policy reforms of the 2000s were made possible by convergences in problem attention and political commitment that created two large policy windows; these are summarized in Table 10.1, in Chapter 10. The 2002 window was the result of the electricity crisis, President Bush's rejection of Kyoto, and the UCS's first report, combined with mobilization by the Bluewater Network and leadership by Fran Pavley and Byron Sher in the legislature. The coincidence of the state's electricity crisis and Bush's policy reversal was crucial. If deregulation and the electricity crisis had occurred a few years earlier or later than Bush's inauguration, or if Al Gore had won the 2000 presidential election, this policy window would not have developed in California. However, once the problem was seen as acute in 2001–02, the political commitment needed to produce major reforms could be created, even though Governor Davis was anything but a leader on the Pavley bill. As in 1967 and 1974, a large problem led to the creation of the political will to address it even in the absence of major independent political developments such as favorable election results. However, in this case, the actors mobilized out of a sense of opportunity to press a contentious policy, rather than in reaction to an intense, widely perceived threat, such as smog or energy supply shortages, to which relatively consensual solutions could be found.

Politics played a much more independent role in creating the 2005–08 policy window, as climate policy advocates helped to shape problem perceptions to a greater degree than the UCS had with its 1999 report. Still, problem events provided them with necessary raw material. The electricity crisis, natural disasters, and the UCS's second report gave energy and climate issues a special prominence in California by the mid-2000s. Then, shifts by business, the news media, and Congress, together with campaigns by environmental organizations, raised the perceived importance of global warming in California and across the United States. In addition, actions by climate policy actors at other levels – congressional advocates of cap and trade, climate scientists publishing reports, post-Kyoto negotiations for international agreements, and the European Union's inauguration of its Emissions Trading System – spilled over into California.

This combination of developments converged with a quite independent political event, Schwarzenegger's defeat of Davis in the 2003 recall election. This, in turn, ultimately led to Schwarzenegger's competing with the Democratic legislature to produce what became by far the strongest climate laws in the United States. As in 1974, an election result made a difference for California climate policy. If the recall effort had failed, climate policy under Davis probably would have been considerably weaker than it turned out to become, and his successor would have had little time to strengthen it before the national economic recession began in 2008.

In these cases, broad advocacy coalitions, the framing of policies in terms of co-benefits and opponents' motivations, and sometimes particular tactics – such as the surprise revival of the Pavley bill with a different bill number – were also important for achieving policy change. These political factors have become more important as the energy and climate policy arena in California (and elsewhere) has become denser, more contentious, and more polarized. As in most places in the United States, party polarization was present in California on this issue in the 2000s, but in this state, it did not block reform, since the legislature had large Democratic party majorities and advocates could afford a handful of Democratic defections.

Path dependence was also evident, since in several ways the policy changes in this period built on previous reforms. First, CARB's administrative capacities, political independence, and credibility were very important. Legislative and public confidence in CARB allowed Fran Pavley to write her motor vehicle legislation and AB 32 as brief bills that delegated much authority to the agency – without alienating moderate Democrats, who understood that CARB was effective and popular in reducing air pollution. As explained in Chapter 6, CARB's capacities and independence had been created by policy reforms in 1967–71, which had their origins in conflicts over air pollution going back to the 1940s.

Second, California's legal authority under the Clean Air Act – to develop more stringent air pollution rules than the federal government (subject to EPA approval) – made the strategy of the Pavley bill possible. The strategy ultimately was successful after the Obama administration ended the EPA's obstruction of those rules. Third, there was positive feedback from earlier energy efficiency, renewable energy, and transportation policies, which had helped to develop a large clean technology sector in California. Since firms in that sector became important advocates of AB 32, business was divided rather than unified in opposition, as shown in Table 7.1, earlier in this chapter.

Finally, the political response to the 1969 Santa Barbara oil spill made the interest group system more favorable to later climate policy making. The intense public and elite concern and environmental mobilization triggered by the spill probably curbed offshore oil development in California and hence reduced the future political clout of the oil industry in the state. It also spurred the growth of environmental organizations and the development of environmental public opinion there and nationally.[138]

Some prominent structural explanations are not supported by the California cases. Although a separation of powers is supposed to prevent strong climate policy, in this case, the legislature and governor competed to be more environmentalist. Although a pluralist interest group system is supposed to hinder climate policy, in highly pluralist California, environmental organizations were able to build broad, successful coalitions. Although a large fossil fuel industry is supposed to prevent strong climate policy, California's large oil production and refining industries were unsuccessful in a series of conflicts over the Pavley rules and the Global Warming Solutions Act – despite mobilizing large resources, getting much help from out-of-state oil and automobile industries, and sometimes using creative tactics.

Education and citizen ideology are somewhat more useful as explanations. California has relatively high levels of education and is more liberal than the average state. These factors probably helped to make the climate policy reforms of the 2000s possible, by undergirding a liberal legislature, encouraging Schwarzenegger to move to the left on climate policy, and promoting environmental organization membership. Polling on Proposition 23 showed that support for retaining the Global Warming Solutions Act was markedly higher among those with at least a college degree and among Democrats and independents than among other groups.[139] But education and citizen ideology do not explain why extraordinary policies were adopted in this state, because California did not have an extraordinarily educated or liberal population. The state's education levels were not much higher than the average state in 2000; California ranked 11th, with an education index of 36%, compared to 32% for the average state. Citizen ideology shows a similar picture. In 2005, California ranked 18th, with a 58% liberal rating on the Berry et al. index, compared to 54% for the average state.

Other structural factors were more important in promoting California climate policy making. The intense smog problem in Southern California, rooted in geographic and climatic conditions, was an essential precursor to CARB's creation and hence to the crucial role that that agency played in the 2000s. The lack of coal reserves and of a large in-state car industry reduced potential resistance.[140] But the latter factor was counterbalanced by California's longstanding dependence on automobile transportation in its economic growth strategy and the hallowed place of car driving in the state's political culture.[141] The massive size of the state's market also encouraged the national car industry and out-of-state oil companies to mobilize against California climate policy.

Finally, California's environmental organizations have been powerful since at least the 1970s, and they played important roles in the 2000s. However, this is not a purely structural factor that flows simply from a deep adherence to environmental protection among the state's residents. Environmental organizations and public support for environmental values grew during decades of conflicts over smog regulation, redwood logging, offshore oil production, and the Diablo Canyon nuclear plant. Nonetheless, public opinion on climate change in California did not lead the rest of the nation in the early 2000s. The high levels of support for climate policy now found in the state are the results, more than the causes, of the legislation that was the focus of this chapter. Indeed, California was not always a leader in explicit climate policy, as it adopted an RPS and greenhouse gas emissions targets later than many other states.[142] That it could catch up with those states and surpass them shows the importance of the processes that converged in the 2000s.

8

New York State: Energy and Climate Policy Making

Those who study state-level climate policies have given New York State much less attention than California, aside from studies of the RGGI emissions trading system that New York spearheaded among the Northeastern states.[1] The relative neglect of New York's broader climate policies has some justification. New York lacks both California's binding targets for greenhouse gas emissions and an economy-wide cap-and-trade program. In the adoption of major climate policies, such as greenhouse gas emission standards for automobiles, renewable portfolio standards, and energy efficiency resource standards, California and other states acted before New York, which still lacks a low-carbon fuel standard. But New York is far from a laggard in climate policy. It was a pioneer alongside California in energy efficiency policies during the 1970s, and it adopted emissions targets and emission trading in the early 2000s, before California. Also, its emissions targets are somewhat tougher than California's, aided by New York's slower population growth. New York State currently has an array of strong sectoral policies in energy efficiency, renewable energy, and transportation.

Climate policy adoption in New York has been concentrated in four episodes.[2*] First, legislation on power plant siting and participation rights for environmental organizations (1972–75) hindered the construction of new power plants and hence indirectly encouraged the energy conservation policies of the next period. Second, during the Carey administration, NYSERDA (1975) and the New York State Energy Office (1976) were created by law, the first appliance standards were adopted (1976), an energy conservation plan calling for a 7% cut in consumption was adopted (1977), and the home energy conservation law was passed (1977). The third period of climate policy reform, under Governor George Pataki from 2001–05, saw the state's current climate policies take shape: greenhouse gas emissions targets (2002), a renewable portfolio standard (2004), and RGGI (2005). The state also adopted goals for energy consumption in state buildings and statewide building codes (2001–02) in this period, as well as the Pavley rules on motor vehicle emissions (2005).

The fourth reform period occurred under Pataki's three Democratic successors (Eliot Spitzer, David Paterson, and Andrew Cuomo) in 2008–14, which saw a significant acceleration in climate policy. The state government adopted an energy efficiency portfolio standard (2008) and emissions targets through 2050 (in 2009), and it tightened the renewable portfolio standard (2010). It also adopted new building codes (2010), power plant standards (2011), new targets for energy consumption in state buildings (2012), and a reduction of the RGGI cap by 2.5% each year from 2014 to 2020 (in 2014).[3*] (See Chapter 2 for details on the policies.)

In the next section, I will place these episodes in the context of the public and political agendas in New York State. Then, in most of the remainder of this chapter, I will show that significant policies were adopted in these four episodes because problems and politics converged to create windows of opportunity that were used by policy entrepreneurs, who often included the state's governors. The conclusions summarize my arguments about why New York's climate policies have been relatively strong and often resemble California's, while also lagging behind that state in crucial respects.

THE PUBLIC AGENDA AND THE POLITICAL AGENDA

At certain times in the last fifty years, events raised public awareness of energy and climate change problems in New York State, which made it possible for policy windows to open. The nationwide energy shortages and price increases of the 1970s put attention on energy problems in the state, with peaks in attention by the *New York Times* in 1974 and 1979, as shown in Figure 9.1, in Chapter 9. Later, public attention on the problem of global warming rose to peaks at four different times, which are shown in Figure 8.1[4*] here: in 1989–92 (due to the scientific discovery of the issue, congressional hearings, and the United Nations conference at Rio); 1997 (due to the adoption of the Kyoto Protocol); 2001 (largely due to criticism of Bush's rejection of the Protocol); and 2007–09 (due to the largely politically driven rise of the issue nationally). These closely correspond to the peaks in attention to the issue in California and nationally, partly because the same national and international events drove the increases in attention on the issue in both states and in the United States as a whole.

These increases in public attention on energy and climate change problems were followed by the rise of the issues on policy makers' agendas. The number of bills mentioning global warming that were introduced in the New York State legislature rose continuously to a peak in the 2009–10 session, with large increases in 2001–02, 2007–08, and 2009–10, as shown in Figure 8.2.[5*] These increases followed increases in public attention to the issue, and they also correspond to the main episodes of climate policy adoption, which were after 2001 and 2007.

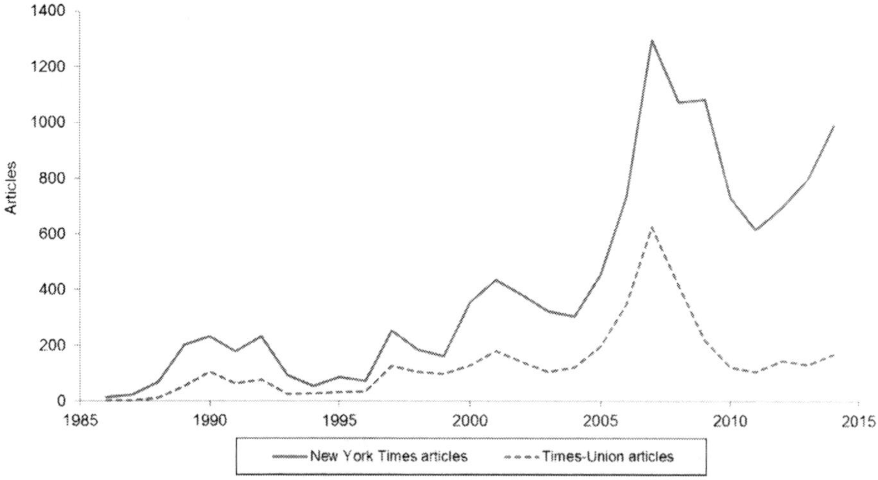

FIGURE 8.1 Public Agenda in New York State: Climate Change in the *New York Times* and the *Albany Times-Union*, 1986–2014
Sources: see note 4.

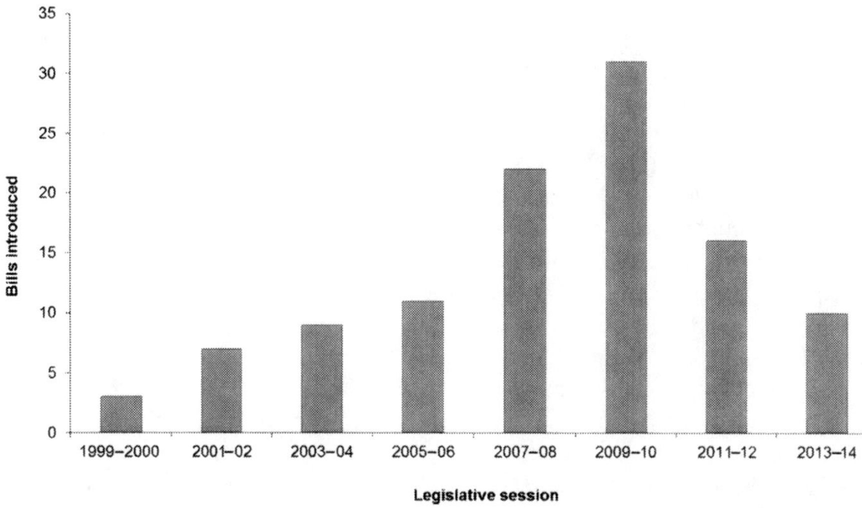

FIGURE 8.2 Legislative Agenda in New York State: Global Warming Bills, 1999–2014
Source: see note 5.

Finally, the governor's agenda in New York, as indicated by the content of annual State of the State speeches, generally shows increases in executive branch attention following the emergence of energy and climate change issues on the public agenda. Figure 8.3[6*] shows that there were major increases in the

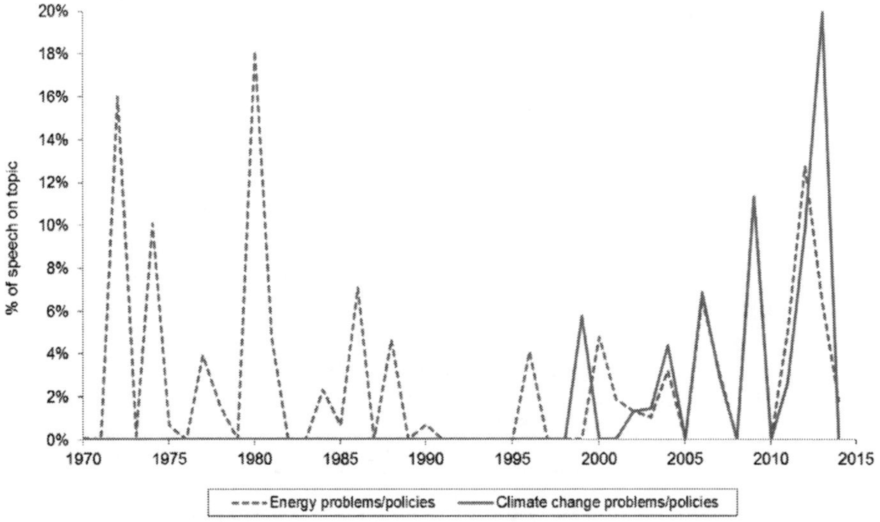

FIGURE 8.3 Governor's Agenda in New York: Energy and Climate Change Issues in State of the State Speeches, 1970—2014
Source: see note 6.

governor's attention to energy issues in 1972–74 and 1980; these were at about the same time that public attention peaked on those issues in the *New York Times*, as shown in Figure 9.1, in Chapter 9. Figure 8.3 also shows that governors' interest in the global warming issue rose in 1999, 2004–06, 2009, and 2013. The first three of these increases occurred a few years after the peaks in public attention that were due to the Kyoto Protocol (1997), Bush's rejection of it (2001), and the mid-2000s shift in business and media positions on climate change (2007). As described in this chapter, political commitment was strong during most of the periods of strong public interest, as the Carey administration was responsive to the energy crisis, Pataki took up the global warming issue when President Bush dropped it, and Pataki's Democratic successors responded to the rise of the global warming issue in the late 2000s.

In sum, the hypotheses in Chapter 3 are generally supported in the New York case. Problem events led to peaks in public attention on energy and climate change problems, which usually were followed by increases in the issue's position on the agendas of the legislature and executive branch; those, in turn, preceded the largest policy changes. The first peak in energy issues helped lead to major energy conservation policies in New York during 1975–77. The increases in attention to global warming in 2000–01 and 2005–09 helped to open policy windows that led to two bursts of policy adoption in New York State under Pataki (2001–05) and under his Democratic successors (2008–14).

However, there were also misfits between agenda changes and policy making. The 1979 peak in public attention to energy led to much more modest reforms than the 1973–74 peak had; these were limited mainly to the adoption of a "six-cent law" for alternative energy in 1981, which had little long-term impact.[7*] The scope for energy policy reform at this time may have been largely exhausted by the time the Iranian revolution occurred. Similarly, two early increases in public attention to global warming, in 1989–90 and 1997, did not lead to climate policy adoption in New York State. At those times, climate policy was seen largely as a federal issue, not one for the states. Partly as a result, the issue had no champion in the governor's office or legislature during Mario Cuomo's last term and George Pataki's first term as governor.

SMOG, INFRASTRUCTURE PROJECTS, AND PROTESTS: POWER PLANT SITING LAWS (1972–75)

Two episodes of policy reform occurred in the 1970s. The first episode, concerning power plant siting and participation rights for environmental organizations (1972–75), set the stage for the second one, which concerned energy conservation (1975–77). Path dependence is evident, as the first set of conflicts created and strengthened environmental organizations and shaped what were seen as the viable policy options for addressing the energy shortages and price increases that began in 1973.

In some ways similar to what happened in California from the 1940s to 1960s, urban smog and other environmental problems converged with environmental mobilization in New York State. In the late 1960s and early 1970s, severe air pollution problems arose in the New York City metropolitan area, despite the area's favorable prevailing wind patterns. A Spring 1966 report by the Mayor's Task Force on Air Pollution said that under the wrong weather conditions, the city "could become a gas chamber," and a statistical analysis estimated that a smog event in 1953 had killed 240 people there.[8] The point was reinforced and public awareness of the air pollution problem greatly increased in November 1966, when a five-day smog event, the most severe to date, coincided with the Thanksgiving weekend; had it been a normal workweek, with offices open and commuters underway, conditions would have been even worse. Reports by the mayor's task force and by federal agencies in 1966–68 found that New York City had the country's highest SO_2 concentrations and the worst overall air quality, and that the utility company Consolidated Edison was the largest contributor to coal smoke and SO_2 emissions.[9]

A second source of problem events during the 1960s consisted of large infrastructure projects with negative environmental consequences that were planned in the New York City area to meet the demand for energy and transportation due to economic growth. In the early 1960s, Consolidated Edison sought to build a pumped storage plant on Storm King Mountain, on a scenic stretch of the Hudson River about fifty miles north of New York City, and to build transmission lines across the river. Also, in 1965, state officials began planning to build an expressway along the east bank of the Hudson River from Croton to New York City, with portions on a landfill that would extend out 1,300 feet into the river.[10] Both projects threatened the scenic character of the Hudson River Valley and jeopardized commercial and recreational fishing. Opponents of these projects also revealed that the Indian Point nuclear power plant was killing large numbers of fish, which Consolidated Edison had been concealing by trucking away the dead fish.[11]

In addition, Con Edison proposed nuclear power plants for several locations in New York City during the 1960s. These included Ravenswood in Queens – across from Manhattan and only two miles from Times Square – as well as under Roosevelt Island and on man-made islands off Coney Island and Staten Island. The proposals raised public awareness of the risks of nuclear power plants, since even the federal Atomic Energy Commission at this time suggested that nuclear plants be located in areas that were unpopulated or had low population density.[12]

At the same time that awareness of the environmental consequences of unregulated economic growth was rising, political developments were increasing the political will to address them in environmental ways – approaches that increased the power of environmental organizations and made energy conservation an attractive alternative to increased production. Primarily, these political developments were bound up with the emergence of the U.S. environmental movement in the late 1960s. They included the creation of new environmental organizations, new spheres of activity for existing organizations, hearings and regulatory efforts by elected officials, and rising media coverage and shifts in public opinion. Although it was to some extent triggered by the specific environmental problems noted earlier, the environmental movement in New York State had complex, varied roots that were also largely independent of those problem events. These causes included the growth of a wide range of environmental problems due to suburbanization, the activism and challenges to authority of the student and antiwar movements, and encouragement from government and politicians.[13]

In the air pollution area, activists began to mobilize in the mid-1960s and got a boost from the 1965 New York City mayoral campaign. A new group, Citizens for Clean Air, and hearings by city councilman Robert Low put air pollution in the center of the campaign, in which John Lindsay emerged as a strong supporter of pollution controls. Lindsay won the election by a narrow, 43% to 39% margin over Abe Beame in a three-way race. Major smog events in 1963 (along the entire Eastern Seaboard) and in 1966 bracketed and reinforced this mobilization. The mayor's report on the issue held a variety of sources responsible, including Consolidated Edison, city buses, and apartment incinerators.[14] Con Edison was a large, visible polluter, and city officials made clear their opposition to its plans for new power plants in the city. The utility, seeking the city government's support for its Storm King plant, offered in 1966 to shut down or downsize eight power plants in the city and to build no new ones there;

it backtracked slightly in 1969 and built one oil-fired plant in the city. After the 1973 oil embargo, the city blocked Con Edison from switching from oil to coal.[15] The results of popular and elite mobilization against air pollution made it virtually impossible for new power plants to be built in New York City and made coal an undesirable fuel for power generation there.

In addition, mobilization by early environmental groups and their allies against the construction of the Storm King plant and the Hudson River Expressway had major ramifications for energy policy. New groups, Scenic Hudson and the Hudson River Fishermen's Association (later called Hudson Riverkeeper) formed to oppose the Storm King plant. They brought together scientists, fishermen, lawyers, and celebrities, and the Sierra Club joined their lawsuits.[16] They immediately gained national attention and formed alliances with New York congressional representatives, especially Democratic Congressman Richard Ottinger, against Consolidated Edison and the Federal Power Commission. After the latter approved the Storm King plant's license in 1970, Scenic Hudson and others continued litigation. Five years later, the New York State attorney general joined their side, and in 1980 the groups finally reached a settlement with Con Edison in which the utility abandoned the plant.[17] A similar ad hoc coalition of groups, which included the Citizens Committee for the Hudson Valley, Scenic Hudson, and the Sierra Club, opposed the construction of the Hudson River Expressway. Joined early by Ottinger and the mayor of Tarrytown, the coalition convinced federal courts to block the project in 1970.[18]

The siting of nuclear plants provoked at least as much environmental opposition as new coal plants or Storm King did. The proposed Ravenswood nuclear plant in Queens was blocked by public opposition in 1963, as were the other nuclear plants proposed for New York City in the late 1960s. Famously, former Atomic Energy Commissioner David Lilienthal told a congressional committee that he "would not dream of living in Queens" if a plant were built there.[19] It was not.

Besides their effects on specific projects, these local environmental conflicts had major consequences for the state's environmental and energy policies. In a form of path dependence, they created new participatory rights, strengthened environmental organizations, and helped to shift public opinion. First, the conflicts led to court decisions that established two new principles. One was that environmental groups had standing to sue in federal court to uphold the public interests in aesthetic and other environmental values; the other was that government agencies were required to develop detailed records showing that the agencies were protecting those public interests. A landmark federal court ruling in 1965 (*Scenic Hudson Preservation Conference v. Federal Power Commission*) was the first to assert these principles. Later, they were affirmed by the federal Environmental Protection Act in 1969 and by the U.S. Supreme Court in its 1972 *Sierra Club v. Morton* decision.

In New York State, these new environmental principles were institutionalized in the state's 1972 power plant siting law (initially in Article 8, later in Article 10) and its 1975 Environmental Quality Review Act, which was modeled on the federal Environmental Protection Act. The latter was passed by the legislature and signed by Governor Hugh Carey despite vigorous opposition from the construction industry, labor unions, and municipalities.[20] The new state laws required that environmental concerns and economic needs be balanced, that environmental impacts be investigated in detail by government agencies and project developers, and that environmental groups be allowed to participate in environmental reviews and even be given funds to facilitate their participation.[21] These legal changes gave environmental organizations much potential leverage over the siting of power plants and other facilities.

The environmental conflicts of the late 1960s also led to the formation of new environmental organizations, while existing organizations shifted from their earlier strategy of lobbying Congress and educating voters to an approach based largely on suing in federal court.[22] The publicity surrounding environmental conflicts, especially the 15 years of controversy about Storm King, helped to durably shift public and elite opinion toward environmentalism. This may help to explain why, since the 1970s, public opinion in New York State has been more environmentalist than in most states.[23*] The accommodation of environmental interests in energy planning became part of the new common sense of New York State politics.

Together, new participatory rights, stronger environmental organizations, and more supportive public opinion together helped to block the construction of coal and nuclear plants. Emboldened by their early successes and empowered by new participatory rights, environmental groups continued to press for limitations on plant siting and air pollution in the 1970s and later. In doing so, they helped to deter or block most new power plants. During the 1970s and 1980s, there was much local opposition to proposed plants along the Hudson River, in New York City, and on Long Island.[24] The federal Clean Water Act of 1972 gave environmentalists additional leverage through arguments about the effects of thermal water discharges.[25]

In New York State, the new plant-siting and environmental laws were intended to allow large projects to be approved while accommodating environmental interests, but they actually made plant siting more difficult, especially for coal and nuclear plants. Article 8 was intended to ease the process of plant siting, by allowing applicants to get all necessary approvals from one agency instead of from multiple state and local agencies. However, only 9 applications for large power plants were submitted during 1972–96, only 6 were approved, and only 1 plant was built, an upstate coal plant. Utilities dropped the others because of environmental opposition, increased costs, and other "changed circumstances" (including reduced energy demand projections) that resulted in part from delays in the approval process.[26] In short, environmentalists helped to force utilities and energy planners to consider conservation as a politically and economically more viable option than new power plants.

There were additional reasons why nuclear energy stalled in New York State, as it did nationally. Citing the waste disposal problem, Governor Carey announced a moratorium on nuclear plant construction a month after taking office in 1975.[27] In 1978, Carey made clear he would use his power as governor to oppose new nuclear plants, while allowing plants under construction on Lake Ontario (Nine Mile Point) and on Long Island (at Shoreham) to proceed.[28] Therefore, during the 1970s, it became increasingly doubtful that nuclear power could make up much of the difference between projected electricity demand and supply in the state, which in turn helped to make conservation more attractive. Rising costs and the 1979 accident at Three Mile Island led the nuclear power industry to shelve its expansion plans, and two other planned nuclear plants were abandoned in New York State in 1979–80.[29] Ultimately, the Shoreham plant was finished in 1984 but never went into operation, because the county legislature and state government refused to approve evacuation plans; the Nine Mile Point plant finally opened in 1988.

In the long term, the series of environmental conflicts in New York State helped to brake efforts to revive coal and hence gave a long-term advantage to natural gas. This helped to limit increases in carbon dioxide emissions even before climate change became an issue.[30] The proportion of electricity production that was coal-fired was halved in New York State from 1965 to 1980, although it increased again after 1985; natural gas and nuclear generation largely replaced it.[31*] After 1998, when power plant siting was made easier, the new plants that were constructed were largely fueled by natural gas.

Furthermore, the growing constraints on plant siting, together with the electricity blackouts that began in 1965 (see the next section), led to a reexamination of the assumption that energy demand would necessarily increase indefinitely. Consolidated Edison, under its new chairman Charles Luce, slowly began to shift its position in May 1971, when the company stopped advertising to encourage electricity and natural gas consumption. Together with Mayor Lindsay, it introduced the Save a Watt program to encourage conservation, in order to address the related problems of peak capacity limitations and the environmental damage from new plants. The utility's efforts included a large advertising campaign encouraging customers to save energy in many areas, shift energy use away from the midday hours, and use more efficient air conditioners. The campaign was modestly successful by 1973, reducing demand by 5% and saving 400 Megawatts in generating capacity compared with the business-as-usual projection, although total demand still rose.[32]

Therefore, a workable solution to New York's energy problem began to emerge in 1971, two years before the Arab oil embargo greatly intensified the energy supply problem and public perceptions of it. What is just as important, Consolidated Edison had begun a long, slow, uneven shift away from opposing conservation toward ambivalence and ultimately support for it; the rest of the state's utility industry began to undergo this shift in the late 1970s.[33*] In short, because of the consequences of local environmental conflicts, when the energy crisis arrived in 1973, policy makers, environmentalists, and even parts of the utility industry were already searching for alternatives to increased energy supply.

BLACKOUTS, ENERGY SHORTAGES, AND THE CAREY ADMINISTRATION: ENERGY POLICIES (1975–77)

The policy window just described, concerning air pollution and power plant siting, overlapped to an extent with a second window of opportunity concerning energy supply and demand. Beginning in the late 1960s and accelerating after 1973, several events increased awareness of energy supply problems and called into question whether continued growth in energy consumption was possible or desirable. Because these problems converged with political developments, New York created NYSERDA and the state energy office, adopted comprehensive energy plans, and became a leader in appliance standards and energy efficiency in buildings beginning in the 1970s.

The first problem was that the electricity industry's decades-long strategy of encouraging increased demand and meeting it with large new power plants ran into internal difficulties. Electricity demand grew 80% in the 1960s and was still increasing at an annual rate of 6% in 1970.[34] But the industry was beginning to confront diminishing returns from its approach of raising steam temperatures and pressures in order to increase output. In response, the utilities rushed the introduction of larger generating turbines, but this decreased the reliability of power plants. The problems hit Consolidated Edison – the provider of power to three million customers in New York City and one of the largest utilities in the United States – especially hard because of its aging power plants and financial problems.[35]

Second, blackouts and brownouts became frequent beginning in the late 1960s, as power plant expansion failed to keep up with growing electricity demand. In November 1965, parts of nine Northeastern states and two Canadian provinces lost power. The entire New York metropolitan area was without electricity for 13 hours, and 800,000 people were stuck in subways during rush hour. In addition, partial blackouts of the city occurred almost every year in the late 1960s and early 1970s, as did preventive brownouts, in which voltage reductions dimmed lights and slowed subway cars.[36*] The blackouts and brownouts underscored the persistent conflict between the growing demand for electricity and the relatively stagnant power plant capacity.

Third, the energy crises of 1973–74 hit New York State, along with the rest of the United States, with shortages of oil, electricity, and natural gas, and with price increases for motor fuels and heating oil.[37] The results included rationing, emergency preparations, and major economic costs. While gasoline lines, rationing, and rising prices harmed ordinary citizens, the oil crisis also affected New York's electric utilities. They had begun switching from coal to oil after 1965, due to declining oil prices and rising environmental action on air pollution. The state's utilities generated 50% of their electricity from oil in 1975.[38] The energy crisis had long-term effects. It created an atmosphere of uncertainty about future energy supplies, prices, and the economic viability of a region that was highly dependent on fossil fuels produced elsewhere, including imported oil.

Fourth, the state's loss of industrial jobs in the 1970s, part of a trend throughout the country's traditional manufacturing centers, also became linked to energy prices. Job losses created economic hardship and threatened the long-term economic development of the state; the official unemployment rate peaked at 10% in 1976, a rate not equaled since.[39] New York tried to compete for factory jobs with other states, many of which had lower energy prices and warmer winters, so officials were under pressure to alleviate energy costs. Hence, in 1977 the state legislature prepared a report, "Operation Bootstrap," authored by Democratic Assemblyman Daniel Haley, that argued for energy efficiency and renewable energy development as sources of job creation.[40]

At the same time that this set of energy supply problems increased public and elite awareness of energy problems, an unrelated political development put officials with a commitment to environmental protection and energy conservation into key elected offices. Due largely to the Republican Party's unpopularity in the wake of the Watergate scandal, the Democrats won the New York governorship in the 1974 elections. They also gained control of the State Assembly with a 90 to 60 seat majority, although Republicans retained a State Senate majority of 35 to 25 seats. If the conservative incumbent Republican, Malcolm Wilson, had defeated Hugh Carey for governor, it is unlikely that conservation would have become a major energy policy emphasis at this time.

Carey had a track record in favor of environmental protection as a U.S. Representative, and he made environmental and energy issues a theme in the 1974 gubernatorial campaign.[41] He strongly defended clean air laws and environmental organizations, calling Wilson's failure to create a plan to cut energy use by 5% "irresponsible." Carey called for a much more vigorous power conservation program, in part by reforming the Public Service Commission and changing its rate structures to stop encouraging increased consumption. Carey gained the endorsement of Political Action for Clean Air, a coalition of the state's leading environmental organizations.[42*]

Wilson took opposing positions, against clean-air requirements, citizen lawsuits, and environment impact statements, and in favor of Storm King and the expansion of nuclear power. Wilson's energy policy was weak on conservation and mainly limited to calling for the creation of a state energy office to deal with emergencies; he also refused to campaign for an ultimately successful mass transit bond act in 1974.[43] Carey, benefiting from a strong national swing to the Democrats in 1974, won easily, by a 57% to 42% margin, making him the first Democratic governor of the state since 1958.

After the 1974 elections, Governor Carey and Democrats in the Assembly[44*] successfully pressed an energy conservation agenda as part of comprehensive energy planning.[45] The combination of acute energy problems, constraints posed by environmentalists, the availability of federal funding, and Carey's rhetoric of resisting OPEC through energy conservation proved difficult for Republicans in the Senate to resist. Hence, the legislature passed a law creating NYSERDA in 1975, sponsored by Haley; it had bipartisan support in the Assembly, although the Republican-led Senate initially opposed it.[46] The new agency was to develop an energy plan for the state that would conserve energy and ensure a dependable and environmentally acceptable energy supply. NYSERDA also would carry out research and development on wind and solar energy, and on energy storage and energy conservation technology.[47] In February 1976, Carey also called on the legislature to create a State Energy Office, and it did so that year.[48] The energy office was to administer energy constraints in emergencies, oversee a comprehensive energy conservation program, and cooperate with NYSERDA.[49] It issued an energy conservation plan in 1977, which called for a 7% reduction in energy use by 1980 and was cited as an early example of integrated energy planning.[50]

Carey showed strong commitment to energy conservation. His State of the State speech in January 1977 challenged OPEC: "Raise your prices, and we lower our consumption," a message that he echoed three years later when he called for ending oil imports through cutting consumption, expanding mass transit, and switching power plants to coal.[51] Besides saving billions of dollars that "would otherwise have gone down the OPEC drain," he predicted that an energy efficient New York would become a magnet for industries wanting to cut fuel and transportation costs.[52] By framing energy policy in terms of energy security and economic development, Carey argued that reducing energy consumption was a viable response to the multiple problems that New York faced. His 1977 address called on the legislature to pass measures to achieve the goals of the forthcoming energy conservation plan. Later that year, the legislature passed the Home Insulation and Energy Conservation Act, which required all major utilities to provide inexpensive energy audits and financing for eligible energy conservation improvements.[53] However, Carey also prioritized increasing energy supply and hence pushed for more coal burning, despite the environmental consequences. Nonetheless, a net switch to coal did not occur in power generation in New York State at this time, because of urban air quality concerns and the strong environmental opposition described in the previous section.[54]

Finally, another political factor that contributed to reform in this period was federal legislation that gave resources to states which adopted energy conservation reforms. The federal 1975 Energy Policy and Conservation Act and the 1976 Energy Conservation and Production Act spurred the creation of a state energy office and the adoption of the 1977 energy plan in New York. However, federal funding for energy efficiency was not generous,[55] and New York State went well beyond what most other states were doing, for example with appliance standards, building code revisions, and later a six-cent feed-in tariff for alternative energy.[56] By 1979, New York was one of the top twelve states in energy efficiency policies.[57]

KYOTO, BUSH, PATAKI, AND COALITIONS: RGGI AND A RENEWABLE PORTFOLIO STANDARD (2001–05)

New York State adopted a number of significant climate policies by executive action during 2001–05: emissions targets, emissions trading in RGGI, a renewable portfolio standard, endorsement of the Pavley standards on motor vehicle emissions, efficiency measures in state buildings, and revised statewide building codes. This burst of policy reform occurred because problems and political developments came together at this time to create policy windows, and because of the presence of specialized administrative capacities and strong environmental organizations, which were legacies of the 1970s policy-making episodes.

First, the problem of anthropogenic climate change received widespread coverage in New York State and the rest of the U.S. beginning in 1997, triggered by the Kyoto climate change conference (see Figure 8.1 earlier in this chapter).[58] Coverage of the issue reached a second peak in 2001, due to President Bush's rejection of U.S. participation in the Kyoto Protocol. Bush's policy reversal triggered criticism by domestic climate policy proponents from both parties and by the leaders of many other countries. Bush's clear signal of federal inaction indicated that the global warming problem would only get more severe in the coming years, which in many states created a sense of urgency to take action without waiting for Washington.[59] Three months after Bush's initial announcement, in June 2001, Pataki announced that he was taking a different approach. He issued an executive order requiring state buildings to cut energy use 35% by 2010 and to obtain 20% of their electricity from renewable sources by 2020. Pataki also announced the creation of the Greenhouse Gas Task Force, which was charged with finding ways to reduce statewide greenhouse gas emissions.[60]

Second, the 2000–01 California electricity crisis and the Bush administration's energy policy, announced in May 2001, increased the perceived importance of energy policy for the public (see Figure 9.1 in Chapter 9). The Bush administration's energy policy downplayed energy efficiency and renewable energy; instead, it sought to increase U.S. fossil fuel supplies by building more pipelines and increasing the extraction of oil and natural gas.

The rising attention to the problems of global warming and energy security took place at a time when political developments in New York also favored action in this area; the convergence of problems and politics created a policy window. First, George Pataki had won the governorship in 1994, in a close, surprising election. He defeated the twelve-year incumbent governor, Democrat Mario Cuomo, by a narrow 49% to 45% margin, in a campaign marked by the issues of tax cuts and capital punishment. Pataki was personally interested in environmental protection and had a strong environmental record in previous offices, and he was re-elected in 1998 and 2002. As the mayor of Peekskill, Pataki had worked to protect open space along the Hudson River in 1982, and as a state legislator, he advocated a landfill cleanup, green space along the river, and wildlife habitat preservation.[61] Pataki's environmental appointees, including John Cahill, his closest aide, shared his environmental interests. Cahill was the Environmental Commissioner from 1997 to 2001 and Pataki's chief of staff from 2001 to 2006; after Pataki left office, Cahill was his partner in the Pataki Cahill group, a business development firm specializing in energy and clean technology.[62]

Despite his environmental values, Pataki's record on environmental issues was mixed at best for the first half of his twelve years in office. He worked to preserve land from development and brokered a $1 billion deal to protect New York City's drinking water supply, but his administration cut spending on mass transit and cleaner buses. It loosened the enforcement of pollution laws, by reducing prosecutions and fines, to the point that the federal EPA threatened to take over enforcement against large-scale polluters. Similar to Schwarzenegger, Pataki's aides were divided between pro-environmental and anti-regulation camps, the latter influenced by his political patron, Alphonse D'Amato, a U.S. senator who retired in 1999.[63]

However, Pataki's desire to be re-elected as governor and to run for president led him to increasingly emphasize environmental issues. His situation was quite similar to that which Schwarzenegger faced in California a decade later. As a Republican governor pressing the legislature for spending cuts and tax cuts in a liberal state, Pataki faced declining approval ratings and found that he could reverse the decline through environmental initiatives. In 1996, he persuaded the legislature to put on the ballot a $1.5 million environmental bond measure, which was carefully crafted to appeal to a very wide range of interests through spending on sewer projects, landfills, and toxic waste cleanups throughout the state.[64] The Pataki administration then led a proactive, carefully designed campaign on the bond act referendum, which included mobilization by environmentalists and labor unions; it led to a 57% to 43% victory. Pataki's advocacy of the environmental bond act led to a boost in his popularity. While Pataki's budgetary and educational policies remained unpopular, the bond act was supported by a 65% to 22% margin after it passed, and Pataki's job approval rating crossed the 50% mark for the first time.[65]

The bond act included a small amount of funding for clean-air initiatives to reduce coal burning in schools and to promote clean-bus technology, which marked the start of Pataki's interest in policies related to climate change. High-profile campaigns by the NRDC and West Harlem Environmental Action against diesel pollution, which began in 1993, played a large role in Pataki's decision to create the Clean-Fuel Bus Program. Pataki endorsed clean bus technology in his January 1997 State of the State speech, in which he called for an immediate switch to natural gas and then the introduction of hybrid electric vehicles. He issued an executive order in 2000 and released a total of $28 million in funding.[66]

After his re-election in 1998 by a 21% margin, Pataki was still a Republican governor trying to implement tax cuts and restrain government spending in a liberal state, so he needed to shore up his public support. Moreover, the Democrat Eliot Spitzer, a former prosecutor and consumer rights lawyer, was elected attorney general in 1998, and Pataki soon found he had to compete with Spitzer for the support of environmentalists. In 1999, Spitzer sued 17 Southern and Midwestern power plants for Clean Air Act violations that caused smog and acid rain in New York State.[67] Spitzer quickly gained a high profile for that lawsuit and for others against polluters, gun makers, drug companies, mortgage companies, and Wall Street firms, and he was seen as a likely future candidate for governor, although he did not actually run until 2006.[68]

Pataki responded to the competitive pressure from Spitzer by seeking to bolster his own reputation and environmental credentials with the state's voters, as he had done quite successfully with the environmental bond act. Therefore, after the issue of climate change rose in the late 1990s and President Bush turned his back on the Kyoto Protocol in 2001, Pataki turned to climate policy. Soon after the September 11, 2001 attacks on the United States by Al Qaeda, Pataki took a different tack from that of the president, one intended to address both energy security and climate change. In his State of the State speech in January 2002, he called for New York to develop renewable energy in order to improve the environment and "reduce our dependence on imported foreign energy," as well as create jobs, diversify energy supplies, and "increase our security."[69] Later that year, the state's energy plan was adopted, including a target of reducing greenhouse gas emissions by 10% over 1990–2010. The Task Force report was clear that the intention was to take a "small but important first step" toward stabilizing atmospheric CO_2 concentrations, similar to the Kyoto Protocol. The task force's report also emphasized the goal of making New York a leader in addressing climate change, building on its leadership role in energy efficiency, renewable energy technology, and transportation policies.[70]

At this time, Pataki also announced the RPS goals, pushed for their adoption by the Public Service Commission, and worked with regulators to ease siting constraints for renewable energy.[71] Sometime after his 2002 re-election, Pataki set his sights on the White House,[72*] which helps to explain why he turned toward high-profile climate policy measures in the next few years, since these helped to differentiate him from President Bush. In 2003, following an initiative by Thomas DiNapoli, a key Assembly Democrat, Pataki called for the state to follow the Pavley rules on motor vehicle greenhouse gas emissions.[73] That year, he also announced the RGGI initiative, and over the next two years, the Pataki administration negotiated with nine other states and ultimately helped to create the first greenhouse gas emissions trading system in the United States.

A second political development that favored climate policy in the early 2000s was the rise and mobilization of a strong coalition of climate policy advocates, which occurred for several reasons. In 1996, electricity restructuring led to altered incentives for politicians and utilities. Utilities were forced to sell off their generating facilities and to purchase power from independent power producers. With rate increases now largely determined by market forces operating on wholesale electricity prices, elected officials were no longer completely responsible for electricity prices. Hence, paradoxically, they became more willing to take actions that would increase rates, as RGGI was expected to do.

Furthermore, the electricity sector became divided between energy producers, who would be required to buy emissions allowances under RGGI, and distribution ("wires-only") utility companies. Once auctioning was adopted, the distribution utilities supported RGGI because the new revenue would help to fund their existing energy efficiency programs. In addition, deregulation created a single wholesale electricity market in the Northeast, which meant that if New York State adopted emissions trading by itself, this would raise wholesale electricity prices throughout the region even if other states did not join the trading system. This gave New York leverage in getting other states to participate in RGGI and to adopt the auctioning of allowances, since the latter would provide them with revenue to help offset the expected price increases.[74]

Finally, environmental and consumer advocates joined with business associations to support climate policy in the early 2000s. Two environmental organizations, NRDC and the Pace Law School Energy Project, began working with wind power interests such as the American Wind Energy Association. They all were part of the Renewable Energy Technology and Environment Coalition, which promoted the renewable portfolio standard beginning in 2003.[75] These groups also helped to form the Alliance for Clean Energy in 2006, which promoted energy efficiency and renewable energy. The alliance grew to 80 members, including environmental, public health, consumer advocate, and educational organizations, as well as private renewable energy and energy efficiency companies.[76]

A major conflict concerned the renewable portfolio standard, in which the advocacy coalition faced an opposing coalition, Multiple Intervenors, which argued that the RPS would increase costs and reduce system reliability. Multiple Intervenors was a decades-old association of large commercial and industrial energy consumers in New York State, including Alcoa, IBM, and Occidental Chemical; it opposed RGGI as well as the RPS.[77*] By contrast, the New York Business Council, a broader business group, registered only moderate opposition, calling for the program to be designed to minimize electricity rate increases and to treat all renewable sources equally. The RPS advocates prevailed in the Public Service Commission in September 2004, after a sixteen-month process involving over one hundred parties and including technical workshops, four comment periods, affidavits, and cost studies.[78]

The formation of the advocacy coalition was facilitated by Pataki personally, by the stakeholder consultation processes that his administration initiated, and by the prior existence of NYSERDA. In June 1996, Pataki convened a key meeting of some of the state's leading environmental organizations, including Environmental Defense, the NRDC, and Hudson Riverkeeper, in order to get their support for the environmental bond act. Within a week, he had gained their backing as well as that of business and labor groups. Pataki also conducted broad stakeholder consultations for a 1997 agreement to protect New York City's upstate watershed, in a 1998 task force on superfund sites (again with business and environmental support), and then for his main climate policy initiatives, the RPS and RGGI.[79] Pataki's Greenhouse Gas Task Force included 4 members from industry, 5 from environmental organizations, and 6 from state government; they overlapped heavily with the renewable energy coalition. In addition, by hosting working groups on the RPS, NYSERDA "provided a neutral, science-based arena" for these groups to meet.[80]

The third political factor was the use of policy design and framing to ease acceptance of the state's new climate policies. Generally, policy makers in New York State did not call for public sacrifice and instead emphasized reducing foreign oil imports, cutting air pollution, and helping local industries. Moreover, RGGI's expected price increases were made more publicly acceptable by directing most auction revenue to support energy efficiency, including for low-income households.[81]

Finally, the policy window that opened in New York in the early 2000s did so in a state where the legacies of the 1960s and 1970s conflicts and energy policies made a strong climate policy response possible. Environmental organizations had grown greatly since the late 1960s, with national organizations such as the EDF and NRDC based in New York City and taking an interest in the state's environmental policies.[82*] Representatives of Environmental Advocates and the NRDC publicly endorsed a staff proposal to cut electricity-sector greenhouse gas emissions by 40%. However, industry and the governor's office strongly opposed the target, and it was scaled back to a 25% reduction in the task force's final report and to a 10% cut in the RGGI program.[83] Also, the environmental organizations' longstanding opposition to new coal plants helped to ensure that the new fossil fuel plants built in New York State after 2004 were fired by natural gas, which made it possible for the state to meet its emissions reductions targets.[84*]

Furthermore, New York had already developed strong institutions for energy forecasting, research and development on energy efficiency, and energy-sector regulation, and it had gained three decades of experience in those areas since the 1970s. NYSERDA, created in 1975, was funded by a tax on utilities, their voluntary contributions, and federal grants – all sources that were independent of state legislative politics. This insulated the agency from partisan politics and made it more resilient.[85] Although NYSERDA's annual budget was initially small, about $15 million, it leveraged these funds by using them to influence research and development projects that were largely funded by private actors or the federal government.

Specialized administrative capacity for climate policy grew further under Pataki. NYSERDA took over the energy efficiency and renewable energy jurisdictions of the State Energy Office (abolished in 1996), including the utilities' energy efficiency programs, mandatory appliance standards, energy conservation building codes, demand forecasting, and supply planning. NYSERDA's budget grew to $220 million in 2005 and to about $600 million in 2011–13. About 55% of the money was spent on energy efficiency programs, 20% on renewable energy incentives, and 15% on technology and market development programs. After federal tax credits for wind production temporarily expired in 2000, the fact that New York was one of only a few states with an agency capable of research and development and technical support in this area was important for attracting wind power developers to the state. In addition, the Public Service Commission has had a large staff, with 500 employees in 2012; although this is less than the peak of 800 in 1994, it still seems adequate. The commission oversees about a thousand utility companies that have a total of $31 billion in operating revenue, as well as municipal electric and water companies.[86]

DEMOCRATIC GOVERNORS AND NEW FEDERAL OPPORTUNITIES: ACCELERATED CLIMATE POLICIES (2008–14)

The next period saw an acceleration of climate policy making in New York State, in which the state government adopted an energy efficiency portfolio standard and a target of reducing greenhouse gas emissions 80% by 2050, and tightened the renewable portfolio standard. It also adopted new building codes, power plant standards, new targets for energy consumption in state buildings, and a reduction of the RGGI cap by 2.5% each year from 2014 to 2020.[87]

These major policies were driven by events in the problem stream, which put attention on the climate change issue nationally and in New York, and by a political shift toward the Democratic Party, which put people into office who were strongly committed to climate change mitigation policy. In the first place, events such as Hurricane Katrina, the business and media shift toward supporting climate science and climate policy, and Al Gore's *An Inconvenient Truth* documentary sharply increased public attention on global warming beginning in 2005, as shown in Figure 8.1, earlier in this chapter.[88*]

However, the major national economic recession that began in December 2007 reduced concern about climate change, as New York State's unemployment rate shot from 5% to 9% in a year and a half during 2008–09 and remained at about 8% through 2012.[89] Nationally, the recession dampened public support for environmental policy, reduced the perceived importance of the climate change issue, and strengthened climate policy opponents in their arguments that climate policy would be too economically costly (see the public agenda section of Chapter 9). In the *Albany Times Union*, the climate change issue faded rapidly beginning in 2008, although it remained relatively prominent in the *New York Times* at least through 2014 (see Figure 8.1).

But in New York State, the recession did not reduce policy makers' interest in climate change. As in previous periods, they emphasized the economic advantages of energy efficiency and renewable energy alongside the benefits for reducing climate change. The 2009 Energy Plan and Andrew Cuomo's 2010 campaign documents put much emphasis on the dangers of climate change and the need to reduce greenhouse gas emissions.[90] The 2009 target was justified by reference to the scientific consensus on anthropogenic climate change and its expected adverse impacts, and secondarily by the job creation expected from energy efficiency and renewable energy projects.[91] A 2013 progress report on Governor Andrew Cuomo's policy goals stated that the first two lessons learned from Hurricanes Irene and Sandy were the need to reduce the RGGI emissions cap and to expand renewable energy.[92] By contrast, when New York governors justified climate policies in State of the State messages in this period, they emphasized the problems of energy costs and the need for local economic development.[93*]

At the same time that the problem stream was highlighting climate change, Democrats took control of the federal government. They won majorities in both houses of Congress in 2006, and they gained the presidency and expanded congressional majorities in 2008. As part of the national political trend, a series of three Democrats committed to climate policy became the governors of New York after Pataki's twelve years in office. The first was Eliot Spitzer, who won the 2006 governor's election by a commanding 69% to 29% margin. Spitzer had a strong record on environmental protection and corporate regulation in his eight years as attorney general, an office he held during the election campaign – during which he was the lead plaintiff in a ten-state federal lawsuit against the EPA for failing to regulate carbon dioxide emissions.[94] Three months after becoming governor, Spitzer announced ambitious energy conservation initiatives to cut state-wide energy consumption by 15% by 2015, including appliance standards, building standards, and conservation in state buildings.[95] The Public Service Commission adopted an energy efficiency portfolio standard in 2008, and a new Climate Change Office was created, with a staff of twelve scientists and policy analysts.[96]

When Spitzer resigned due to a sex scandal in March 2008, he was succeeded by the lieutenant governor, David Paterson, a veteran, liberal Democratic assemblyman who continued Spitzer's climate policy course during his three years in office. In his first State of the State speech, Paterson called for a 15% reduction in energy consumption through increased efficiency, as well as a new, 30% target for the RPS; the Public Service Commission adopted the latter in 2010.[97] Under Paterson, the state government adopted a number of new policies, mainly through executive action. It set a target of reducing emissions 80% by 2050, commissioned the 2010 climate action plan, implemented the first RGGI auctions, adopted the energy efficiency portfolio standard initiated by Spitzer, tightened state building codes and requirements for new government buildings, and expanded net metering to small-scale renewable energy systems.[98] Environmental organizations were also involved in New York State climate policy at this time, with twelve different groups participating in the working groups that drafted the climate action plan, including the National Wildlife Federation, the Nature Conservancy, the Alliance for Clean Energy New York, and the NRDC.[99]

Plagued by scandals and very unpopular, Paterson ultimately did not run for election to continue as governor in 2010. Andrew Cuomo won the Democratic nomination for governor and the general election by a huge 63% to 33% margin, took office in January 2011, and won re-election in 2014. Cuomo's 2010 election campaign included a detailed energy policy program, which was broadly similar to the climate action plan released in the last months of the Paterson administration.[100] However, Cuomo's plan criticized the state for not moving quickly enough to achieve its goals for energy efficiency and renewable energy.[101] During his first three years in office, Cuomo adopted new power plant standards and new targets for energy consumption in state buildings, obtained the agreement of the other participating states to lower RGGI's emissions cap in 2014,[102] and used $100 million in RGGI money for smart growth programs in communities. His administration also created a solar initiative for households and businesses (with $800 million in investment planned) and eased permitting requirements for small-scale renewable energy facilities.[103]

CONCLUSIONS

Windows of opportunity developed in New York in ways that were broadly similar to California; they are summarized in Table 10.1, in Chapter 10. The two policy windows in the 1970s were products of growing air pollution problems, environmentally threatening infrastructure projects, blackouts, and energy supply crises – together with political constraints and incentives posed by environmental mobilization and the elections of Lindsay and Carey. The early 2000s window was a product of problem awareness created by international action on climate change, Bush's retreat from the Kyoto Protocol, his promotion of fossil fuels, and the September 11 attacks – together with the political commitment provided by Pataki's election, his strategic shift toward the climate issue, and the development of a broad advocacy coalition for climate policy. The late 2000s window was a product of the national shift in media attention to the climate issue, which had many causes, combined with the election of Democrats committed to climate policy in Washington and Albany. The latter two periods of major policy change also were made possible by the administrative capacities and environmental organizations created in the previous period of reform, in the late 1960s and 1970s.

As in California, the problem stream and the political stream were not always independent of each other. The intense, concrete problems of air pollution and energy supply crises were capable of generating a degree of political commitment on their own in both states, although they did not determine the nature of the policy solutions adopted. Recurring smog events in Los Angeles and New York City in the 1960s helped to convince policy makers that action was urgently needed – especially given the risk of mass deaths due to smog, as had occurred in Donora, Pennsylvania in 1948. So did the rapid fuel price increases, power outages, gas lines, energy rationing, and emergency plans in 1973–74.[104] These problems posed severe and obvious threats to dominant values, since they threatened economic activity, long-term economic growth, and sometimes loss of life or civil disorder. In addition, the political parties were not highly polarized through the early 1970s, which helped to make relatively consensual policy solutions possible.

By contrast, the more abstract, distant threat of climate change usually lacked the power to drive the political stream, especially given the much more polarized party landscape of the 2000s. Therefore, the explicit climate policies adopted in this period were only possible because independent political developments converged with problem events. One such political development was the election in 1994 (and re-election in 1998) of a politician, George Pataki, who had strong environmental interests. Those elections put him in office during the late 1990s and early 2000s, when the Kyoto Climate Change Conference took place and Bush rejected the Protocol. The Democrats' mid-2000s election victories, at a time of peaking fossil fuel prices and public attention to climate change, comprised another key political development. By this time, the energy policy field had become more complex than in the 1970s, with more actors, agencies, and policy areas involved, which created opportunities either for broadly consensual solutions or for larger scale conflicts.

The New York case also shows partial path dependency, through the institutionalization of aspects of the political stream that had appeared in earlier rounds of policy change. Specialized administrative capacity, created and built up in the 1970s, played a major role in the 2000s reforms. NYSERDA and the State Energy Office were created in the mid-1970s in response to that decade's energy supply crisis, the growing interest in energy conservation, federal legislation, and the political commitment of Governor Carey and legislative Democrats. The energy office was crucial in implementing energy efficiency policies in the subsequent decades, while NYSERDA took the lead role in renewable energy and climate policy in the 2000s, in a cooperative relationship with the Public Service Commission, an older agency.[105] Finally, the 1977 Clean Air Act gave New York and other states the legal authority to adopt California's zero-emission vehicle policy and its Pavley rules, once the EPA approved them.

In addition, environmental organizations were formed and environmental public opinion became crystallized through the environmental conflicts of the late 1960s and 1970s. In that period, grassroots activists mobilized against major new projects such as the Storm King pumped storage plant and the worsening problem of air pollution in New York City. Their efforts helped to raise environmental consciousness among the public and elites, to build durable environmental organizations, and to challenge the prevailing assumption that energy demand and supply necessarily would grow indefinitely. Environmental organizations remained factors motivating elected officials and helping to construct advocacy coalitions in the 2000s.

New York-California Comparison

Although I have emphasized the processes that led to policy changes, some structural features were important in New York, as they were in California. New York State's high education levels undergirded the strength of its environmental organizations, and the generally liberal values of its population made vigorous government action possible with less risk of electoral backlash than in more conservative states. Additionally, the size and density of New York City made its smog problems severe and potentially alarming, especially given its national importance as a media and economic center. Indeed, New York ranked 8th highest among all states in education level in 2000,[106*] was the 5th most liberal state in 2005, and had the 3rd worst air pollution during 1990–2002.

New York also differs structurally from California in several ways that are of importance for climate policy. New York lacks a significant fossil fuel industry,[107*] and hence has a greater reliance on fossil fuel imports and somewhat higher electricity prices than California. This presumably helped to focus attention in New York on conservation rather than increased production in the 1970s and later. New York's political institutions also concentrate power to a greater extent in the hands of the governor and other executive branch officials than is the case in California, as discussed in Chapter 4. Pataki was able to develop the RPS and emissions trading system through agency actions alone; a lawsuit challenging RGGI as an illegal tax was dismissed in 2012.[108] Executive authority, of the governor and agency heads at NYSERDA and the Public Service Commission, was used to develop and implement almost all the policies described in this chapter. New York's lack of a fossil fuel industry and its stronger executive branch powers should make for stronger climate policy there than in California – and yet the opposite is the case.

Indeed, comparison of the two states shows that relatively strong climate policies could be reached through different paths, one involving competition between the governor and legislature (in California) and the other in which the governor largely sidelined the legislature (in New York). Other features of political institutions had little effect on climate policy making. California has a highly plural executive, while New York has a more unified executive with only four statewide elected officials. Nonetheless, Pataki found himself in competition with the attorney general on environmental issues after 1998. Both states have highly professional legislatures that compete intensely with the executive on issues like the budget, but in California the legislature also participated in energy and climate policy making (led by Pavley and Sher), while in New York it did relatively little on these issues after the 1970s, especially during the Pataki administration.

By contrast, differences in political events and in problem events in the two states can help to explain why New York's policies are weaker than California's in several areas. In the 1970s, New York mustered less political commitment for reform. Hugh Carey was not as committed to environmental protection and renewable energy as Jerry Brown was,[109*] and this helps to explain why New York's renewable energy policies were not strong enough to encourage a boom in that sector in the 1980s.

In the areas of enforceable targets, emissions trading, and low-carbon fuel standards, New York's policies are weaker than California's not because Pataki was personally less interested in climate policy than Schwarzenegger, but because the executive branch strategy favored by Pataki has inherent limitations. After Pataki won the 1996 environmental bond referendum campaign, effectively circumventing the legislature, he continued to pursue a largely unilateral strategy in climate policy. Pataki was not spurred by competition from a legislature that was active on the issue, and although he had competition from the attorney general, Spitzer could not block or enhance Pataki's proposals on state buildings, the renewable portfolio standard, or RGGI. However, there are legal and political limits on what the governor can do on his own. Legally binding targets or an economy-wide cap-and-trade system probably would require legislative approval.

Schwarzenegger, on the other hand, lost on the four ballot measures he had proposed in an attempt to evade the California legislature in 2005, and this led him to change his strategy and work with legislators on climate policy in the following years. If Schwarzenegger had chosen to implement his climate policy through executive orders and actions by the California Environmental Protection Agency rather than through legislation, California's policy would probably more closely resemble New York's, lacking enforceable targets and an economy-wide cap-and-trade system. Timing also may have played a role in making Schwarzenegger more flexible in his strategy than Pataki had been. Pataki had had seven years of conflict with the New York legislature before his climate policy push in the wake of Bush's 2001 decision. By contrast, Schwarzenegger took office after 2001 and had had only two years of conflict with the California legislature before he initiated his climate policy via executive order in 2005. Presumably, Schwarzenegger did not need to overcome as much bad blood with the legislature as Pataki did.

Another difference between the two states in the early 2000s is that the problems relevant to climate policy, and the public attention on them, were smaller in New York than in California. There was no major electricity crisis in the wake of deregulation in New York State, and no major report on climate change impacts in that state until 2011, seven years after the highly influential UCS report in California. In other ways, the two states had probably equally large, but different problems related to climate policy: California had droughts, wildfires, and a relatively large, vulnerable agricultural sector, while New York had the blackout of a major urban area in 2003 and the unique, though somewhat peripheral, event of September 11. Moreover, the weakness of legislative involvement made the impacts of focusing events smaller in New York, because the path from the event to the news media to the public was also less important for policy making there than in California.

9

Federal Government: Energy and Climate Policy Making

As described in Chapter 2, the climate policies of the U.S. federal government are weak compared to those of many other industrialized democracies, with one climate policy index ranking the United States 19th out of 24 industrialized democratic countries in 2005.[1] I will argue here that while the structural barriers examined in Chapter 4 clearly have contributed to the relative weakness of federal climate policy, they did not determine this outcome. Rather, when national policies have failed, this also has been due to the mistiming of problem events and political developments, and to flaws in advocacy coalition formation and strategy.

Moreover, the federal government sometimes has adopted moderately strong national climate policies – in the late 1970s, in the early 1990s, and since 2009. Each of these episodes of relative success can be traced to a convergence of attention on the problem of climate change and political commitment to address it. At the same time, the increasing polarization of the political parties on environmental policy, combined with the institutional separation of powers and elections that often have produced divided government, have limited the expansion of federal climate policy, and they continue to do so.

The episodes of relative success have been interrupted by periods of stagnation or retrenchment. Therefore, in this chapter I will focus on eight subcases of attempted U.S. national energy and climate policy making: the adoption of major energy efficiency and renewable energy policies under President Jimmy Carter (1975–78); the reversal of those policies under President Ronald Reagan (1981–85); the major energy and climate policies of the George H. W. Bush administration (1990–92); two subcases involving the stagnation of climate policy under President Bill Clinton (1993 and 1995–2000); the climate policy stalemate and focus on fossil fuel production during the George W. Bush administration (2001–05); and two subcases that center on the legislative failures (2009–10) and executive branch successes (2009–present) of the Obama administration. (See Chapter 2 for details on these policies.)

In the next section, I will analyze developments in the public and political agendas on energy problems and global warming, showing how they correlate with these eight policy-making episodes. Then, the episodes will be explained in terms of the convergence or non-convergence of problems and politics. The conclusions will relate the process explanations developed here to structural features, describe several aspects of path dependence, and examine how recent trends affect the prospects for federal climate policy making.

THE PUBLIC AGENDA AND THE POLITICAL AGENDA

At several points since 1970, problem events led to peaks in public attention to the energy and global warming issues, which created opportunities for policy entrepreneurs to put these issues on the political decision-making agenda. First, the energy supply crises of the 1970s, which involved massive price increases and sometimes shortages and rationing of oil, gasoline, and natural gas, thrust energy problems onto the public agenda.[2] Figure 9.1[3*] shows that public attention to energy issues peaked in 1973–74, during the oil embargo by the Arab members of OPEC, and again in 1979, after the Iranian Revolution disrupted oil supplies from Iran. This figure also shows three later peaks in attention to energy issues, which played mostly secondary roles in climate policy making during the periods in which they occurred: briefly in 1990 (due to Iraq's invasion of Kuwait), in 2001 (due to the electricity crisis in California), and in 2005–07 (due to fossil fuel price increases, as global energy demand grew).

Second, global warming emerged as a problem in the late 1980s. Attention on the issue has peaked a total of four times in newspapers since then, which is also shown in this figure.[4*] The first peak in attention, in 1989–92, was driven by the emergence of a scientific consensus that anthropogenic global warming was a serious problem and by hearings on the issue that Congress held during Summer 1988, while heat waves and extreme drought struck the United States. This period of heightened attention was prolonged and reinforced by the negotiation of the United Nations Framework Convention on Climate Change at the Earth Summit, in Rio, in 1992. Next, the negotiation of the Kyoto Protocol created a second, briefer attention peak in 1997. Attention rose to a third peak in 2001, when President George W. Bush rejected the Protocol and came under fire from domestic and international critics.

The fourth peak, by far the largest, was in 2007–09, with a sharp increase in attention beginning in 2005. As analyzed later in this chapter, it was produced by a confluence of hurricanes (in 2005), media and business shifts, mobilization by advocacy organizations (especially Al Gore's 2006 documentary), energy price increases, scientific developments (the 2007 report by the UN's Intergovernmental Panel on Climate Change, IPCC), and international negotiations (at Bali in 2007 and Copenhagen in 2009).[5] Although it has declined since 2007,

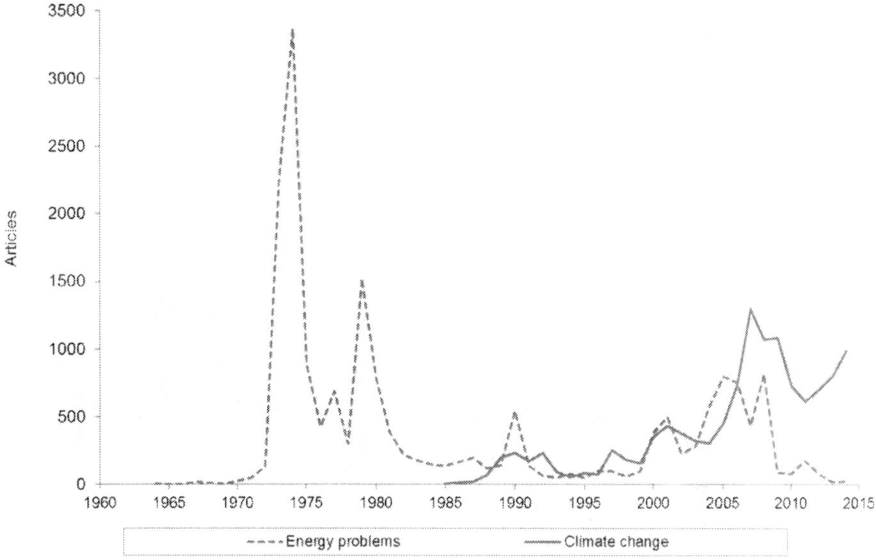

FIGURE 9.1 Public Agenda in the United States: Energy Problems (1964–2014) and Climate Change (1985–2014) in the *New York Times* and TV News Broadcasts
Source: see notes 3–4.

public attention on climate change, at least as measured in newspapers, has remained very high, well above that of the pre-2005 period.[6*] Figure 9.2[7*] shows that during each of the peaks, attention to scientific evidence and climate events increased, while attention to domestic and international government action increased even more. Hence, most of the variation in public attention over time was due to political actions, which may have either responded to or shaped the climate and scientific events.

Recent opinion surveys also give information about the perceived importance of the energy and global warming issues for the public, which largely accords with the newspaper data. Polls show a stable or increasing interest in energy and global warming problems from 2003 to 2008, a sharp decline to a trough around 2010, and a partial recovery over the next few years. The percentage who considered the energy situation to be "very serious" hit three peaks, in 1979 (Iranian oil crisis), 2001 (California electricity crisis), and 2008 (high gasoline prices).[8] Those who said that high gas prices posed a financial hardship rose from 2001 to a peak in 2008, before declining through 2012.[9] The percentage of those who "worry a great deal" about global warming hit peaks in 2000 and 2007,[10] and other surveys also recorded maximum values in 2007 for those who considered the issue to be very serious or who thought that immediate action was needed.[11] Those giving the environment priority over economic growth also peaked in 2007, though at a lower level than in the 1980s and 1990s.[12]

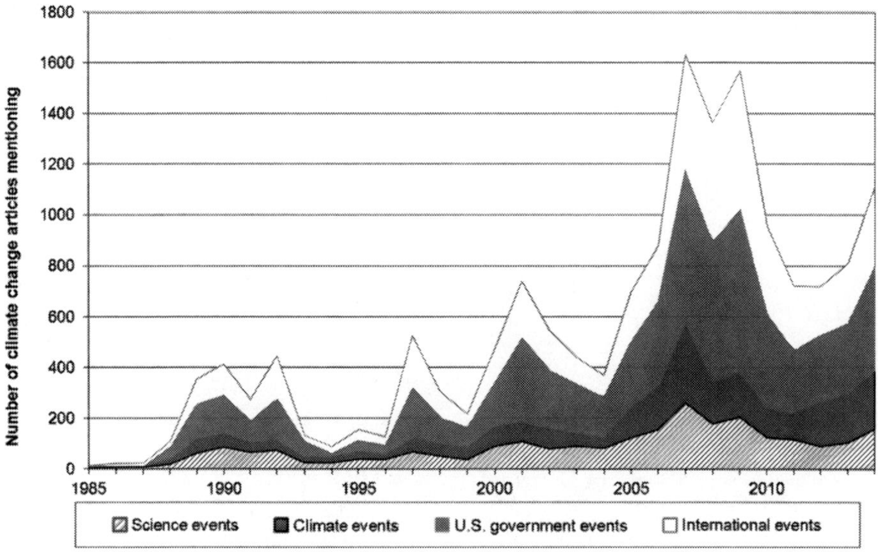

FIGURE 9.2 Drivers of Public Attention on Climate Change in the *New York Times*, 1985–2014
Source: see note 7.

The peaks in attention to energy problems and global warming on the public agenda were often, but not always, followed by peaks in these issues on the political agendas of Congress and the president – because political commitment to these issues was usually, though not always, available at those times. Figure 9.3[13*] shows that in Congress, the numbers of committee hearings and bill introductions explicitly addressing global warming or climate change peaked in 1989 and again, at a much higher level, in 2007–11. However, congressional activity on global warming showed only a modest response to the peaks in public attention to this issue in 1997 and 2001.

Congressional interest in energy policy innovations rose to a greater extent after increases in public attention to energy problems or climate change, with peaks in 1975–79, 1991, and 2001, and then a fairly steady rise to a large peak in 2007–11. By the early 1990s, these energy policy innovations were understood in Congress as a form of climate policy. Especially in 2001–08, Democrats and centrist Republicans in Congress pursued them as a response to President George W. Bush's retreat on climate policy. Congressional attention also showed momentum; it tended to remain high or decline slowly for several years after peaks occurred.

Presidential attention to climate policy correlates somewhat less strongly with the public agenda than does congressional attention.[14*] In response to the 1970s energy crisis, presidential interest in energy issues, measured in annual State of the Union messages,[15*] peaked sharply in 1975 and 1977, due to the energy policy proposals of Gerald Ford and Jimmy Carter, respectively.

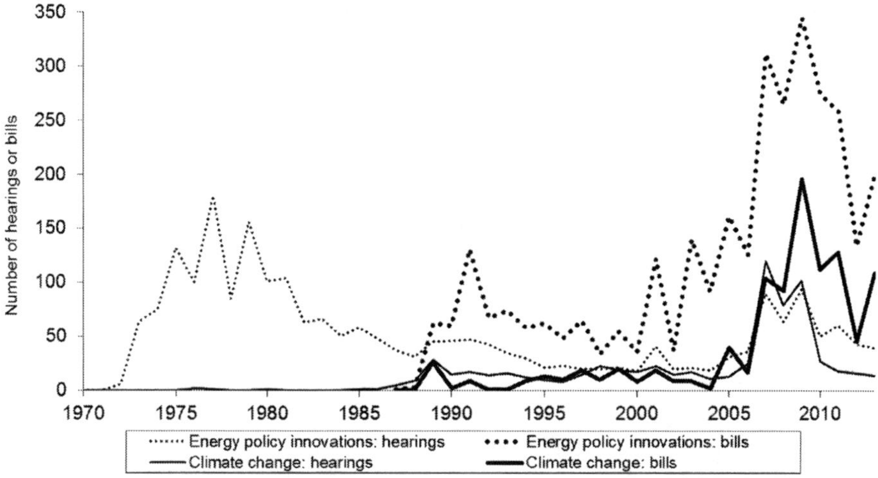

FIGURE 9.3 Congressional Agenda in the United States: Hearings (1970–2013) and Bills (1987–2013) on Energy Policy Innovations and Climate Change
Source: see note 13.

A small spike in presidential attention to energy issues (due to Clinton's BTU tax proposal) occurred in 1993, just after the rise in public attention to climate change during the Rio Earth Summit.

Figure 9.4[16*] shows that peaks in presidential attention to climate policies occurred in 1998–2000 (by Clinton, following the Kyoto peak in public attention), in 2007 (by Bush),[17*] and in 2009–13 (by Obama). The latter two peaks followed the major rise in public attention to global warming that began in 2005 and the Democrats' election victories in 2006 and 2008. However, presidents responded only selectively to energy and climate issues, and were more likely to do so at the start of a new administration, as Ford, Carter, Clinton, and Obama did. For their part, the last two Republican presidents (George H. W. Bush and George W. Bush) responded only weakly to the climate issues that were in the public eye during the first years of their administrations, in 1989–91 and 2001.

Comparing Figures 9.1 through 9.4 leads to several conclusions. First, as expected theoretically (see Chapter 3), attention on energy and climate problems tends to cluster in response to problem events, notably energy shortages, price increases, and international negotiations and agreements. However, at times the drivers have a strong political component, such as President Bush's rejection of the Kyoto Protocol and the complex of factors behind the rise of the climate change issue in the mid-2000s. Second, issues peaked on the public agenda before or at the same time as they peaked on the political agenda, but the latter also depends on independent political events such as elections. Therefore, while the agendas of the president and Congress are responsive to the public agenda, they do not simply mirror it. Third, the periods of greatest attention by policy makers were also those with the largest policy changes – the

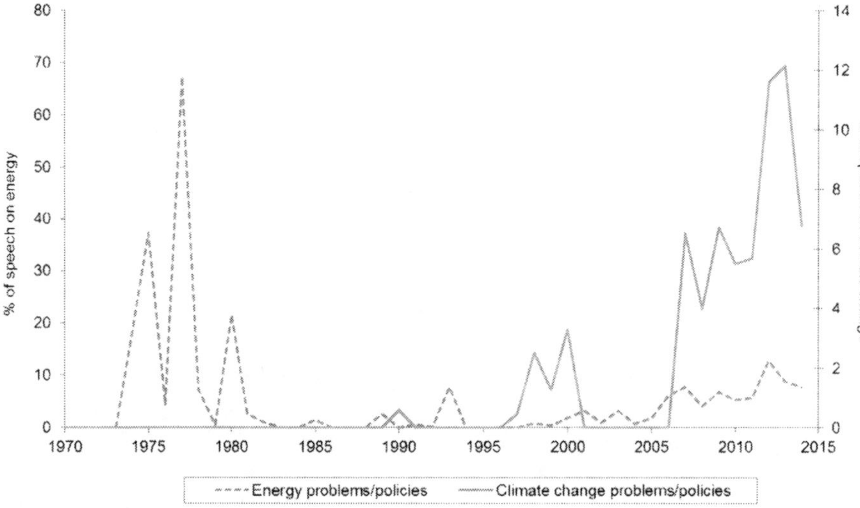

FIGURE 9.4 Presidential Agenda in the United States: Energy and Climate Change Issues in State of the Union Speeches, 1970–2014
Source: see note 16.

1970s (with attention by Congress and the president), early 1990s (by Congress), and 2009 to the present (by Congress and the president).

In these processes, public opinion is mainly an intervening variable rather than a driver of change in climate policy. The perceived importance of the global warming issue has been generally low, even during periods of peak public attention to it. During 2007–13, only 25% to 40% thought it should be a top priority for the Bush or Obama administration, ranking behind 21 other issues.[18] Even in 2007, when global warming was briefly named the most important *environmental* problem, only 7% considered the environment or global warming to be the most important priority for presidential candidates, ranking it 7th out of 8 issues.[19]

As in California (see Chapter 7), public opinion on climate change responds to the initiatives of political actors on the issue, as well as to economic conditions.[20] The percentage who "worry a great deal" about global warming peaked at about 40% in 2000 and 2007, which were both high points in economic expansion, fell sharply to 25% during the recession in 2011, and then rebounded to 33% in 2013, as the economy recovered.[21] These shifts in public opinion were also affected by political activity, which itself was affected by changing economic conditions: the Bush administration's opposition to climate policy around the time of the 2000-01 recession; environmental, business, and Democratic party mobilization on climate issues during the economic growth period of the mid-2000s; and the Tea Party's anti-government campaigns that began during the recession at the end of that decade.[22] At the same time, partisan polarization on the issue means that there is an entrenched minority who doubt the reality of anthropogenic global warming. Since 2006, between 16% and 34% have thought that the earth has not been warming, and an average of 42% have either denied the warming trend or thought it is mostly due to natural causes.[23]

THE ENERGY CRISIS, CONGRESSIONAL DEMOCRATS, AND JIMMY CARTER: ENERGY CONSERVATION AND RENEWABLE ENERGY POLICIES (1975–78)

The energy crisis of the 1970s converged with the election of large Democratic majorities in Congress and of Jimmy Carter to the presidency, which created a policy window for national energy conservation and renewable energy policy in the late 1970s. The results included significant federal policies for energy conservation and renewable energy promotion, as well as support for such policies implemented through state governments.

The energy crisis had many underlying causes. These include rising consumption in the United States and other industrialized countries, and supply constraints due to domestic price controls and the growing power of OPEC. The crisis was triggered by the 1973–74 oil embargo by the Arab members of OPEC in retaliation for the U.S. support of Israel during the Yom Kippur War.[24] World oil prices quadrupled virtually overnight, and shortages in oil, gasoline, and natural gas appeared in the United States. Electricity shortages also increased during the 1970s, due to rising demand, worsening urban air quality, and increasing environmental regulation of power plants, as described in Chapter 8.[25] A second oil supply shock occurred in 1979, when the Iranian Revolution cut oil production and exports from Iran. For the public, the energy crises involved long lines and rationing at gasoline stations, shortages and price increases for heating fuels, and electricity blackouts and brownouts.[26] Retail gasoline prices rose continuously from 1974 to 1982, almost doubling in real terms.[27] Because of these events, public attention to energy shortages and rising prices in the *New York Times* skyrocketed in 1973–74, and they peaked again in 1979, as shown in Figure 9.1 in the previous section.

When the energy crisis began, ensuring adequate energy supply and reducing dependence on imported oil became primary policy goals of the federal government. Promoting conservation and renewables were potential paths to those goals, but political commitment to them grew only in stages as other approaches were tried first. Initially, President Richard Nixon continued price controls on oil and natural gas while trying to break the oil embargo.[28]

After Nixon's resignation in August 1974 due to the Watergate scandal, the political context for mandatory energy conservation policies began to improve. While President Gerald Ford favored energy production and voluntary conservation measures, congressional Democrats favored a regulatory approach to energy conservation. During the November 1974 elections, the Democrats benefited from public revulsion concerning Watergate; they increased their majorities to 291 to 144 seats in the House of Representatives and 61 to 38 in the Senate. Ford emphasized increased oil, coal, and nuclear energy production, in part by delaying new environmental rules, as well as the decontrol of prices for domestic oil and new natural gas in order to promote conservation.[29] Congressional Democrats, led by Jim Wright, Jim Pastore, and Lloyd Ullman, preferred to increase gasoline taxes and to pursue energy conservation through mandatory auto efficiency standards, home insulation programs, appliance labeling, and utility rate restructuring. Democrats also insisted on maintaining energy price controls, in order to reduce the short-term burden on consumers. However, since retail prices reflected a combination of domestic and world prices, controls also had the effect of subsidizing the price of imported oil and encouraging its consumption, along with that of domestic oil and natural gas.[30]

Congressional Democrats largely shaped the resulting legislation: the National Energy Policy and Conservation Act and the National Energy Production and Conservation Act in 1975. Signed reluctantly by Ford, the 1975 laws, among other things,[31*] created the Corporate Average Fuel Economy (CAFE) program, which was to increase standards for new cars to 20 miles per gallon (mpg) for the 1980 model year and to 27.5 mpg by 1985. This program enjoyed bipartisan support, as few Republican senators responded to Ford's lobbying against it.[32] The initial version of the program had been weakened by the opposition of carmakers, who achieved a reduction in the penalties for noncompliance.[33] The legislation also gave the federal government authority to create voluntary targets and if necessary mandatory standards for 13 consumer appliances, with the aim of achieving at least a 20% gain in efficiency by 1980; state governments were permitted to set higher standards and received federal funding to promote energy conservation.[34*]

Much stronger political commitment to energy conservation and renewables arrived with the surprising election of Jimmy Carter. A virtual unknown two years earlier, Carter won the 1976 presidential election by only 50% to 48%. The margin of victory probably was provided by Ford's unpopular pardon of former President Nixon and a gaffe that Ford made concerning Poland during a presidential debate. Carter, trained as a nuclear engineer, had a strong interest in energy policy and had had good relations with the environmental community as governor of Georgia.[35] He advocated for the creation of an energy department during the presidential campaign,[36] and once elected, he appointed environmental and consumer advocates to high positions in his administration.

Indeed, Carter made energy policy the primary focus of his first two years in office and declared the problem to be "the moral equivalent of war."[37] He put the former defense secretary James Schlesinger in charge of a task force that would develop a national energy policy within three months. The task force's proposals emphasized government actions more than market forces and conservation more than increased production.[38] Carter announced his energy plan in a televised national address on April 18, 1977, in which he declared that the

United States was running out of oil and natural gas. The country needed to shift to "strict conservation" and the increased use of coal and renewable energy sources.[39]

President Carter obtained major energy policy reforms from Congress. With significant bipartisan support, it created the cabinet-level Department of Energy in 1977 and charged it with developing efficiency standards for cars and appliances, building codes, and tax incentives for homeowners and small businesses engaged in energy conservation.[40] Then, in 1978, Congress passed legislation on public utility regulation (giving access to renewable energy producers), energy tax credits (including for renewable energy), and energy conservation. The laws also mandated efficiency standards for appliances, replacing the voluntary targets of the 1975 act, although implementation of these was delayed until after 1987. In addition, this legislation mandated state energy conservation plans and provided grants for low-income households, schools, hospitals, and local governments. It also created programs to increase efficiency and solar energy use in federal buildings, increased the penalties in the CAFE program, and added a gas-guzzler tax to give consumers incentives to buy more fuel-efficient cars.[41]

Furthermore, the Carter administration vigorously promoted renewable energy. Within the new Energy Department was an office for renewable energy, led by an assistant secretary. Research and development funding was sharply increased for solar energy, wind energy, and energy conservation, as well as for synthetic fuels from coal and oil shale.[42] In addition, the 1978 legislation included provisions allowing state governments to grant generous feed-in tariffs for independent power producers, which spurred renewable energy and cogeneration development in some states. The federal laws also granted a 20% tax credit for business investments in wind or solar energy and a 20% to 30% tax credit for residential installations. Finally, they exempted small power producers and cogeneration facilities from federal and state energy regulation while guaranteeing them access to electricity grids.[43]

However, Congress failed to pass many of the measures that Carter had called for.[44] The policies adopted could have been stronger if energy shortages and public concern had persisted during the first years of the Carter administration and if the administration had taken steps to develop an advocacy coalition to counter the opposition of fossil fuel, utility, and automobile manufacturing interests. The Senate rejected increased taxes on crude oil, gasoline, and natural gas, due to lobbying by the oil and natural gas industries, electric utilities, carmakers, labor unions, and consumer groups.[45] Carter might have gotten more out of Congress if he had not rushed the policy (which was developed in just three months) and if he had worked to build support from Congress and interest groups (his task force worked largely in secret). Also, Carter might have had more success had he framed his policies in ways that did not rely heavily on calls to sacrifice and the denunciation of waste in personal consumption habits.[46]

Another, perhaps larger, problem was that Carter's election occurred several years after the peak in public perceptions of energy problems. Since gasoline and heating oil supplies had returned to normal before Carter took office, public concern about energy had dropped markedly. In this context, Carter's attempt to appeal to public opinion fell flat. In 1979, after two years of addressing the issue as a high priority, the share of the public accepting his view of the seriousness of the energy problem remained unchanged, at only 33%.[47]

FALLING FOSSIL FUEL PRICES AND RONALD REAGAN: ENERGY POLICY REVERSAL (1981–85)

After 1980, a policy window for the reversal of energy efficiency and renewable energy policies opened, and the Reagan administration used it to retrench or terminate many of the policies adopted under Carter. A major event in the problem stream was the large decline in fossil fuel prices during the 1980s, led by a sharp drop in oil prices beginning in 1981. The price decline was caused by an "oil glut," which itself had multiple causes: energy conservation and fuel switching away from oil in the United States and other industrialized countries; the 1980 and 1981–82 recessions; the expansion of oil production outside the OPEC countries; and the increasing ineffectiveness of the OPEC cartel's ability to restrict production. After peaking in 1980, crude oil prices fell for the next six years, declining by 70% in real terms. Coal and natural gas prices also fell sharply during the 1980s.[48*] Public awareness of energy problems, as measured in news coverage, declined even more sharply than did prices, reaching levels in 1982–89 that were only about one-eighth of their levels during 1973–80 (see Figure 9.1). Public and elite interest in energy conservation dropped dramatically, while energy consumption and oil imports rose.[49]

At almost the same time, political commitment to energy conservation and renewables was drastically reduced when Reagan became president in the 1980 election. He defeated Jimmy Carter by a 50% to 41% margin, aided by the Iranian hostage crisis, rising fuel prices, and the recession that began in January 1980, all of which made Carter's administration appear ineffective. Reagan ran on a promise to reduce the size and scope of government and to allow private enterprise and market forces to restore prosperity. A deregulatory approach to energy policy was part of the Republican platform, and Reagan sounded the theme in campaign speeches.[50] Reagan's disinterest in environmental protection was well known, and environmental organizations had backed Carter in the election.

In office, Reagan remained committed to deregulation, especially in energy policy. In a July 1981 message to Congress, he rejected any "rigid set of production and conservation goals dictated by government" in favor of free markets and individual choices.[51] Reagan accelerated the deregulation of oil and natural gas prices begun under Carter, and otherwise his energy policy focused on "strategic reserves and strategic forces" in order to maintain a secure and inexpensive oil supply, according to his budget director David Stockman.[52] His administration quickly sought to reverse the energy conservation and renewable energy policies of the 1970s, by sharply cutting federal budgets for research and development and allowing tax credits to expire.[53]

As energy prices fell, Congress largely accepted Reagan's proposed cuts to these programs. Federal research and development spending for wind power fell by 50% from 1981 to 1984 and kept declining through the early 1990s. Total spending for renewable energy and energy conservation dropped about 90% during the 1980s.[54] Federal tax credits for wind and solar energy were allowed to expire in 1985, dealing a severe blow to the industry.[55] Fuel economy standards for cars and light trucks were relaxed slightly, ending seven years of increases. After Reagan tried unsuccessfully to have the Energy Department abolished, he reoriented it toward nuclear weapons production during the 1980s, as part of a military buildup.[56] The Energy Department refused to issue new appliance standards, despite losing two federal lawsuits brought by the NRDC and the Consumers Union. However, the adoption of appliance standards in six states led appliance manufacturers to negotiate a deal with environmental organizations. This resulted in Congress twice passing a bill that mandated new, uniform appliance standards. Although Reagan vetoed it in 1986, he finally signed the law the next year.[57]

GLOBAL WARMING AND GEORGE H. W. BUSH: EXPLICIT CLIMATE POLICY EMERGES (1990–92)

Just as Reagan was leaving office, a convergence of problem and political events favorable to climate policy began in the late 1980s. There were four major developments on the problem side. First, James Hansen's testimony at congressional hearings in June 1988 coincided with an extreme heat wave in Washington, D.C., with seven straight days of temperatures about 100 degrees Fahrenheit. That summer also brought severe droughts to the United States, which harmed agriculture, stilled barge traffic on the Mississippi River, and led to forest fires in Yellowstone National Park. News media durably expanded their coverage of global warming, and the EDF, Greenpeace, and the Sierra Club began to make the issue a top priority.[58]

Second, a scientific consensus emerged that global warming was a serious problem, with a large spike in scientific publications in 1990 and 1991, including the First Assessment Report by the IPCC, which was formed in 1988.[59] Third, the international negotiations that produced the United Nations Framework Convention on Climate Change, which was signed at Rio in 1992, attracted renewed attention to the problem.[60] Fourth, Iraq's invasion of Iraq in August 1990 led to concerns about oil supplies and to a doubling of oil prices.[61] The Bush administration's National Energy Strategy, under development since 1989, had emphasized energy security, along with energy efficiency and respect for the environment. The invasion of Kuwait put energy independence back on the political agenda and gave the development of energy legislation new urgency for a time.[62] As a result of these problem events, both global warming and energy problems gained public attention in 1989–92, as shown in Figure 9.1.

The emergence of the global warming issue, as well as the brief return of the energy supply issue, coincided with a major increase in political commitment to environmental issues in the White House. In his campaign for president in 1988, George H. W. Bush emphasized his concern for environmental protection and promised to discuss global warming at an international environmental conference at the White House.[63] In office, Bush was certainly much more favorable to environmental policy than Reagan had been. After eight years of Reagan, the Democratic-controlled Congress responded very favorably to the new global warming issue and the opportunities afforded by Bush's presence in the White House. Hearings and bills on global warming increased sharply in 1989, and continued at elevated rates in the 1990s, as shown in Figure 9.3. However, Bush's commitment to environmentalism weakened after he took office, because of the rise of business opposition, including that organized by the Global Climate Coalition, and because of the 1990–91 recession.[64] When Bush issued his Action Agenda on Climate Change in February 1991, it lacked binding targets for reducing greenhouse gas emissions; as his administration developed its National Energy Strategy, its initial emphasis on energy efficiency and renewable energy was replaced by a focus on economic growth and fossil fuel production.[65]

Nonetheless, this period saw the adoption of the first explicit climate policies in the United States, as well as several important implicit ones. First, Congress passed air pollution and transportation bills in 1990–91 that curbed greenhouse gas emissions. The amended Clean Air Act (1990) was a response to the discovery of the acid rain issue in the 1980s, combined with the opportunity to finally pass environmental legislation under a more favorable administration after the long drought of the Reagan administration.[66] The act limited SO_2 emissions from power plants and provided for emissions trading in allowances for that pollutant. Hence, it potentially encouraged energy conservation and fuel switching from coal to natural gas, both of which would tend to reduce CO_2 emissions.[67]

The Clean Air Act amendments also contained a major, explicit climate policy by creating the Significant New Alternatives Policy program, implemented by the EPA. This program regulates substitutes for ozone-depleting substances, which are being phased out under the 1987 Montreal Protocol; it limits the use of hydrofluorocarbons (HFCs), which are potent greenhouse gases that are used as substitutes for chlorofluorocarbons (CFCs) and hydrochlorofluorocarbons (HCFCs). The act also directed the EPA to develop regulations on air pollution from landfills and a voluntary program to identify and label energy efficient products and buildings; these two programs, which later had major effects on greenhouse gas emissions, were implemented by the Clinton administration (see the next section). In addition, the landmark Intermodal Surface Transportation Efficiency Act, which Congress passed in 1991, helped to reorient transportation planning away from increased car travel. It permitted states to use federal transportation funds for mass transit projects and encouraged the use of funds for bicycling and walking infrastructure; this set a precedent for future transportation acts.[68]

Second, the Bush administration signed the United Nations Framework Convention on Climate Change at the Rio Earth Summit in June 1992, and the Senate ratified it in October. The convention requires industrialized ("Annex I") countries to produce emissions inventories and non-binding national climate action plans, and the United States has complied with these requirements in great detail.[69] However, at the urging of the United States, the treaty did not include specific greenhouse gas targets, let alone binding targets, which had been favored by the European Community and Japan.[70] The treaty ultimately included only a call to try to "prevent dangerous anthropogenic interference with Earth's climate system."

Third, the Energy Policy Act of 1992 established a number of mostly[71*] implicit climate policies, which were not linked explicitly to the problem of greenhouse gas emissions, but which tended to reduce those emissions or at least their rate of growth. Hence, in practice, the act helped to implement the United Nations Framework Convention's intention of curbing greenhouse gas emissions.

The Energy Policy Act promoted energy efficiency and renewable energy in a variety of ways. It directed the Energy Department to require states to update their commercial building codes and to consider updating their residential codes; states needed to do so by 1994 and in future years when the department determined revisions were needed.[72] It also directed the Energy Department to adopt standards for lamps, motors, and commercial heating, cooling, and plumbing equipment.[73*] To promote renewable energy, the legislation established a Production Tax Credit for wind and closed-loop biomass facilities (1.5 cents per KWh, indexed to inflation).[74] It required federal and state agencies to buy alternative-fuel vehicles and the federal government to save energy in its buildings and vehicles.[75] The act also authorized a ten-year budget for the Energy Department, including increases in research and development spending for energy efficiency, renewable energy, and clean coal technology, increases that were implemented during the Clinton administration.[76]

Moreover, the Energy Policy Act changed the regulatory context for utilities in ways that tended to reduce greenhouse gas emissions. It funded state energy offices and required natural gas and electricity utilities to price their products to encourage conservation and to undertake integrated resource planning, which would include renewables and conservation measures, evaluated at the same marginal costs as new plant capacity.[77] The act also opened electricity providers to national competition from a new class of "exempt wholesale generators," which ultimately led state governments to deregulate electricity markets in the late 1990s; this was expected to give natural gas advantages over coal.

After the bill was intensely debated and negotiated within Congress and between it and the White House, the Energy Policy Act passed by large bipartisan majorities and was signed in October 1992. The energy efficiency and renewable energy provisions of the act passed Congress largely because of broad advocacy coalitions and congressional Democrats' support for the reform course that the measures represented. Congressional Democrats pushed for the enhanced energy efficiency provisions, as well as for additional limits on fossil fuels and increased CAFE standards, which they failed to get.[78] The policy changes also were made possible by the initial reform impetus from the George H. W. Bush administration and Bush's ultimate willingness (in contrast to Reagan) to accept a bill that had strong provisions for energy efficiency and renewable energy. The natural gas industry, the energy efficiency industry,[79*] and the solar industry lobbied together on provisions that benefited all of those sectors; consumer and environmental organizations joined them on energy efficiency measures. A joint report projected that 200,000 jobs could be created by promoting natural gas, renewable energy, and energy efficiency. The advocates' success also depended in part on divisions among the traditional energy industries (natural gas, coal, oil, and nuclear).[80]

However, as an example of climate policy, the Energy Policy Act was much weaker than it could have been, largely due to a lack of commitment from the Bush administration. The effects on energy use and greenhouse gas emissions were not integrated into an overall plan or linked to national targets. Moreover, they fell far short of the measures that the National Academy of Sciences and other advocates had called for, which they had said would have achieved a 10% to 40% reduction in greenhouse gas emissions at little or no economic cost.[81] Despite his environmental leanings, Bush was not firmly committed to a particular energy policy course, and he was already on the defensive against the conservatives in his party because of tax increases he had accepted in 1991.[82] Therefore, Bush allowed others[83*] in his administration to move its position to the right during the energy policy debates.[84] Bush used veto threats to weaken the bill, by blocking any greenhouse gas emissions targets (which had been proposed by the Democrat Henry Waxman but not pushed to a floor vote by him), tax increases, or bans on offshore oil drilling. Efforts to increase CAFE standards were dropped in the Senate in exchange for the removal of a provision permitting drilling in the Arctic National Wildlife Reserve, which had been a Bush administration priority.[85]

BILL CLINTON AND CONGRESSIONAL RESISTANCE: THE BTU TAX PROPOSAL (1993)

Attention on global warming was high at times during the Clinton administration, and President Clinton initially was committed to creating a vigorous climate policy. But major new policies with reduction commitments did not result, largely because of a mismatch between problem awareness and political constraints. During the first two years of his administration, Clinton showed strong political commitment to the global warming issue, but it was at a low ebb in public and elite awareness in 1993–94, as shown in Figure 9.1 earlier in this chapter. This, combined with congressional ambivalence and strategic mistakes, led to the defeat of Clinton's proposed energy tax (the BTU tax) and to the end of serious legislative efforts at climate policy during the rest of his administration. However, executive branch action continued, including successfully pressing Congress to increase funding for existing programs and vigorously implementing existing policies.

Bill Clinton's election in 1992 brought strong climate policy proponents into the White House, including Vice President Al Gore, who had published a best-selling book on global warming and environmentalism during the election campaign. Energy tax proposals fit well into Clinton's market-oriented approach to public policy, and the new administration took the view that green energy sources could be a source of economic growth.[86] In the first months of his administration, Clinton appointed a climate policy advocate, Hazel O'Leary, as Energy Secretary, and he called for the development of a plan to reduce emissions to 1990 levels by 2000. Clinton also called for a relatively small energy tax (the "BTU tax"), of about 3% to 7% on fossil fuels, in order to reduce the federal budget deficit and curb pollution.[87]

However, Clinton's efforts to pass an energy tax failed, due to a combination of low issue salience at the time, poor framing, and lack of advocacy coalition mobilization. After twelve years of Republican administrations, the Democrats' allies in environmental organizations were ready for legislation with environmental goals, but Clinton argued for the energy tax largely as a deficit reduction measure. He also undercut the support of environmental groups by rapidly making concessions that moved the tax collection point closer to consumers.[88] Interest group lobbying against the tax, led by the National Association of Manufacturers and American Petroleum Institute, became intense when industry saw the administration begin to make concessions.[89] Although the BTU tax passed the House of Representatives, it failed in the Senate Finance Committee, despite having been modified with 13 exemptions. The Senate ultimately replaced the BTU tax with a small, 4.3-cent gasoline tax increase.[90] After the White House accepted that result, the conference committee approved the Senate version.[91]

Although two Democratic senators from fossil fuel producing states blocked the BTU tax, this result was not foreordained by the size of the fossil fuel industry and the structure of the Senate. Several oil companies initially supported the tax because they viewed it as unstoppable and saw good chances to pass the tax on to consumers.[92] The BTU tax might have fared better had it been sold as an environmental measure or if the revenue had been dedicated to popular programs such as Social Security.[93] Chances of passage would have been higher if the Clinton administration had had a clearer strategy and had at the outset communicated with the public about the preferability of energy taxes compared with income or consumption taxes.[94] Above all, the occurrence of an event that raised the profile of the global warming issue – such as the 1992 Earth Summit[95*] or the 1997 Kyoto Protocol – would have given the administration the opportunity to make strong environmental arguments and to sustain the momentum that it had in the early months of the proposal.

The defeat of the BTU tax in Congress ended Clinton's attempts to pass legislation involving mandatory greenhouse gas reductions. His October 1993 climate action plan simply drew on existing law and called for new voluntary programs with business.[96] The blocking role of Congress was reinforced when Newt Gingrich led Republicans to victory in the 1994 elections on a platform of smaller government, which was expressed in his party's Contract with America. In environmental policy, the new congressional majority party was committed to deregulation, and in its intense budget battles with the president, it sought to reduce spending on energy efficiency measures and the administration's climate action plan.[97]

CLINTON AND CONGRESSIONAL OPPOSITION: THE KYOTO PROTOCOL AND VOLUNTARY MEASURES (1995–2000)

In 1997, Clinton's engagement in the global warming issue and public awareness of it both rose during the negotiation of the Kyoto Protocol, which included a conference of experts convened by the White House in October.[98] Public attention to the issue rose sharply that year, approximately matching the 1992 peak in news coverage, as shown in Figure 9.1.[99*] Moreover, compared to 1993, the administration communicated better with its supporters about the nature of its climate policy strategy and it worked harder to gain public support.

However, this favorable constellation was counteracted by new, highly unfavorable political factors. The Republicans took control of Congress in January 1995, party polarization on environmental issues increased sharply,[100*] and the Lewinsky sex scandal distracted President Clinton from the climate change issue beginning in January 1998. Because of the mismatch between high problem awareness and a political context that was negative in important respects, this was a period of only minor climate policy reform, in which the policies initiated under President George H. W. Bush were consolidated and expanded incrementally, rather than a period of major expansion.

The Clinton administration helped to negotiate the Kyoto Protocol. In 1996, it called for binding emissions targets, despite domestic industry opposition, while at the same time pressing other industrialized countries to accept provisions for joint implementation and emissions trading. However, the Senate preemptively rejected the Protocol in a symbolic 95 to 0 vote on the Byrd-Hagel resolution in July 1997. This resolution was the result of heavy lobbying by the energy industry and congressional concerns about trade competitiveness with China and India. International negotiations on the Protocol concluded in Kyoto, Japan, in December 1997; the Clinton administration had agreed to a

7% cut in U.S. greenhouse gas emissions. But one month later, in January 1998, at a time when Clinton might have called on Congress to pass implementing legislation, the Lewinsky scandal broke. It preoccupied Clinton for the next year.[101]

Congressional opposition to the Kyoto Protocol also increased after the Energy Department conducted an analysis projecting that compliance would increase gasoline and other energy prices and would cost tens of thousands of coal miners their jobs. Even energy efficiency programs became suspect as measures "preparing for the implementation" of the Protocol, and they were banned by congressional budget amendments in 1999 and 2000.[102] Congress also blocked any increases to CAFE standards for light trucks by preventing the Department of Transportation from spending money to develop them from 1996 to 2000.[103]

This left Clinton with executive actions, international diplomacy, and budget battles with Congress as the main tools of his climate policy. The Clinton administration used those tools well. It continued and reinforced the energy efficiency and renewable energy policies in the 1992 Energy Policy Act, used its executive authority and leverage over the budget, and created a host of small domestic and international initiatives that modestly strengthened climate policy. As part of its Kyoto strategy, the administration's Climate Change Technology Initiative initiated or expanded existing cooperative programs with electric utilities, carmakers, and industry. These cost $6.3 billion over five years in tax credits for energy efficiency and renewable energy and in funds for research and development. Spending for energy efficiency, renewables, and alternative-fuel vehicles rose 50% over the 1997–2001 fiscal years, to $1.2 billion, while spending for coal and nuclear power fell 20%.[104] Clinton also used executive orders to require federal agencies to reduce vehicle fuel consumption by 20% and to shift toward alternative fuels.[105]

Furthermore, the Clinton administration completed the implementation of two major policies, which proved to be highly effective in reducing greenhouse gas emissions, as shown in the last section of Chapter 2: the expansion of the Energy Star program and the adoption of landfill methane regulations. Both policies were based on the authority of the Clean Air Act and were initiated by the Bush administration, but they were brought to completion under Clinton without much political conflict. These policies lay in the discretion of the EPA, and hence depended on the political commitment of the executive branch, including Clinton's EPA administrator Carol Browner, and not on Congress. The EPA under Browner greatly expanded the voluntary Energy Star program in 1993 and 1995 to include commercial buildings, office equipment, industry, electric utilities, and consumer products.[106] In addition, the agency initiated the Combined Heat and Power Partnership and the Green Power Partnership around 2000.[107] These voluntary programs represented a continuation of the Bush administration's market-oriented approach to climate policy, and benefited from policy feedback during the 1990s as businesses saw the benefits of increased efficiency.

By contrast, the methane regulations were a regulatory measure authorized by the 1990 Clean Air Act. Their primary intention was to reduce smog formation and toxic air pollution from non-methane organic chemicals, but the Bush administration, in issuing proposed rules in 1991, already had noted that a side effect would be the reduction of emissions of an important greenhouse gas, methane. In March 1996, the EPA issued the final rules on landfill air emissions, which required landfill operators to collect and burn landfill gas. The regulations were projected to reduce methane emissions from this major source by 40%.[108]

KYOTO REJECTED, GEORGE W. BUSH, AND CONGRESSIONAL DEMOCRATS: STALEMATE (2001–05)

After George W. Bush's controversial election to the presidency in 2000, problem awareness and political commitment concerning climate change once again diverged. While the global warming problem reached a new high in public attention, Bush attempted to roll back the previous twelve years of climate policy. However, resistance from congressional Democrats prevented a policy reversal, and the result was a stalemate in which only incremental advances in some policy areas occurred.

Public attention to climate change reached a peak in 2000–01, which was even higher than in 1997. The increase in attention was due partly to the 2000 presidential election campaign between Gore and Bush, and to international efforts to revise the Kyoto Protocol, which were completed in July 2001. But the increased attention was also due to the incoming Bush administration's announcement of new energy and environmental policies – especially its rejection of Kyoto – and the criticism of those policies by the governments of other countries, congressional representatives, and some U.S. states. Public opinion also showed increases from 1997 to 2001 in those worrying about climate change, as well as majority support for the Kyoto Protocol.[109] However, public and congressional attention on global warming peaked in 2001, and they both declined for the next three years.

The incoming Bush-Cheney administration was strongly committed to fossil fuel development and had little interest in a strong climate policy. The latter, at least, was somewhat surprising to environmentalists and U.S. allies abroad. In the 2000 presidential election campaign, Bush had said he would address global warming through new power plant standards for carbon dioxide and other pollutants, and through legislation that would clarify that the Clean Air Act applied to carbon dioxide. As late as March 10, 2001, Bush indicated he would pursue regulation of greenhouse gas emissions along with other pollutants. However, four days later, he completely reversed his position in the face of opposition from conservative Republicans and lobbyists from the oil, gas, and electric utility industries. At this time, the business coalition in favor of a cap-and-trade program was growing, but it was not actively lobbying for climate policy.[110] In a letter to Republican senators, Bush justified his reversal in terms of the "incomplete state of scientific knowledge" about the causes of climate change, as well as energy shortages (the California electricity crisis was ongoing) and the need to protect consumers.[111] Bush's publicly expressed doubts about the scientific consensus helped to fuel increasing partisan divisions on the issue in public opinion during the 2000s.[112]

After Bush's announcement, his administration attempted to carry out a climate policy reversal. Bush's EPA administrator, Christine Whitman, announced that the United States would reject the Kyoto Protocol, and Bush affirmed this during a trip to Europe in June 2001, arguing that it was a flawed treaty due to the exclusion of developing countries and that it would harm the U.S. economy.[113] The rejection was of crucial symbolic importance, even though nothing had been done during the previous four years to get the agreement and its 7% reduction target approved in the Senate. Bush tried to mollify his opponents with his February 2002 Global Climate Change Initiative, in which the United States adopted the non-binding goal of reducing the greenhouse gas intensity of the U.S. economy by 17.5% from 2002 to 2012, with $4.6 billion in business tax credits over five years proposed. However, meeting this target would merely have continued the trend in greenhouse gas intensity improvement already seen in the previous ten years.[114] Even with falling emissions intensity, total U.S. emissions were projected to grow by 14% over this period.[115]

However, Bush's efforts at a climate policy reversal met with congressional resistance, which led to a kind of stalemate. Congressional opposition to the administration's new policy course included the introduction of bipartisan cap-and-trade legislation sponsored by Senators John McCain and Joe Lieberman in 2003.[116] The U.S. Senate also passed legislation establishing a national renewable portfolio standard several times. Despite strong lobbying by environmental organizations and support from some utilities, it did not advance further because of opposition by most electric utilities and by politicians linked to the coal and oil industries.[117] In 2005, Bush initiated efforts to aid developing countries to reduce their greenhouse gas intensity through the new Asia Pacific Partnership on Clean Development, which included China and India and addressed technology development and transfer; the partnership ended in 2011.[118]

But the Bush administration's opposition to domestic greenhouse gas regulation remained steadfast, even after the Supreme Court ruled in April 2007 (in *Massachusetts v. EPA*) that the EPA has the authority to regulate carbon dioxide under the Clean Air Act. The EPA rejected California's request for a waiver for its Pavley rules in December 2007, and Bush ordered the agency to retract the first draft of an endangerment finding about the effects of carbon dioxide. The EPA issued a report in 2008 concluding that greenhouse gases would be difficult and costly to regulate, without drawing any conclusions about their threats to human health and welfare.[119] U.S. negotiators firmly opposed even provisional targets at the international climate meetings in Bali in December 2007.

The Bush administration's attempted climate policy reversal is also evident in its energy policy, led by Vice President Dick Cheney and announced in May 2001. This policy sought to undermine efficiency standards and downplay conservation, while promoting the development of fossil fuel resources through weakened environmental regulation and drilling in the Arctic National Wildlife Refuge. The energy plan, drawn up by a secret task force dominated by representatives of the fossil fuel industry, called for easing environmental restrictions on energy production and for building 1,300 new power plants by 2020.[120] Bush's first budget also proposed cutting spending on energy efficiency programs by 15%.[121]

But this attempted reversal was only partially successful, as the Democratic-controlled Senate blocked Bush's 2002 energy bill, despite Bush's popularity in the wake of the attacks by Al Qaeda on September 11, 2001.[122] The Energy Policy Act, finally adopted in 2005, was a compromise that primarily reflected the Bush administration's commitment to fossil fuels and nuclear power. It granted energy companies $12.3 billion in subsidies (mainly for oil exploration and coal extraction) over a ten-year period and gave the oil industry easier access to drilling permits. It also exempted hydraulic fracturing from existing clean air and clean water legislation. The administration dropped the controversial provisions for Arctic oil production while Democrats gave up on raising of CAFE standards.[123] This legislation probably helped lead to an expansion in domestic natural gas production (which rose 36% during 2005–14) and ultimately in domestic oil production (which rose 74% during 2008–14), the latter after more than 35 years of decline.[124]

But the 2005 Energy Policy Act also reflected the political commitment to energy efficiency and renewable energy by congressional Democrats, such as Senator Jeff Bingaman.[125] Therefore, it included continuations or incremental improvements to several of these implicit climate policies. It required a commercial building code update and new standards for consumer products and commercial equipment, which were again urged by manufacturers in order to preempt differing state standards that had been adopted in recent years.[126] The law also established a new renewable fuel standard, which was expanded in 2007 and then curtailed by the EPA in 2013; moreover, it extended the Production Tax Credit for renewable-source electricity and expanded the credit to include additional sources.[127] Finally, the law created a new type of bond (initially funded at $800 million) to finance renewable energy projects by municipalities and cooperatives, and it increased tax credits for residential and business use of alternative energy, for contractors, and for hybrid vehicle purchasers.[128] Therefore, as a result of congressional Democrats' resistance, on the whole there was stagnation in climate policy rather than retrenchment during the George W. Bush administration.

CLIMATE CHANGE, CONGRESSIONAL DEMOCRATS, AND BARACK OBAMA: LEGISLATIVE FAILURE AND REGULATORY PROGRESS (2009–PRESENT)

Timing and Mistiming

More than the other periods studied here, the Obama administration presents a mixed picture of climate policy failure and successes. The failure of national climate legislation has been accompanied by the first direct regulation of greenhouse gas emissions, through executive actions on road vehicles, power plants, landfills, aircraft, and the oil and natural gas industry. Hence, the period includes two subcases that overlap in time. The major failure occurred in 2009–10, when comprehensive climate legislation, including emissions trading and a national renewable portfolio standard, died in the Senate. The successful attempts to adopt major reforms actually began in 2007, after the Democratic victories in the 2006 elections gave them control of both houses of Congress and led to the passage of legislation raising CAFE standards the next year. Climate policy expansion continued during the Obama administration with regulatory standards for greenhouse gas emissions by motor vehicles (in 2009 and 2011) and by power plants (beginning in 2014).[129*]

The failure to get major climate policy legislation adopted in this period was due partly to a mistiming between problem and political events. There was only a brief period (2009–10) in which high awareness and high political commitment (with unified Democratic control of Congress and the Presidency) coincided, and this was also a period of intense economic concern due to the recession.

The problem context for climate policy was very favorable in 2007–08, but political commitment to climate policy was still relatively low, though increasing, at that time. After the Democrats gained control of Congress in January 2007, the new Speaker, Nancy Pelosi, made passage of energy legislation to promote efficiency and renewable energy a priority. This led to the passage of the Energy Independence and Security Act, which raised CAFE standards and promoted biofuels, that year. However, provisions for a national renewable electricity standard on utilities and increased oil taxes, which passed in the House of Representatives, were dropped after utilities, the Chamber of Commerce, the National Association of Manufacturers, and the oil industry successfully lobbied Republican senators and the Bush administration threatened to veto the bill.[130] In addition, Senator Joe Lieberman (by then an Independent) revamped previous emissions trading proposals to include for the first time a declining cap, but the bill was blocked by a Republican-led filibuster. The Bush administration, which steadfastly opposed greenhouse gas restrictions, opposed congressional efforts on cap and trade.[131]

Then, the political context improved dramatically for two years beginning in January 2009, when the Democrats increased their congressional majorities (to 256 to 178 in the House of Representatives and 58 to 41 in the Senate). In addition, Henry Waxman of California took over as chair of the House Energy Committee (in place of John Dingell of Michigan), and Barack Obama became president with a strong commitment to climate policy. But by this time the problem context was becoming less favorable to climate policy, as a severe economic recession was underway and unemployment was rising rapidly. As a result, the climate issue was on the verge of a decline in terms of public attention and its perceived importance in public opinion, which undercut the prospects for legislative progress. Those were reduced to approximately nil after the Republicans gained control of the House of Representatives in the November 2010 congressional elections. During the remainder of the Obama administration, conditions have been very mixed. The problem context has remained favorable: news media attention to the climate issue has remained very high, international diplomacy revived in the run up to the 2015 Paris agreement, and public opinion on climate change has recovered to an extent. Although Congress has not acted, the executive branch has showed strong commitment in pursuing greenhouse gas regulation on the authority of the Clean Air Act.

Problem Awareness: A Conjuncture of Causes

Events in the problem stream and the political stream caused public attention on climate change to increase beginning in 2005, leading to what was by far the largest peak in attention to the issue in 2007–09, as shown in Figure 9.1 earlier in this chapter. Public opinion measures also show a peak in concern about global warming in 2007–08. These trends were due to a powerful combination of climate, business, media, advocacy, economic, scientific, and international events. Some of the shifts began gradually in the years prior to 2005, and they reinforced each other until a critical mass or tipping point was reached. Hence, it would be very difficult to identify the most important drivers of the change.

First, the strongest Atlantic hurricane season in over 150 years occurred in 2005. Seven major hurricanes developed that year, including the devastating Hurricane Katrina and three others that struck the United States.[132*] Causing an estimated 1,800 deaths and $108 billion in property damage, Katrina was the economically most destructive storm in U.S. history. Although its connection to global warming was controversial, climate policy advocates and journalists tried to make the connection.

Second, around 2005, major corporations began to shift rapidly toward accepting climate change and adjusting to the apparent inevitability of climate policy. Some major U.S. corporations had begun to accept climate policy in 2001, after the IPCC's Third Assessment Report. The anti-regulatory Global Climate Coalition continued to lose major corporate members, and it disbanded in 2002.[133] In 2005, U.S. companies began to shift more noticeably, and *Time* magazine called 2006 the turning point for many corporations, as General Electric, Walmart, and other firms pledged to limit their greenhouse gas emissions. Many others began to take account in their business plans of the risks of climate change and the prospects of future carbon pricing.

There were many reasons for this shift. Large institutional investors applied pressure on companies because they saw regulation as inevitable, given developments in the European Union and efforts in Congress, and hence they had come to see a company's carbon dioxide emissions as a liability. Insurance companies were raising rates and changing terms of policies concerning damages from storms, fires, and droughts. Litigation was beginning, and more was feared. Some top executives cited pressure from their children, and businesses that shifted their positions improved their images and staff morale.[134]

Third, during 2004–06, the news media shifted sharply toward reporting mainly the views of mainstream scientists, rather than giving equal space and credibility to climate change deniers; the media had been doing the latter since about 1995.[135] The news media apparently were responding to the business shift, as well as the strong Atlantic Hurricane season in 2005 and Al Gore's 2006 documentary. A *Time* magazine cover story in October 2005 asked "Are We Making Hurricanes Worse?",[136] and the *Boston Globe* and Fox News also linked the hurricanes to global warming.[137] Newspapers announced that the debate on global warming was over, television networks broadcast special reports, and the Weather Channel on cable television added climate change to its coverage.[138] The share of U.S. newspaper articles that gave credence to the denial of anthropogenic climate change dropped from 37% in 2003 to only 3% in 2006.[139]

Fourth, national environmental organizations increased their campaigns on climate change. While some organizations had been active on the issue since the 1990s, such as the EDF, NRDC, Sierra Club, and Friends of the Earth, other organizations began to give the issue attention in the early or mid-2000s, such as the Nature Conservancy, the Wilderness Society, and the National Audubon Society.[140] Private foundations more than doubled their annual spending on climate and energy issues from 2004 to 2006, to $120 million, although they increased it even more in the next three years, to $600 million per year in 2009-10, largely in an effort to pass national cap-and-trade legislation.[141] In particular, the May 2006 release of Al Gore's documentary, *An Inconvenient Truth*, had a large influence on elites and the public in the middle of the decade. The film was a box-office success – at the time the third highest grossing documentary in U.S. history – and won two Academy Awards. It was accompanied by a number one best-selling book and copious press coverage that helped to create a large spike in news coverage of climate change, as shown in Figure 9.1.[142]

International events also helped to drive increased attention to global warming in the United States. Oil, gasoline, and natural gas prices rose rapidly after 2002, due to increased demand for oil from China, India, and other emerging economies, as well as strong demand from industrialized countries and lags in production increases. Oil prices rose from $25 to nearly $150 per barrel, natural gas prices almost tripled to about $8 per thousand cubic feet in 2008, and retail gasoline prices doubled in real terms from 2002 to 2008.[143] Consumers in the United States began to move away from large vehicles such as sport utility vehicles, and carmakers announced plans to produce a greater proportion of relatively fuel-efficient cars, including the highly publicized hybrid electric vehicles.

Finally, major scientific reports and international meetings drew attention to the issue. The Stern review, released in October 2006, had a major impact by comparing the costs of climate change with the costs of climate change mitigation policy. In advance of the release of the IPCC's Fourth Assessment Report in February 2007, scientific bodies, such as the American Geophysical Union and the American Physical Society, endorsed the consensus view of global warming and its causes. The Fourth Assessment Report concluded that human-caused climate change is now apparent throughout the world and that worse is to come; it received much more media attention than the 2001 report had.[144] In addition, high-profile international climate meetings were held at Bali in 2007 and at Copenhagen in 2009. However, the Copenhagen meeting was preceded a few weeks earlier by the release of stolen emails purporting to show scientific fraud by leading climate scientists. This created much publicity on the climate change issue as the media, to an extent, shifted back toward giving credibility to climate change skeptics. The proportion of mainstream news coverage promoting climate change denial or a "balanced" view rose from 5% at the start of 2009 to 25% in the last quarter of the year, and remained at about 15% in the second half of 2010.[145*]

Public attention to climate change remained high and stable even in the recession year 2009, when over 1,000 articles were published in the *New York Times*, still more than double the peak in 2001. Interest also remained very high after 2009, as the annual number of articles during 2010–14 averaged more than 700, still more than three times as many as during 1990–2005 (see Figure 9.1).

However, a key component of the problem stream turned negative as the economy went into recession in 2008 and the unemployment rate rose to nearly 10% by the end of 2009. In 2008, public concern with the economy shot up, as 35% to 55% named the economy or jobs as the country's most important problem, eclipsing the Iraq War; the economy remained the dominant problem through the end of 2012.[146*] In March 2009, a majority gave economic growth priority over environmental protection for the first time in 25 years of polling on the question.[147] Hence, given the opposing trends in the problem stream and the political stream, the early months of 2009 marked the greatest opportunity to get global warming onto the congressional agenda and to get a bill passed. In fact, hearings and bills on climate change peaked that year, although congressional and presidential attention remained high in 2010–13.

Political Conditions: Favorable but Unstable

The political context for climate policy was very favorable when Obama took office. On the one hand, the 2007 Supreme Court decision in *Massachusetts v. EPA* had given the EPA major authority to regulate greenhouse gas emissions without new congressional legislation, and Obama had a strong commitment to climate policy. The Democrats had gained control of Congress in the November 2006 elections and aggressively promoted cap-and-trade and renewable energy legislation. Immediately after the Democratic electoral victories, President Bush agreed to raise CAFE standards, which previously had been a Democratic priority opposed by Republican lawmakers and not supported by the White House.[148] In 2007, Congress passed, and Bush signed, legislation raising CAFE standards and expanding a biofuel mandate that mainly benefited corn ethanol. The Bush administration did not begin to develop the new fuel economy rules, which were left to the new administration.

Moreover, major U.S. business corporations, including energy companies (e.g., BP America and Duke Energy) and manufacturers (e.g., Alcoa, General Electric, and Pacific Gas and Electric) formed the U.S. Climate Action Partnership with major environmental organizations (e.g., the NRDC, EDF, the Pew Center on Global Climate Change, and the World Resources Institute) in order to press for cap-and-trade legislation in Congress. Sixty leading institutional investors joined their call to action.[149] During the 2008 election campaign, both Obama and McCain supported caps on greenhouse gas emissions, and Obama's energy plan included a national renewable portfolio standard with a requirement of 25% renewable-source electricity by 2025.[150] President Obama immediately appointed as Lisa Jackson, a former New Jersey environmental official with much experience in air pollution regulation and a strong advocate of greenhouse gas regulation, to be EPA director. Moreover, within days of taking office, Obama directed the EPA to review its previous rejection of California's vehicle rules.[151*]

But political events during the Obama administration soon posed new constraints on climate policy. Rising party polarization after 2008 made legislation more difficult, as Republicans in Congress who had favored a cap-and-trade program reversed their positions when Obama endorsed the approach. They now considered it to involve too much intervention by the federal government.[152] After about 2008, partisan divisions increased in public opinion on climate change and in congressional voting on environmental policy.[153] Then, the rise of the Tea Party faction within the Republican party and the Republicans' capture of the House of Representatives in 2010 scuttled any prospects for legislative approaches to climate change.

Partly because of the mismatch between problems and politics in this period, only one major climate policy measure passed Congress during the first six years of the Obama administration. That consisted of about $51 billion in support for renewable energy, energy efficiency, home weatherization, and electricity grid improvement, which was included in the 2009 economic stimulus bill.[154] That bill passed in the first one hundred days of Obama's administration, when his popularity was still very high and the sense of crisis posed by the rapidly deepening recession made arguments for stimulus and job creation effective.

Legislative Failure

The major failure of the Obama administration's climate policy efforts was the blockage of the proposed American Clean Energy and Security Act (the Waxman-Markey bill) in Congress. This comprehensive climate policy bill included a national renewable portfolio standard (20% by 2020), subsidies for energy efficiency and renewable energy, detailed energy efficiency standards, and a cap-and-trade system to meet a target of a 17% reduction in emissions over 2005–20.[155] It also would have limited the EPA's regulation of greenhouse gas emissions, and it would have preempted state-level cap-and-trade programs (such as California's) for five years.[156] The broad alliance of businesses and environmental organizations organized in the U.S. Climate Action Partnership supported the Waxman-Markey bill.[157] Although it passed the House of Representatives by 7 votes in June 2009, with 8 Republicans voting for it and 44 Democrats from coal and oil producing states opposed, Democratic leaders did not try to bring the bill to a vote in the Senate.[158]

Resistance from Republicans was a factor in this failure, but not a decisive one, since major health care reform and financial reform bills were passed in 2010 despite strong Republican resistance.[159] The failure to pass major climate policy legislation was partly due to concerns about the weak economy. However, the Waxman-Markey bill went a long way toward addressing those concerns by issuing 85% of allowances for free and providing for generous offsets in agriculture and forestry, to be overseen by the business-friendly Department of Agriculture rather than the EPA. By one estimate, loose definitions of offsets meant that the bill would not require any actual emissions reductions for 18 years.[160]

While the recession made the passage of comprehensive climate legislation more difficult, what doomed the Waxman-Markey bill was the strong mobilization against it by the fossil fuel industries and the strategic mistakes of climate policy proponents. The proponents began the fight for the bill with a long-term disadvantage in spending on lobbying. In 2007–08, the electric utilities and fossil fuel industry outspent environmental advocates by a margin of $492 million to $35 million. Fossil fuel industries spent $120 million on lobbying and advertising against the bill in the first four months of 2009. By contrast, the main coalition favoring the law, the Alliance for Climate Protection, spent only $19 million, although other organizations later spent at least $100 million trying to get passage in the Senate. [161] However, a shortage of money was not the advocacy coalition's main problem; rather, strategic and tactical decisions hampered their efforts.

First, most foundation funding on climate change went to organizations that were legally barred from lobbying because of their tax-exempt status, which weakened the coalition's efforts once Congress began to debate the bill. Second, from the start, the alliance relied heavily on Washington-based environmental organizations (the EDF, NRDC, Pew Center, World Resources Institute, and Nature Conservancy), without the involvement of state and local groups or other potential allies. Al Gore's Alliance for Climate Protection made a similar strategic mistake when it planned to spend $300 million in 2008–11, mostly on television advertising, to try to recruit millions of new activists, rather than spending money on grassroots organization building.[162]

Third, the coalition began its public campaign for the legislation late, only weeks before the House committee vote, in contrast to the more successful campaign by advocates for health care reform, which began in mid-2008. The climate policy coalition was also unprepared for the Senate campaign and lacked clear allies there. It did not expect the strong industry backlash that occurred and relied on the White House to try to deliver the essential Republican support in the Senate. Fourth, the initial bill was relatively unambitious, with little auctioning of allowances, and when it was weakened further by the addition of offsets and other provisions, it lost support from some advocates.[163]

Indeed, after the House vote, the climate policy advocates began to split just as the opposition counterattacked. Three members of the U.S. Climate Action Partnership (BP America, Conoco Phillips, and Caterpillar) dropped out of it; the oil companies were unhappy with their failure to gain free emissions allowances, as the coal and utilities industries had.[164] Moreover, environmentalists split to some extent between those who advocated the bill and others who rallied around a simpler, more rigorous proposal by Senators Maria Cantwell (a Democrat) and Susan Collins (a Republican). They proposed a cap-and-dividend plan that would auction all allowances and return an estimated average $1,100 per year to each family to compensate for higher energy costs. Unlike the Waxman-Markey bill and similar to what Obama initially preferred, this plan was designed to increase energy costs and create substantial revenue. It gained the support of smaller environmental organizations and the Rockefeller Family Foundation.[165]

Given these strategic mistakes and divisions – which came on top of the industry backlash, the weak economy, and Republican opposition – the Waxman-Markey bill had little chance in the Senate. Even the Deepwater Horizon oil spill in April 2010, which gained a very high media profile and cast the oil industry and offshore drilling in a bad light, did not revive the bill's chances. EPA officials wanted Obama to use the spill to vilify the oil companies in order to advance his environmental agenda.[166] Obama may have been reluctant to do so because the nature of the accident cut against the Senate version of cap and trade then under consideration, which included an expansion of offshore oil drilling as part of the package in order to gain support.[167]

Regulatory Progress

But the Obama administration also has had important climate policy successes. As in the Clinton administration, yet to a greater degree, executive authority under existing legislation made progress possible even with an uncooperative Congress. Indeed, even before the Waxman-Markey bill failed in Congress, the Obama administration initiated major climate policies through executive action. Within Obama's first few months in office, the EPA began explicit regulation of greenhouse gas emissions on the basis of the Supreme Court's 2007 *Massachusetts v. EPA* decision, which had not been implemented by the Bush administration. In his first days in office, Obama directed the EPA to reconsider California's request for a waiver to allow its Pavley rules to take effect, and the agency granted the waiver later that year.[168]

Also in 2009, the EPA issued an endangerment finding concerning six greenhouse gases; it officially began to regulate them in January 2011, starting with motor vehicle emissions. The agency effectively nationalized California's Pavley approach on regulating vehicle greenhouse gas emissions, through an agreement between the administration, carmakers, and California and other states in May 2009. Under the agreement, the EPA created a nationwide standard for CO_2 emissions per mile and required a 22% decrease in emissions over the 2011–16 model years.[169*]

On the international front, Obama attended the climate negotiations at Copenhagen in December 2009, where he said the United States would cut its greenhouse gas emissions by 17% over 2005–20, in line with the Waxman-Markey bill. This helped lead to an international agreement to work toward binding commitments for both industrialized and developing countries by 2016, with the goal of limiting global warming to two degrees Celsius.[170] In March 2015, the Obama administration submitted to the United Nations a new target of a 26% to 28% reduction by 2025. It had announced this target the previous year as part of an agreement between the United States and China, in which the latter committed to stop the rise in its emissions growth in 2030.[171] The relatively ambitious U.S. target allowed the Obama administration to take a leading role in negotiating the historic Paris agreement, in which all countries agreed to limit their greenhouse gas emissions, in December 2015.

However, U.S. credibility internationally also has depended on the regulatory progress made by the Obama administration. In 2011, two years after the first deal with carmakers, the EPA reached another agreement with 13 carmakers, the autoworkers union, and the state of California. This agreement extend the EPA's vehicle emissions standards, requiring an additional 35%

emissions cut during the 2017–25 model years.[172] The carmakers agreed to the reductions because they preferred meeting one national standard and because they gained concessions concerning light trucks.[173] The 2009 and 2011 agreements were promptly implemented by the EPA and Transportation Department, under the authority of existing legislation; the relevant laws were the 2007 Energy Independence and Security Act and the Clean Air Act, the latter in light of the Supreme Court ruling in *Massachusetts v. EPA*. In 2011, the EPA also issued the first-ever federal standards for medium and heavy-duty trucks and other large vehicles, requiring reductions in greenhouse gas emissions that amount to a 15% improvement in fuel economy by 2018.[174]

As the EPA continued to implement greenhouse gas regulations after 2010, the administration confronted a mixture of favorable and unfavorable factors for climate policy. The unemployment rate began to decline after November 2010, and public attention to climate policy, as measured in newspapers[175*] and shown in Figure 9.1, remained relatively high during 2010–14. But with sluggish growth in incomes and jobs in the recession's aftermath, public support for climate policy has recovered only partially. It is still below the 2007–08 peak in the percentage who accept the scientific consensus, see climate change as already occurring, and consider the problem serious. Congress has remained a strong constraint on major advances in climate policy, but executive action on power plant emissions has continued to unfold, subject to the Supreme Court's further interpretation of the EPA's rule making authority under the Clean Air Act.

The pace of executive action was slowed somewhat due to the weak economy, a strategic interruption for Obama's 2012 re-election campaign, and the EPA's caution in the face of federal litigation. Despite the *Massachusetts v. EPA* decision, the EPA's authority to regulate greenhouse gas emissions has been vigorously challenged in the federal courts by several states, major business associations, and congressional Republicans.[176] After a federal appeals court rejected the challenges in June 2012, the Supreme Court refused to hear the case, and in June 2014, the Court largely upheld the EPA's authority in this area.[177*]

In this context, the EPA slowly has begun to adopt greenhouse gas emissions standards for power plants. It issued the first rules, for new power plants, in April 2012, and after the agency received more than 2.5 million public comments, it issued revised rules in September 2013. The rules require that new power plants have greenhouse gas emissions no greater than 1,000 to 1,100 pounds of CO_2 per Megawatt-hour. This would permit combined cycle natural gas plants and would effectively bar new coal-burning power plants unless they had carbon capture-and-storage technology in place. Draft rules for greenhouse gas emissions by existing power plants were issued in June 2014, with final rules issued in August 2015 (the Clean Power Plan); implementation by states had been scheduled to begin a year later, until the Supreme Court issued an injunction that suspended the plan in February 2016.[178] In 2013, President Obama also announced that his administration would use its direct lending and influence with multinational lenders to block the funding of coal power plants in developing countries, except under narrow circumstances in the least developed countries, a move that was joined by the United Kingdom and the Nordic countries.[179]

CONCLUSIONS

The United States is a laggard in climate policy among the industrialized countries, and yet it has adopted some relatively strong policies at certain times during the last forty years. In several periods, a fortunate convergence of problems and politics made the adoption of major climate policies possible. In the 1970s, the converging factors were the energy crisis and the election of Democrats to Congress and the presidency. In the late 1980s and early '90s, they were the hot, dry summer of 1988 in the United States, the emerging scientific and international concern about global warming, and George H. W. Bush's election. And in the late 2000s, the converging factors were damaging hurricanes, energy price increases, shifts in corporate and news media attitudes, and the election of Democratic congressional majorities and Barack Obama. As a result of these convergences, the United States adopted significant federal policies promoting energy efficiency and renewable energy, adopted regulations on substitutes for ozone-depleting substances and landfill methane emissions, and finally began to directly regulate greenhouse gas emissions from motor vehicles and power plants. These episodes are summarized in Table 10.1, in Chapter 10.

Nonetheless, these periods and others could have produced stronger policies had events unfolded somewhat differently. Table 10.2 shows that the failures occurred in the context of mismatches between the problem stream and the political stream. If the United Nations Framework Convention on Climate Change had been hammered out at a conference in 1993, after Clinton had become president, he might have used the threat of global warming to help justify his energy tax proposal and might have maintained the initial momentum that the tax had in Congress. If Al Gore had been elected president in 2000 – instead of losing by four electoral votes due to a narrow Supreme Court decision that halted the recounting of Florida's popular votes – a Gore administration probably would have been strongly committed to climate policy. It might have responded to the September 11, 2001 attacks in part with a renewed focus on energy efficiency and renewable energy policies, in order to reduce U.S. dependence on foreign oil, as even some Republicans such as George Pataki did. If the Great Recession had begun a year after Obama's inauguration, instead of a year before it, some kind of cap-and-trade legislation and a national renewable portfolio standard would have had much better chances of passing Congress. In addition, at many points where congressional battles over legislation occurred, such as in 1978, 1992, 1993, and 2010, if policy advocates had been more united, strongly mobilized, and strategically savvy, they probably would have achieved better results.

To an extent, earlier climate policy successes laid the groundwork for later expansions. Although many of the Carter-era reforms were reversed under Reagan, the creation of the Energy Department established specialized capacities for research on energy efficiency technology and standards, which were used repeatedly beginning in the late 1980s. The Energy Star program, which began with a lighting program (Green Lights) and consumer products (computers and monitors) in 1991–92, proved so popular with business that it was easy to expand to buildings, office equipment, utilities, and industry, with much business cooperation, in the mid-1990s.

Positive feedback involving economic actors also occurred to a degree in renewable energy and energy efficiency. In these areas, economic sectors and their interest groups, such as the American Council for an Energy-Efficient Economy, grew with the help of the 1992 Production Tax Credit and, beginning in the late 1980s, federal energy efficiency standards. In the latter example, positive feedback also ran through policy making by state governments, as state-level efficiency standards drove appliance manufacturers into a reluctant alliance with environmental and efficiency advocates, which mobilized together in favor of national appliance standards. As the renewable energy industry has grown, it has continued lobbying for extensions to the Production Tax Credit, usually with bipartisan support made possible by the industry's strong presence in Republican- as well as Democratic-governed states, and ultimately it has gotten them.

Legal authority is another aspect of path dependence that sometimes has operated at long time lags. The congressional granting of authority over mandatory appliance standards to the Energy Department in 1978 finally became a basis for the adoption of standards after business opposition to them eased in the late 1980s. The Clean Air Act, adopted at the height of environmentalism, had much later consequences for greenhouse gas regulation due to the authority it granted the federal government in section 202a (to regulate pollution that endangers public health or welfare) and in section 111d (to regulate air pollutants not anticipated at the time of the legislation). However, it required decades of accumulating scientific evidence, the landmark 2007 Supreme Court decision, and the election of a Democratic president before this authority would be applied to greenhouse gases. Other provisions of the Clean Air Act, which allow California to apply for exemptions from federal preemption of its standards and which allow other states to copy California, made possible that state's role in pushing vehicle emission regulation, through the Pavley law. That California law, in turn, helped lead to both the *Massachusetts v. EPA* decision and the nationalization of the Pavley approach by the Obama administration. The new federal policy preserved California's program and its legal right to adopt standards stricter than the federal government, as described in the sections on transportation policy in Chapter 2. This makes it possible for California and states allied with it to press the federal government to adopt more stringent standards in the future.

Structural conditions played mixed roles in these policy-making episodes. Some favorable structural conditions, identified in Chapter 5, were important for the successes. Compared with other industrialized democracies, the United States is a relatively affluent country, with an approximately average proportion of college graduates, and these in turn probably have promoted relatively strong environmental organizations in the United States.[180] Air pollution problems created by economic growth in an urbanizing society were severe enough in the 1960s to lead to the federal Clean Air Act, which has become the legal basis for federal regulation of greenhouse gas emissions. Despite high levels of fossil fuel production, the United States' dependence on oil imports rose to a peak of 60% of oil consumption in 2005, and it was still 49% in 2010.[181] The relatively high dependence on oil imports created a national interest in reducing energy consumption in transportation and power generation, and it made arguments about energy independence or energy security potentially effective.

In addition, the institutional separation of powers guaranteed by the Constitution often has been a barrier to climate policy, by making mismatches between problem events and political commitment more likely. But the effects of this institutional feature also have depended on election results, such as the Republican congressional victories in 1994 and 2010 (see Chapter 4). Therefore, the separation of powers did not prevent the convergence of problem awareness and political commitment by both the president and congressional majorities in several periods: in the late 1970s (with commitment supplied by a Democratic president and Congress); in the early 1990s (an environmentalist Republican president and Democratic Congress); and in 2009–10 (a Democratic president and Congress). The separation of powers also did not prevent a negative convergence in the early 1980s, when a Republican president allied with Republicans and conservative Democrats in Congress.

The pluralist interest group system, on the whole, has also braked climate policy making, but its effects, too, have depended on other factors. The fragmentation of economic interest groups, their adversarial relation to the federal government, and their many access points in both houses of Congress, the executive branch, and the federal courts certainly have given certain interests – especially the fossil fuel, electric power, and energy-intensive manufacturing industries – large opportunities to block climate policy, as they did in Congress in 1993 and 2009. These forces have operated to limit what policy makers will propose in Congress or in the form of regulations, such as the Obama administration's power plant rules.

However, these industries were successful in blocking legislation only because they mobilized strongly and effectively, while the advocates often did not. Beginning in 1989, with the formation of the Global Climate Coalition by oil, car, and other industrial companies, certain business interests, conservative foundations, think tanks, and media outlets used what proved to be a very effective public relations strategy.[182] They funded scientists who doubted climate change, got them access to news media, and hence helped to shift public opinion toward skepticism about climate change and the human role in it.[183] Although the news media generally has given less credence to climate-change deniers after 2005, the "Climategate" email controversy in 2009 had some lasting effects on public opinion, helping to raise the proportion who doubt the scientific evidence and the media reporting of it. Climate change denial organizations are remarkably influential in the United States, compared to other developed countries, although similar groups have grown in Australia, Canada, and the United Kingdom.[184]

While the opponents used effective methods, the advocates often did not. In 1993, the Clinton administration did little to build support for the BTU tax, and it inadvertently fed the lobbyists' appetite for exemptions by granting concessions early and easily. In 2001, climate policy advocates did little while the fossil fuel industries were persuading President Bush to reverse his position on greenhouse gas regulation. In 2009, the advocates were slow to start their campaign, ill-prepared for the industries' counter-offensive, and increasingly divided. But the pluralist system does not just benefit the fossil fuel lobbies. It also gives opportunities for climate policy advocates to create or fund new groups and coalitions, which activists and foundations began to use in the mid-2000s, although they may need more time to learn effective strategies and tactics.

Probably more than they have been harmed by structural barriers or strategic mistakes, the chances for climate policy have been undermined by party polarization since the early 1990s, which has strongly affected climate policy and all other kinds of environmental policy. A degree of bipartisan cooperation is crucial for bridging the United States' fragmented institutions, and it was an important element in the reforms of the 1970s and early 1990s, as well as in attempts at emissions trading since the early 2000s. But cooperation across party lines on energy and climate policy has grown increasingly uncommon since Reagan became president. The Republican Party shifted strongly toward environmental deregulation under Reagan and experienced the rise of an intensely anti-government faction with the party's 1994 congressional victories, a faction that became even more influential after Obama's 2008 election. The Republicans' conservative wing blocked any revision to CAFE standards under Clinton, helped to push President George W. Bush away from greenhouse gas regulation, and made it nearly impossible for congressional Republicans to support climate policy under Obama.

From 1992 to 2012, partisan divisions in public opinion grew to a greater degree on environmental protection than on 14 other issues, apparently the result of the shift by Republican elites and conservative media outlets.[185] In the 2012 elections, no major Republican presidential candidate[186*] endorsed cap and trade, and previous congressional backers of the policy, such as John McCain, had become silent due to fears of a primary challenge backed by Tea Party groups. Fifty-eight percent of Republicans in Congress (and half of all Republican governors) in 2013 had publicly denied the existence of anthropogenic climate change, while only 3% of congressional Republicans had gone on the record as accepting climate change science in recent years.[187]

Yet several other recent trends have improved the prospects for climate policy and may be able to counteract the effects of party polarization. Scientific knowledge and public awareness of the impacts of climate change is growing, the Supreme Court has repeatedly endorsed greenhouse gas regulation by the EPA under the Clean Air Act, and business is adjusting to climate policy in Europe and the leading states. California and the RGGI states have a chance to show the economic feasibility of a lower-carbon economy, and the sectors that benefit from climate policy continue to grow. Given this mixture of promoting and inhibiting factors, the future of national climate policy making is quite open, a point to which I will return in the concluding chapter.

10

Conclusions: Political Opportunities for Climate Policy in the United States

Most of this chapter will assess the structural, process, and path dependence theories of climate policy in light of the empirical evidence presented in the previous six chapters. After doing so, I will consider the prospects for the adoption of stronger climate policies by the states and the federal government over the next ten years.

STRUCTURAL EXPLANATIONS

The structural explanations receive mixed support from the case studies and the state-level statistical analysis, as shown in Tables 4.1 and 5.1, in earlier chapters. The major political institutions that help to explain the weakness of federal policy making are actually conditioned by political processes; moreover, they are of little help in explaining state-level policy adoption.[1*] However, other political and socioeconomic structures *are* important factors, with varying abilities to explain federal and state policies.

The separation of powers clearly helps to explain the failure of federal climate policy making at important junctures. Congress weakened President Carter's energy legislation in 1978, whittled down President Clinton's BTU tax proposal into a small gasoline tax, blocked the ratification and implementation of the Kyoto Protocol in the late 1990s, and allowed comprehensive climate legislation to die in 2009–10. But the effects of the separation of powers have also depended on the Republican Party's growing opposition to environmental policy, including that party's denial of climate change – and on certain election results, especially Reagan's election to the presidency in 1980 and the Republican congressional victories in 1994 and 2010. Party polarization and divided government have made the separation of powers a major barrier in recent years.

Moreover, variation in the separation of powers cannot explain differences in climate policy adoption in the states. In fact, the evidence there lends more support to the theory that separated institutions offer alternative paths to reform and sometimes compete to bid up environmental standards. The statistical analysis in Chapter 5 and the case studies in Chapters 7 and 8 suggest that weaker institutional powers for the governor and more intense competition between executive and legislature, as in California, may promote climate policy more than dominant executives can in states such as New York.

One reason for this difference between the state and federal levels is that many states are relatively homogeneous ideologically, dominated by either liberal or conservative publics. By contrast, Congress represents a large diversity of public opinion, from liberal, conservative, and mixed states. In addition, the U.S. Senate's filibuster rules, which allow a minority of 41 members to block legislation and have been used increasingly aggressively in recent decades, do not have counterparts in most state legislatures.

But the importance of these institutional differences should not be exaggerated. At the federal level, the increasing scope for unilateral executive action, based on previous legislation and encouraged by favorable court rulings, has weakened the separation of powers as a barrier to climate policy making. Facing congressional opposition in the 1990s, President Clinton was able to implement important landfill methane regulations and voluntary programs for energy efficiency, and Obama's EPA has taken far-reaching regulatory actions. Advocates seek alternative paths to reform, and recently have found them in the Supreme Court and EPA, as shown in Chapter 9.[2] And sometimes the separation of powers has supported climate policy, by helping to prevent the undoing of previously adopted policies. President Clinton resisted the attempts of a Republican-led Congress to undermine energy efficiency and renewable energy policy in the 1990s, and Congress resisted similar efforts by President George W. Bush in the 2000s.

The United States' pluralist interest group system has been another structural barrier to federal climate policy making, but only to an extent. Private campaign finance, business access to elected officials, and professional lobbying together enhance the ability of energy and manufacturing interests to block legislation and deter executive action. However, pluralism's effects have been ambivalent, since the interest group system also has encouraged the growth of powerful environmental organizations and the development of broad advocacy coalitions. Fossil fuel and manufacturing industries were successful in blocking legislation partly because they mobilized strongly and effectively, while climate policy advocates often did not. The Global Climate Coalition was influential in promoting the denial of anthropogenic climate change, and fossil fuel and manufacturing interests mobilized promptly, professionally, and vigorously to block the Kyoto Protocol and domestic climate legislation. By contrast, climate policy proponents were slow to mobilize resources, poorly prepared for political conflict, and sometimes divided, as shown in Chapter 9.

Moreover, the extent of pluralism does not help to explain the state-level differences analyzed in Chapters 5 through 8. Pluralism was not a significant variable in the statistical analysis, and it was not an insurmountable barrier in the leading states. California's politics include a highly pluralist interest group system, and its interest groups generated intense, well-funded opposition that made climate policy making in that state a highly public, conflictual process – and yet this process ultimately produced strong climate policies. Nor did pluralist contention over New York's RPS and over RGGI block those policies.

Administrative capacity emerged as a key factor in explaining state-level climate policy both in the case studies and the statistical analysis. States historically have had agencies regulating energy and other areas related to climate policy, such as economic development. Since the 1960s, they have created new administrative agencies, such as CARB and NYSERDA, with the expertise needed to implement energy and climate policies and to develop new proposals. As large states,[3*] California and New York may have had advantages in creating and building up new agencies, as discussed in Chapter 4. The creation of new administrative capacity is an important aspect of path dependence, which is discussed in the relevant section below.

Political culture receives qualified support as an explanation. National public opinion in the United States is less favorable toward environmental protection and climate policy than in most European countries. However, majorities in the U.S. support the scientific consensus on global warming, as well as a wide variety of specific climate policies based on regulations or subsidies, though not tax increases. Nonetheless, a liberal (pro-intervention) citizen ideology was a significant predictor of climate policies in the fifty states. In California and New York, relatively liberal citizens helped to put relatively liberal officials into office, especially in their state legislatures, although New York has been the more liberal of the two states since 1960. Facing liberal legislatures and liberal voters, even Republican Governors Pataki and Schwarzenegger pursued climate policies as part of their efforts to appeal to the political center in their states.

Education levels were good predictors of state climate policies, and can help to explain the successes in California and New York. Both states had education levels that were higher than the national average in the 1960s and in 2000, although they were higher in New York than in California. In those cases, and at the national level,[4*] highly educated populations supported the growth of environmental organizations, which have drawn their members disproportionately from professionals and other groups with large amounts of education.[5] Those organizations played vital roles at many stages in the development of air pollution, energy efficiency, renewable energy, and explicit climate policies in the leading states and at the federal level. This point is solidly supported by the statistical analysis of state policies and by the case studies.

There is much evidence that a large domestic fossil fuel industry adversely affects climate policy, both because the industry is a powerful political force that can deter or mobilize to defeat climate policies and because fossil fuel production reduces the incentives for energy conservation and supply diversification. Fossil fuel interests were important in opposing, weakening, delaying, or defeating federal climate policies. They lobbied or litigated against increased oil and natural gas taxes in the 1970s, binding targets in the United Nations Framework Convention on Climate Change, Clinton's BTU tax, cap-and-trade legislation, and the EPA's power plant rules, as described in Chapter 9. They also funded the climate change denial movement, which has been very successful in influencing media coverage (especially during 1995–2005), politicians' positions, and public opinion.

The statistical analysis also shows that states with less fossil fuel production have stronger climate policies. New York State fits that pattern. It has virtually no fossil fuel production, along with 21 other states, many of which also have strong climate policies. Hence, the coalition opposing New York's climate policies did not include fossil fuel interests, but rather centered on energy-intensive industries and coal-burning utilities. Trying to reduce dependence on oil imports was a major justification for energy efficiency policies in the 1970s, and energy security was also a significant part of the rationale for Pataki's climate policy initiatives in the early 2000s.

However, the California case shows that a strong fossil fuel industry does not necessarily doom the prospects for climate policy. California has had significant oil production and refining capacity, still ranking third in the nation in both, and its total per-capita production of all fossil fuels was higher than that of 28 other states in the late 1990s. Oil production and refining interests, both in-state and out-of-state, have been important members of the coalition opposed to climate policy in California. However, they were defeated at important junctures, such as the passage of AB 32 and the rejection of Proposition 23, by an even more vigorous and larger mobilization by environmental organizations and their allies. The opposite occurred at the federal level, where the fossil fuel industries were able to block climate legislation partly because they mobilized more effectively than the advocates did.

Finally, severe smog problems, which were conditioned by urban density and geographic features, were major spurs to climate policy making because they led to early action on air pollution, including the creation of administrative and legal capacities. In Southern California and New York City, urban smog was an unusually serious problem in objective terms. Moreover, by the late 1960s, environmental groups and local officials had heavily politicized it, and their efforts led to important reforms at that time. CARB was created in California and became a powerful, capable regulator of air pollution, while in New York State, coal-fired generation was curbed, environmental law reforms empowered environmental organizations, and increased energy efficiency became a preferred approach for meeting the state's energy demand. The air pollution issue continued to resonate in California in later decades. Legislators justified the Pavley bill partly in terms of health impacts, and CARB earned a reputation as an effective agency, and hence the legislature expanded its authority, because of its demonstrated successes in reducing air pollution from the 1970s to the 1990s. More generally, air pollution problems were a significant predictor of climate policies in the statistical analysis of all 50 states, and California and New York ranked 3rd and 11th in air pollution during 1990–2002.

Finally, the response to urban air pollution had crucial implications for national climate policy, in two ways. The 1970 Clean Air Act, adopted under the pressure of reform movements in several urban states, empowered California to continue to lead other states and the federal government toward more stringent air pollution standards for motor vehicles, starting with the Pavley law in 2002. In addition, the Clean Air Act gave the EPA the authority, confirmed by the Supreme Court in 2007, to regulate greenhouse gas emissions as pollutants that the agency found to be threats to public health and the public welfare of current and future generations.

STREAM CONVERGENCE AND WINDOWS OF OPPORTUNITY

Convergence and Non-Convergence

The windows of opportunity theory is also well supported by the three case studies, which include a total of 22 policy-making episodes. In each of the 15 episodes of major policy reform, problem events led to increases in public attention to energy and global warming issues, which preceded or coincided with increases in those issues on policy makers' agendas. This is shown in the figures presented early in each case study chapter and discussed there. Most of these problem events were related to severe smog occurrences, energy supply crises, scientific information about human-caused global warming, and international agreements. In the major reform episodes, political commitment and mobilization were also high enough for the issue to rise on policy makers' agendas – as measured by counts of legislative bills and hearings and by content analysis of the speeches of presidents and governors – and for significant policies to be adopted. Table 10.1 summarizes the convergences of problem events and political events in these 15 cases.

But political commitment and public attention to an issue did not always converge. In the 7 episodes that did not result in major reforms, several different patterns can be observed, as shown in Table 10.2. There were two cases of policy reversal in the early 1980s, in California and the federal government. In these cases, renewable energy policies were reversed and federal energy efficiency policies were stalled due to nearly simultaneous reversals in the problem stream (falling fossil fuel prices) and in the political stream (the elections of President Reagan in 1980 and Governor Deukmejian in 1982). By contrast, the attempted reversal of California climate policy through Proposition 23 in 2010 failed because the political stream and the problem stream diverged, since the

TABLE 10.1 *Fifteen Episodes of Major Climate Policy Making*

Period of policy making	Problem events that increased (decreased) public awareness	Political events that increased (decreased) political commitment	Outcome	Degree of convergence of problem events and political events	Type of outcome
California cases					
1947	Economic and population growth; Smog events.	Public protests; Campaigns by *Los Angeles Times*, business.	State law permitting county air pollution control boards	High	Major policy adoption
1960	Smog events; Inadequate policies; Research findings on car exhaust.	Public protests; Campaigns by Air Pollution Control District, Air Pollution Foundation.	State law establishing Motor Vehicle Pollution Control Board	High	Major policy adoption
1967–68	Smog events; Inadequate air pollution policies; Emerging federal air pollution policy.	Campaigns by Air Pollution Control Board; Al Wiman's "Breath of Death" series.	Mulford-Carrell Act, creating CARB; Pure Air Act; California's exemption from federal air pollution standards.	High	Major policy adoption
1970	Conflicts over nuclear power plant siting, redwood forests; Santa Barbara oil spill; NEPA.	Mobilization by Sierra Club; Sponsorship by John Knox; Support from Attorney General Younger; California Supreme Court ruling in *Friends of Mammoth*.	California Environmental Quality Act (CEQA) enacted and implemented	High	Major policy adoption

1974	Utilities' power plant plans vs. CEQA requirements; Threats of blackouts; Arab oil embargo.	Anti-nuclear movement; Sierra Club alliance with Rand Corporation and Charles Warren; Reagan's switch in position.	Warren-Alquist Act	High	Major policy adoption
1976–82	Energy crises; Utilities' financial crisis.	Jerry Brown's election; Anti-nuclear movement and Brown's shift on nuclear power; Environmental organizations' mobilization; Initiatives of Democrats in legislature.	Energy efficiency and renewable energy policies	High	Major policy adoption
2002	Electricity crisis; Kyoto Protocol and Bush's rejection of it; First UCS report on climate change effects in California.	Mobilization by Bluewater Network, other environmental organizations; Pavley's leadership; Existence of CARB.	Pavley law (AB 1493); Renewable portfolio standard.	High	Major policy adoption
2005–08	Electricity crisis after-effects; Drought, wildfires; Second UCS report; EU's emissions trading system; National rise of global warming issue.	Schwarzenegger's election and shift toward climate policy; Initiatives by Pavley and other legislative Democrats; Actions by Attorneys General Lockyer and Brown.	Emissions targets; Global Warming Solutions Act (AB 32); Low-carbon electricity and fuel standards; RPS acceleration; Transportation planning law (SB 375).	High	Major policy adoption

(*continued*)

TABLE 10.1 (*continued*)

Period of policy making	Problem events that increased (decreased) public awareness	Political events that increased (decreased) political commitment	Outcome	Degree of convergence of problem events and political events	Type of outcome
New York cases					
1972–75	Air pollution; Storm King plant and Hudson Expressway proposals; Nuclear power plant proposals.	Environmental mobilization; Lindsay's election.	Power plant siting law; Environmental Quality Review Act.	High	Major policy adoption
1975–77	Growth of energy demand; Blackouts and brownouts; Arab oil embargo; Loss of industrial jobs.	Carey's and Assembly Democrats' elections; Environmental mobilization against coal and nuclear energy.	NYSERDA and State Energy Office created; Appliance standards; Energy efficiency policies.	High	Major policy adoption
2001–05	Kyoto Protocol and Bush's rejection of it; Bush's energy policy and 9/11 attacks.	Pataki's election and shift to environmental policy; Spitzer's election and actions as attorney general; Broad advocacy coalition mobilization.	RGGI; RPS; Energy efficiency policies (buildings); Emissions targets.	High	Major policy adoption
2008–14	National rise of global warming issue, due to multiple factors; Rising fuel prices; (Economic recession).	Democrats' national election victories; Spitzer's and Andrew Cuomo's elections.	Energy efficiency portfolio standard; Emissions targets; Building codes; Power plant standards; RGGI cap lowered.	High	Major policy adoption

U.S. federal government cases					
1975–78	Oil embargo, shortages, price increases	Congressional Democrats' and Carter's elections	EPCA 1975, PURPA 1978: energy efficiency and renewable energy legislation.	High	Major policy adoption
1990–92	Hansen's testimony, East Coast heat wave, drought; Scientific discovery of global warming; UN Framework Convention; Gulf War; (Economic recession).	George H. W. Bush's election and his environmental interest	Clean Air Act 1990 and EPAct 1992: energy efficiency policies; Production Tax Credit; SNAP, methane regulations, Energy Star programs begun.	Moderate	Major policy adoption
2007–14	Strong Atlantic hurricanes; Business and media shifts; Environmental campaigns; Rising fossil fuel prices; IPCC reports and UN meetings; (Economic recession).	Democrats' election to congressional majorities in 2006; Obama's election; *Massachusetts v. EPA* decision; Public opinion shift toward (then away from) climate policy; (Rise of Tea Party); (Republicans' election to House majority in 2010).	CAFE increases; EPA's endangerment finding; Greenhouse gas standards for motor vehicles; Proposed standards for existing power plants and other sources.	Moderate	Major policy adoption

TABLE 10.2 *Seven Episodes of Failed Climate Policy Making or Policy Reversal*

Period of policy making	Problem events that increased (decreased) public awareness	Political events that increased (decreased) political commitment	Outcome	Degree of convergence of problem events and political events	Type of outcome
California cases					
1983–85	(Oil price decline)	(Deukmejian's election)	End of renewable energy policies	High (both negative)	Major reversal of policies
2010	(Economic recession and high unemployment)	(Mobilization by coalition of climate policy opponents, led by out-of-state oil companies); Most of business neutral or supportive of climate policy.	Defeat of Proposition 23	Low	Failure of attempted policy reversal
U.S. cases					
1981–85	(Fossil fuel price decline)	(Reagan's election)	Expiration or blockage of energy efficiency and renewable energy policies	High (both negative)	Major reversal of policies
1993	(None)	Clinton's election	Failure of BTU tax; Very small gasoline tax increase.	Low	Very little policy adoption

1995–2000	Negotiation of Kyoto Protocol	(Election of Republicans to majorities in Congress)	R & D spending on energy efficiency and renewable energy; Landfill regulations and Energy Star implemented; But: Kyoto rejected by Senate.	Low	Minor policy adoption; Significant implementation of previously adopted policies.
2001–05	Bush's rejection of Kyoto Protocol and ensuing criticism of that; Bush's energy policy proposals.	(George W. Bush's and Dick Cheney's election); Congressional support for renewable energy and energy efficiency.	Some energy efficiency and renewable energy policies renewed and funded; But: EPA's blockage of Pavley rules.	Low	Very little policy adoption
2009–10	Strong Atlantic hurricanes; Business and media shifts on climate change; Environmental campaigns; Rising fossil fuel prices; IPCC reports and UN meetings; (Economic recession and high unemployment).	Democrats' election to congressional majorities in 2006; Obama's election; *Massachusetts v. EPA* decision; Public opinion shift toward (then away from) climate policy; (Rise of Tea Party).	Failure of comprehensive climate policy legislation (Waxman-Markey bill)	Moderate	Failure to adopt policy

climate policy coalition had become stronger relative to the opposing coalition even though unemployment rose in the aftermath of the 2008–09 recession.

In four other episodes, all at the federal level, major reforms were attempted but failed, in 1993, 1995–2000, 2001–05, and 2009–10. In the first three of these, problem awareness and political commitment by the president and Congress were out of sync. In 1993, Clinton's initial political commitment to a BTU tax was stymied by a low level of public attention to energy and climate issues and a lack of interest by the Democratic-controlled Congress. In the late 1990s, public attention was high, but intense presidential interest was met with congressional inattention and resistance after the Republicans won control of Congress. Then, the reverse occurred, when the Bush administration resisted congressional efforts on climate change in the early 2000s, at another time when attention on the issue was high.

The 2009–10 case was somewhat different, with a moderate degree of convergence between problem events and political events. Hence, the public, the president, and Congress were all focused on climate change at this time, and major legislation did pass the House of Representatives, but it failed in the Senate. The reasons for failure lay partly in organizational weaknesses and strategic errors of the advocacy coalition, as well as in the worsening problem context due to the recession and the related shift in public opinion and rise of the Tea Party.

Elections and Environmental Organizations

Another lesson from the case studies is that elections sometimes matter a great deal for climate policy making, when they affect the amount of political commitment provided by governors, presidents, or new legislative majorities. Elected officials' potential interest in environmental policy can be assessed from their previous records, campaign statements, early relationships with environmental organizations, and key advisors' leanings. Based on these indicators, Jerry Brown, Hugh Carey, Jimmy Carter, George Pataki, Arnold Schwarzenegger, and Barack Obama were much more interested in energy and environmental policies and hence potentially more committed to climate policy than their electoral opponents – Houston Flournoy, Malcolm Wilson, Gerald Ford, Mario Cuomo, Gray Davis, John McCain, and Mitt Romney.

The election of the officials on the first list, who became climate policy advocates, was not inevitable. All of them were elected either by narrow margins (Brown by 3%, Carter by 2%, Pataki by 3%, Obama re-elected by 4%) or under extraordinary circumstances. Brown, Carter, Carey, and the Democrats in Congress and the New York State Assembly benefited from the Watergate scandal in the mid-1970s, and Schwarzenegger defeated Davis through an unprecedented recall campaign in 2003. Their elections also depended on unrelated issues such as the Iraq War, which helped the

Democrats to gain control of Congress in 2006, or taxation and spending, which aided Pataki's and Schwarzenegger's electoral successes.

Any of these elections easily could have turned out differently, probably with major implications for climate policy making. The same holds in spades for the 2000 presidential election. If Al Gore had prevailed in Florida and hence in the national electoral vote, he very likely would have pursued climate policies to implement the United States' Kyoto commitment, and the September 11, 2001 attacks would have given him an opportunity to advance them because of their co-benefits for energy security. As party polarization has grown, election results have become even more portentous for climate policy, which now often depends on the success of Democratic candidates. Finally, particular legislators, sometimes newly elected to office or newly chosen for key committee assignments – such as Charles Warren and Fran Pavley in California – also provided important political resources by framing solutions and organizing advocacy coalitions.

Mobilization by environmental groups has been the other major source of increased political commitment to the adoption of climate policies or their precursors. These include many episodes in California: smog control policy (grassroots protests), the Environmental Quality Review Act (Sierra Club), the Warren-Alquist Act (anti-nuclear groups), the utilities' adoption of energy efficiency programs (EDF and NRDC), the Pavley bill (Bluewater Network), and SB 375 on transportation planning (Center for Biodiversity). In New York State, the shift toward energy efficiency instead of continued power plant construction was largely spurred by local protest groups against air pollution, the Storm King pumped storage facility, and the Hudson Expressway (Scenic Hudson, Hudson River Fishermen's Association, and Sierra Club). Moreover, environmental organizations (NRDC, Pace Law School, and Alliance for Clean Energy) were important backers of the state's renewable portfolio standard, RGGI, and hybrid-electric bus program.

In federal policy making, environmental organizations also were important at many junctures. They advocated for mandatory appliance standards in the 1980s, lobbied for energy efficiency measures in the 1992 Energy Policy Act, publicized concerns about global warming in the 2000s, and advocated for cap-and-trade legislation in the first Obama administration. Other elite allies of environmentalists were sometimes important for reforms, too, such as the *Los Angeles Times* and radio station KLAC for smog control policies in California, as well as several state attorneys general: Evelle Younger in California in the early 1970s, for implementing the state's environmental quality act; Eliot Spitzer in New York beginning in 1999, for spurring Pataki to stronger environmental positions; and, in California in the late 2000s, Bill Lockyer for defending the Pavley rules and Jerry Brown for applying environmental quality reviews to greenhouse gas emissions.

The strong role for these two political factors – the election of politicians with environmental interests and mobilization by environmental organizations within advocacy coalitions – is also supported by the statistical analysis in Chapter 5. Government officials' ideology and environmental organization membership were the two process variables that were significant predictors of climate policy strength across the fifty states.

Since problem awareness and political commitment must converge if major reforms are to occur, timing matters. Sometimes the timing favored climate policy. Jerry Brown, Hugh Carey, large Democratic majorities in Congress and the New York State Assembly, and Jimmy Carter were elected soon after the Arab oil embargo, largely because it coincided with the Watergate scandal. The hot summer of 1988 in Washington, D.C. occurred while scientific concern with global warming was accelerating and a few months before George H. W. Bush, a self-styled environmental president, was elected. Arnold Schwarzenegger was elected soon after George W. Bush helped to put the global warming ball in the states' court in 2001 and at a time when droughts and wildfires were bringing the issue home to California. George Pataki was elected and re-elected before Bush's 2001 rejection of the Kyoto Protocol and the September 11 attacks the same year made global warming and energy policy more salient in New York State. The global warming issue peaked in 2007–09, driven by a multitude of forces, just as the Democrats were winning control of Congress and Obama was being elected president.

In other episodes, timing was not favorable, but a small difference in when key events occurred could have made a big difference for climate policy making. The United Nations environmental conference in Rio in 1992 probably would have led to a stronger climate policy in the United States had it been held two years earlier, before a recession began. Al Qaeda's attacks on September 11, 2001 created a large window for energy policy. If they had occurred in the Clinton administration, or if a President Gore had been in office in 2001 rather than President Bush, the EPA might have begun regulating greenhouse gases eight years earlier than it did. If the Great Recession of 2008–09 had occurred two years later, Obama and the Democrats in Congress probably could have passed a national RPS and a cap-and-trade program before public opinion had shifted to prioritize economic recovery over environmental protection.

Focusing Events

Focusing events are potentially very important for climate policy, because climate change probably will produce an increasing number and severity of extreme weather and climate events involving economic damage and the loss of life. But how effective have focusing events been in helping to produce climate policies so far? In the episodes analyzed in this book,[6*] increases in problem awareness often were due to focusing events, both concerning air pollution and energy problems through the 1970s and concerning climate change since the late 1980s. Urban smog events, the oil supply crises of the 1970s, New York City's blackouts and brownouts in the 1960s and '70s, California's electricity crisis in 2000–01, that state's droughts and wildfires in the early 2000s, and the 2005 hurricane season all helped to open policy windows.

However, there was not a simple one-to-one correspondence between focusing events and policy windows. There was often a lag between the events and policy changes. Twenty years of intermittent smog events preceded the 1967 Mulford-Carrell Act that created CARB, and the 1973–74 oil embargo still affected energy policy in 1976–78, both in leading states and at the federal level. The 2000–01 electricity crisis influenced the California Global Warming Solutions Act and other climate policy reforms that were adopted in 2006–08 and justified in good measure by reference to energy security.

Furthermore, the objective size of a focusing event was not always related to the size of its political effects. Ironically, several large climate-related events occurred at times and in places where the scope for policy reform had largely been exhausted before they occurred: the droughts in California in 2007–09 and from 2012 to the present, which were much more severe than the politically influential 2001–03 droughts, and Hurricane Sandy, which devastated New York City in 2012. These events had little influence on climate change mitigation policy because so much had already been adopted in the immediately preceding years; the scope for major reform had been exhausted for the time being.[7*] However, the latter did spur climate change adaptation measures in New York State, including a $16 billion federal-state program for infrastructure upgrading and protection.[8]

Moreover, focusing events sometimes did not have much effect on climate policy, because key political actors did not take advantage of them for that purpose. For example, Pataki did not vigorously use the September 11 attacks to argue for increased energy efficiency in order to increase energy security. He also made no such use of the 2003 blackout, which affected 55 million people in eight Northeastern states just a few months after his Greenhouse Gas Task Force issued its report. By contrast, Pataki's counterpart in New Jersey, Governor James McGreevey, did use the blackout to argue for renewable energy and emissions trading as a path for energy supply diversification.[9] Similarly, Obama did not link the Deepwater Horizon oil spill, which was the top national news story for four months in 2010,[10] to programs to reduce oil consumption through vehicle efficiency, alternative fuels, or reductions in vehicle miles traveled.

Interdependence and Coordination of Problem Awareness and Political Commitment

The case studies also suggest the need to modify and elaborate upon the windows of opportunity theory. In Kingdon's theory, the events of the problem stream and the political stream are largely independent of each other, as is the stream of events that produces policy solutions. Similar to the garbage can model of organizational decision making, policy changes depend on mostly accidental or random encounters between salient problems, the will to address them, and feasible solutions. Policy entrepreneurs are important in making the links between problems, solutions, and political commitment, but they do so only when a policy window has opened.[11] Indeed, as described in the section on convergence, the book's case studies show important examples where problem awareness and political commitment developed independently of each other and happened to converge to create major policy windows.

However, the case studies also show that problem awareness and political mobilization often were not independent of each other, a point that others also have noted.[12] Three patterns can be observed here. First, problems sometimes are so severe that they largely produce the political commitment to address them, while usually leaving open the content of the policies to be adopted.[13*] Smog in Southern California and New York City in the 1960s, and energy shortages in all three cases in the 1970s, were such problems. They threatened economic prosperity and growth, and in the case of smog, they directly threatened health, lives, and property values, too. Therefore, these problems were urgent ones for the public and business elites, and were very hard for policy makers to ignore. Most candidates in both parties competed to reduce smog, ensure energy availability, and curb price increases. Hence, election results did not play a large role in the *opening* of policy windows before the 1980s, although they had major effects on how those windows were used and on the nature of the reforms that were adopted in the 1970s.

However, this pattern did not hold for the explicit climate policies of the 2000s and 2010s. Problems could drive political commitment to address them only if the problems were severe, immediate, and concrete, so that business and political elites were united in seeing an urgent need to address the problem. But global warming is a different kind of problem than smog or energy shortages. The most visible, concrete effects of global warming are intermittent, such as storms, droughts, and wildfires. Moreover, such events cannot be definitely and mainly traced to greenhouse gas emissions, because the events also have natural causes and the losses from them are influenced by factors such as water usage, land management, and settlement patterns. The largest effects of global warming are in the future, making the problem abstract for people living in the present, unlike a choking haze of smog, a power outage, or a long line at the gasoline station.

As a result, it is not clear whether and when the effects of global warming will be large enough to lead to an elite and societal consensus that action is urgently needed. Moreover, if a consensus on the need for action does emerge, the proposed solutions probably will be controversial. For the foreseeable future, Republicans probably would prefer an expansion of nuclear energy, carbon capture and storage for coal plants, adaptation measures (e.g., sea walls and genetic engineering), and perhaps geo-engineering. By contrast, Democrats are likely to emphasize energy efficiency, renewable energy, transportation policies to reduce car travel, and regulatory measures in power generation, industry, and agriculture.

The potential of the global warming problem to drive the political stream is weakened further by the polarized party landscape that has existed since the 1990s, in which one party has come to deny the importance or even the existence of global warming. Therefore, in the last few decades, the federal government's adoption of major climate policies has depended on the independent development of adequate political commitment, largely through election results. Indeed, the continued relatively high position of the climate change issue on the public and political agendas since 2007 *may* mean that problem awareness now is normally sufficient and that reforms now can result whenever political commitment emerges.

The second pattern occurs when political events and actors influence, or even drive, problem perception. It should be noted that some events in the problem stream, as conceived by Kingdon, are actually initiated by political actors (reports on problem indicators), while the others are at least potentially subject to significant interpretation by them (focusing events and feedback from existing policies). Also, given the multi-level governance context of state climate policy making, I have conceptualized policy feedback to include policy decisions made at other levels of government. Thus, politics potentially permeates the problem stream, rather than the problem stream being a purely objective and separate set of events. This has major practical implications. It means that climate policy advocates can help to create their own opportunities, by trying to affect both problem perception and political commitment, rather than needing to wait for extraordinary problem events to occur.

The politicization of problems can be seen in many examples described in this book. In California in the early 2000s, influential scientific reports on climate change impacts were produced by a political actor (the Union of Concerned Scientists), and the CEC and CPUC used the state's electricity crisis to keep energy supply problems, and hence energy efficiency and renewable energy, on policy makers' agendas. The sharp and sustained rise of the global warming issue on public and political agendas in all three cases after 2005 owes more to political mobilization by environmentalists and to shifts in position by business corporations and the news media than to change in the objective problem or in scientists' understanding of it. Generally, there is much scope for the interpretation of the severity and urgency of the problems involved in global warming.[14]

In short, the relationship between the streams of problem events and political events that impinge on climate policy making has changed over time, as the relevant problems shifted from air pollution and energy to include global warming in the late 1980s, and as the political parties became polarized about the latter issue beginning in the 1990s. From the 1940s to the 1970s, during the period in which implicit climate policies and their precursors were adopted, large new problems created both the political commitment to address them and a search for solutions. Bipartisan consensus was often possible. Since the 1980s, the political stream has been more independent of the problems. Elections and leadership have mattered more, political actors have influenced problem perception to a significant degree, and partisan conflict has become intense.

The third pattern occurs when events or actors coordinate the problem stream and the political stream by affecting them both. International events affected problem awareness in the United States, California, and New York State and at the same time motivated political actors to press for climate policy. The IPCC's reports and the international negotiations concerning the United Nations Framework Convention on Climate Change and the Kyoto Protocol led to peaks in public attention in the early 1990s and in 1997, as did the international backlash against Bush's policy reversal in 2001. The 2007 IPCC report and United Nations conferences in Bali and Copenhagen contributed to the 2007–09 peak in attention. So did policy feedback from the European Union's climate policies, especially the inauguration of emissions trading in 2005, which led to increased business acceptance of climate change and climate policy. These developments from outside the United States helped to motivate actors such as Pataki, Schwarzenegger, and Obama to invest political capital in the pursuit of climate policy.

Window Duration

Another area where the windows of opportunity theory requires some modification concerns the duration of policy windows. According to Kingdon's theory, policy windows are brief, and bursts of major policy change are separated from long periods of relative stability.[15] Climate policy making generally does follow this pattern, but a major caveat is needed. The periods of reform became relatively long and difficult to delineate during the last decade. Major reforms were adopted in California during three years in 2005–08, but they were followed soon afterwards by another major enhancement of the RPS in 2011 and a successful ballot initiative in 2012 that committed $550 million per year in new out-of-state corporate taxes to fund clean energy programs.[16*] Similarly, major reforms most recently occurred over at least a six-year period, from 2008 to 2014, in New York State. And at the federal level, the EPA has announced major policies regulating greenhouse gas emissions from motor vehicles (2009, 2011), power plants (2014), landfills (2014), and the oil and gas industry (2015), and the agency also has similar plans for aircraft. These regulatory measures have continued throughout the Obama administration, even though one might think that the window for federal climate policy making had closed when the Republicans captured the House of Representatives in 2010.

As climate policy has become more established and the problem remains high on public and elite agendas, a new pattern of policy making may be emerging. In this pattern, major policies are adopted or significantly enhanced every few years, rather than reforms occurring in short bursts followed by long periods of stability. Moreover, if important decisions concerning the implementation of previously adopted policies are considered, policy making appears even more incremental than the policy windows and critical junctures models suggest. Important examples include the landfill methane rules and Energy Star programs implemented throughout the 1990s, CARB's issuing of the Pavley rules in 2004, the Supreme Court's 2007 ruling on greenhouse gas regulation by the EPA, the California emissions trading system (legislated in 2008 and implemented in 2013 and 2015), and the EPA's rules on power plants and industrial sources (developed beginning in 2012 and continuing today).

Policy windows have become elongated in the climate policy area not only because policy implementation can take years, but also for other reasons. One quite general consideration is that political commitment may be sustained because of policy feedback or spillover from one policy decision to the next; that is, the successful adoption of a policy may demonstrate the political viability of new, similar measures. Also, advocacy coalitions may remain active and transfer their attention to the new measures.[17] This helps to explain why there is inertia in the attention of policy makers, which tends to remain high even after public attention has peaked and declined, as shown in the figures depicting public and political agendas in Chapters 6 through 9.[18]

In addition, the accumulation of statutory authority and agency capacity makes major policy reform through unilateral executive action more likely.[19*] Executive action depends mainly on political will and is less likely to require fortuitous increases in problem awareness than is legislation. Unilateral executive action has been important in federal climate policy making when the executive and legislative branches are deadlocked. This was the case when the Clinton administration issued methane regulations and expanded the Energy Star program in the 1990s and when the EPA during the Obama administration acted to regulate vehicle efficiency and power plant emissions. In New York State, too, the executive has acted on its own to initiate major reforms, most recently in expanding the state's RPS (2010) and tightening RGGI's cap (2013).

Finally, climate change issues have remained prominent on the public and political agendas ever since 2005 (see Figures 9.2 through 9.4 in Chapter 9). While this partly reflects the actions and statements of the Obama administration's strong political commitment to climate policy, it also results from increasing problem pressure from outside the policy system. Evidence of harmful anthropogenic climate change is growing, the concerns of the scientific community are accelerating,[20] and leading industrialized countries persistently have been trying to forge an international agreement.

ADVOCACY COALITIONS, POLICY ENTREPRENEURS, FRAMING, AND STRATEGIC DECISIONS

Many of the cases of major climate policy change involved mobilization by broad advocacy coalitions, with environmental and business organizations at their core. Depending on the period and the policy, the environmental groups varied from new, local, grassroots groups (against smog or power plants) to formal state- or national-level organizations such as the Sierra Club, EDF, and NRDC. The business side of these coalitions usually included only segments of business, with clean energy and high-technology industries most likely to be found in favor of climate polity – and the fossil fuel, automobile, and other manufacturing industries usually on the opposing side. Electricity utilities and general business associations were more ambivalent, and their switch from opposition to neutrality or support was a hallmark of successful cases, such as the passage of AB 32, the defeat of Proposition 23 in California, and the adoption of RGGI in New York. The importance of business interests to the advocacy coalitions helped to ensure that the climate policies adopted would minimize costs to broad business sectors and the general economy, and business also shaped specific policies, such as the adoption of emissions trading in RGGI and California. Advocacy coalitions often were broadened further by the inclusion of other interests, such as health, local government, and labor unions.

Elected officials were also crucial members of the advocacy coalitions. These were usually Democrats (Carter, Clinton, Obama, Brown, Spitzer, Paterson, Andrew Cuomo, and key Democrats in the legislatures of all three cases). But Republicans also played crucial roles in New York and California (Pataki, Schwarzenegger), as did George H. W. Bush to an extent at the federal level during the relatively consensual early 1990s. Schwarzenegger had to deal with a Democratic-controlled legislature, while Pataki appealed to voters in order to evade a lower chamber controlled by Democrats; both were running for re-election as Republicans in liberal states where the Democratic Party was dominant. Therefore, they used climate policy and other environmental policies as means to shore up support while pursuing other, less popular policies, such as those reducing government spending or restricting labor unions.

Elected officials often play the role of policy entrepreneurs, who bring together political commitment, a framing of the problem, and politically viable solutions. They have several advantages over other actors in doing so. They can supply a good portion of political commitment themselves, by using the authority of their executive or legislative offices and mobilizing their networks of political support. In addition, their access to the news media allows them to influence perceptions of problems and solutions. This role for policy entrepreneurs can be seen in public statements in which they interpret focusing events to mean that their preferred policy should be adopted. The case studies show that elected officials often play this role visibly, while their aides and executive agency appointees may do so in less visible settings.

Advocates' arguments about the co-benefits of climate policies often were important in gaining broad support for them. Reducing air pollution was crucial for a series of major reforms: the creation of CARB in California; the federal Clean Air Act, with its special role for California in developing more stringent standards; and New York State's participatory rights for environmental groups, which effectively limited new coal and nuclear plants and hence shaped later energy policy. Air pollution remained salient in California, where it was a major factor in the passage of the Pavley law. Energy security, in terms of reducing reliance on imports and ensuring adequate supply, was the primary motivation for the energy efficiency and renewable energy reforms in all three cases during the 1970s. It also was a major justification for the federal Energy Policy Act in 1992, Pataki's RGGI program, the California climate policies of the 2000s, and the federal cap-and-trade proposals in 2009. Economic development and job creation were salient arguments in New York in the 1970s, in New York and California in the 2000s, and in the 2009 federal stimulus package. They were used to justify climate policies that promoted the domestic renewable energy sector and energy efficiency businesses.

Finally, strategy and tactics also affected the chances of success. Schwarzenegger's decision to deal with the California state legislature helped to produce stronger climate policy than Pataki's strategy of unilateral executive action. Advocates' attacks on the oil and car industries were effective moves in California in the 1947, 1967, 2002, and 2010 conflicts there. The Democrats' last-minute relabeling of the Pavley law to a different bill number helped to convince a crucial handful of legislators to change their votes. Pataki's systematic construction of environmental-business advocacy coalitions began early and undergirded support for his climate policies.

Advocates at the federal level were often clumsier than those in the leading states, especially in comparison to the skill of their opponents. President Carter did a poor job of building public, congressional, and interest group support for his energy policies. President Clinton failed to frame the BTU tax as environmental policy and to mobilize its potential supporters, and he made concessions to industry too readily. He later became distracted by the Lewinsky sex scandal to the point that legislation to implement the Kyoto Protocol was not even submitted to Congress. President George W. Bush quickly chose to give in to the pressure from conservative Republicans and industrial lobbyists, rather than defend his earlier commitment to greenhouse gas regulation. The environmental organizations backing cap-and-trade legislation failed to create or connect with grassroots mobilization and were ill-prepared for the lobbying process and their opponents' counterattack.

PATH DEPENDENCE AND POSITIVE FEEDBACK

The case studies also show that there is partial path dependence in climate policy making in the United States. Processes, including political conflict over policy making, sometimes create new structural conditions that make later climate policy making more likely. This positive feedback may continue iteratively, progressively strengthening both climate policies and the structures that promote them.

Five kinds of new structural conditions were created by the processes analyzed in the case studies. First, new, specialized administrative agencies were created during early policy-making episodes, and later they were used to implement climate policies and sometimes influenced policy making. The main agencies for climate policy in California have been CARB (for the Pavley rules and AB 32 implementation) and the CEC (for energy efficiency and renewable energy); these were created in the 1960s and 1970s policy-making episodes described in Chapter 6. In New York State, the major agency for climate policy has been NYSERDA (for energy efficiency and renewable energy), which was created in 1975 as part of Carey's push for energy conservation.[21*] In the federal government, a key agency, the Energy Department (for appliance standards), was created in 1977, as part of the burst of energy policy reforms under Carter.[22*]

CARB and NYSERDA have significant scientific expertise and funding sources independent of their states' legislatures. CARB gained a reputation for successful reduction of air pollution, which gave it political credibility and the capacities needed to undertake comprehensive regulatory measures in order to implement AB 32. It gained the confidence of the public and state legislators, easing the legislature's decision to allow the agency to decide on important features, such as the Pavley fuel economy rules and the emissions trading system. The prior existence of competent administrative capacity is an important factor for legislators considering new policies.[23] In addition, the state agencies went beyond implementation to develop and press for new policies, such as the stronger efficiency and renewable energy policies that CARB and the CEC advocated in the early 2000s, or the measures that CARB recommended in its 2008 scoping plan. However, administrative capacity must be coupled to political will in order to be effective at addressing contentious problems. For example, the CPUC had 900 staff in the late 1970s, but it was still ineffective in countering the utilities' plans for new power plant construction, due to the agency's heavy, decades-long dependence on the utilities for information and cooperation.[24]

Furthermore, administrative capacity in air pollution, energy, and climate policy, once created or harnessed to alternative policies, became a structural force driving new policies, which, in turn often increased administrative capacity. For example, the implementation of AB 32 was estimated to require an additional 160 full-time CARB employees.[25] In this and other ways, administrative capacity is an element in a process of positive feedback rather than a structural factor that is independent of political processes.

A second structural condition was created when early legislation gave states and agencies legal authority, which they later used to expand and strengthen policies. At the federal level, beginning in the late 1980s, Congress gave the Energy Department the authority to develop new standards for currently regulated appliances as well as standards for new appliances. Congress reinforced the department's role in 2007, by directing it to regularly update standards. The same year, Congress gave the Transportation Department the authority to raise fuel economy standards for cars and light trucks in the CAFE program without further legislative approval, which the department did jointly with the EPA in 2009 and 2011. A crucial section of the 1970 Clean Air Act (section 111d) gives the EPA authority to regulate newly recognized air pollutants, which the agency used to develop rules for carbon dioxide emissions by power plants. At the state level, the Clean Air Act was crucial because it permitted California to develop stricter standards in transportation emissions and allowed other states to copy California's standards. This made possible the adoption and diffusion of the zero-emission vehicle program and the Pavley standards for greenhouse gas emissions.

Third, climate policies shifted the economic landscape in ways that strengthened advocacy coalitions, especially in the leading states. In California, the clean technology sector that supported AB 32 was partly a product of earlier energy efficiency, renewable energy, and transportation policies.[26] A similar process occurred in all three cases when renewable energy associations grew as a result of earlier climate policies and then pressed for new or expanded policies, as in the advocacy coalition for an RPS in New York. Positive feedback was also important in stabilizing and expanding the broad advocacy coalitions that favored emissions trading in RGGI and California, which included clean technology firms, environmental organizations, regulators, utilities, and their customers. In contrast, the advocacy coalition failed to reach the critical mass needed to pass a cap-and-trade program at the federal level.[27] Renewable energy and energy efficiency industries have grown nationally and are important lobbies for the preservation of the Production Tax Credit and the expansion of appliance standards.

Advocacy coalitions also expanded as businesses adjusted to the prospect or reality of greenhouse gas regulation. Sometimes they shifted from opposition to neutrality, as the California Chamber of Commerce did on Proposition 23. At other times, former opponents became supporters, though often grudgingly. Appliance manufacturers shifted to support federal appliance standards, some utilities did so on state-level energy efficiency policies and in all three cases on emissions trading, and carmakers did so on national greenhouse gas standards for motor vehicles.

A fourth new structural condition consists of procedural rights for environmental organizations to intervene in siting decisions concerning power plants and other large projects, which were created by earlier legislation. The environmental quality acts in California (1970) and New York (1972) gave opponents of nuclear and coal plants leverage to delay or block particular projects. In turn, this gave policy makers and utilities additional incentives to pursue energy efficiency, cogeneration, and renewable energy as alternatives to new power plants. More than three decades later, California's environmental quality law, as used by the Center for Biological Diversity and later by the state attorney general, became the basis for extending AB 32's greenhouse gas reduction targets to land-use planning decisions.

Fifth, earlier conflicts in the 1960s and early 1970s, and media coverage of them, strengthened environmental organizations and public opinion in favor of environmental protection. In California, mobilization against urban smog, redwood logging, oil development in the Santa Barbara channel, and nuclear power plants played this role. In New York, mobilization against urban smog, the Storm King power plant, and the Hudson River Expressway did so. Given the size and prominence of these states, their links to federal policy making, and the engagement of elite allies for these environmental causes, these early conflicts also helped to build environmental organizations and opinion at the national level at this time.

However, the California case study suggests that generally favorable environmental public opinion is not a stable feature of a state's political system, and that it does not necessarily translate into support for climate policy specifically. In California, strong support for environmental protection coexisted with weak support for climate policy in the early 2000s. Public support for the latter grew only after the Democratic legislature and Schwarzenegger began to press the issue and environmental organizations mobilized.

OPPORTUNITIES FOR FURTHER CLIMATE POLICY MAKING IN THE STATES

At this writing, the prospects for further climate policy development in California, New York, and other states with relatively strong policies are quite open. There is much potential to continue and strengthen the policies, but also some chance of stagnation. In California, there are many favorable conditions in place. AB 32 provides the authority and CARB the administrative capacity for pursuing the state's emissions target, which requires a 13% decrease in net emissions over 2004–20. California's accumulated stock of regulatory instruments, together with an emissions trading system that covers 85% of statewide emissions, shows promise in meeting the target. For over a decade, the state's legislature and the governor both have been heavily engaged in explicit climate policy making, and after intense public debates, they have adopted a series of major measures, including the Pavley law and AB 32. California politics is dominated by the Democratic Party, which currently holds all eight executive offices and large majorities in both legislative houses, and leading Democrats, especially Governor Brown, continue to support climate policy in the state.[28] Oil production and refining are major industries in California, and out-of-state fossil fuel and automotive interests are also likely to oppose an expansion of California climate policy, as they have in the past. But environmental organizations have been strong forces in the state's politics for decades and have succeeded in publicly vilifying those opposing interests. Positive feedback between climate policies and the clean technology sector has developed, and AB 32 survived a crucial test in the defeat of Proposition 23, which demonstrated the law's political viability.

However, California faces major challenges, too. AB 32 provides binding targets only through 2020, although Schwarzenegger's executive order had announced a target of an 80% reduction by 2050. A bill to direct CARB to begin planning for the 80% reduction failed in 2014.[29] Any effort to require decreases in emissions of that scale will encounter the need to accelerate reductions sharply, from 0.8% per year during 2004–20 to 5.2% per year during 2020–50 – even in the face of further increases in population, which is projected to grow 30% by 2050. Given the population growth challenge, a CARB-sponsored study concluded that even the aggressive adoption of "available" technology, defined to include carbon capture and storage, nuclear power, and the electrification of transportation and heating, could lead to an emissions reduction of only 60% below 1990 levels by 2050.[30]

The lack of a regional emissions trading agreement has been a weakness for California's climate policy. However, California, Oregon, Washington, and British Columbia signed the Pacific Coast Action Plan on Climate and Energy in 2013, pledging cooperation on linking carbon-pricing programs and low-carbon fuel standards.[31] California made a similar agreement with Mexico and began to integrate its emissions trading system with Quebec's in 2014. Although the Washington state legislature considered a comprehensive cap-and-trade system for its economy in 2015, it rejected the plan, leading Governor Inslee to pursue the direct regulation of greenhouse gas emissions in his state through executive action.

New York State also has a number of favorable conditions for continued climate policy making. RGGI has proved durable because it is popular with the state governments that are its members and with the electricity utilities; it has lost only one state, New Jersey, since its inception. After its initially rather unambitious cap was overwhelmed by the effects of fuel switching and economic recession, RGGI is now implementing a declining cap that should produce emissions reductions. Moreover, RGGI might gain new members, such as Pennsylvania, as a result of the EPA's emerging power plant rules, which as currently drafted will permit states to create or join existing emissions trading systems as one option for meeting the new federal standards.

In addition, New York State has reasonably ambitious greenhouse gas targets and probably can achieve them through existing state and federal climate policies, given slow population growth and fuel switching away from coal; this could help to build momentum for steeper reductions later. New York also has large, influential environmental organizations that probably could reactivate the broad advocacy coalitions on climate policy that they formed with parts of business in the past. Finally, New York enjoys the legacies of executive authority exercised by its governors and administrative agencies, with continued activity on energy efficiency, renewable energy, and power plant standards by NYSERDA and the Public Service Commission. The episode of climate policy making that began in 2008 has produced a steady flow of reforms up to the present.

However, New York also faces some major challenges in expanding its explicit climate policies. Its governors have preferred executive action, without involving the legislature very much so far. This limits the scope of policies that can be adopted; in particular, New York has no binding emissions target or economy-wide emissions trading system. It also means that the degree of public support for climate policy has not been tested in the way that it has in California. Moreover, the state government's commitment to its greenhouse gas targets may be flagging, after the target of a 10% reduction by 2020 was nearly achieved in 2011 with the help of the economic recession and low natural gas prices. The climate action plan still has not been formalized as of 2015, five years after it was drafted and seven years after CARB's adopted its scoping plan. In the near future, the political response to Hurricane Sandy may direct the state's energies toward adaptation, such as the hardening of mass transit and utility systems, rather than to climate change mitigation.

The leading states have the advantage of having policies in place and, for the most part, the political and structural conditions that helped to produce them. But what are the prospects for stronger climate policies in other states? The fifty states can be roughly grouped into relatively strong, middling, and weak climate policy states, as shown in Figure 2.1, in Chapter 2. In this book's statistical analysis, lower levels of climate policy in the weak and middling states were partially explained by structural factors. Although these factors change only slowly, recent trends toward greater administrative capacity and education generally bode well for climate policy making, as detailed in the conclusions to Chapter 5. The other significant structural variables – air pollution from ground-level ozone, citizen ideology, and fossil fuel production – show stable or mixed trends. In addition, the statistical analysis also suggests that there is scope for political processes to make a difference in these states, since government ideology and environmental organizations were significantly correlated with climate policies even when controlling for the structural factors. According to the regression analysis, if the average state were to double its environmental organization membership and elect a governor and legislature that are 16 percentage points more liberal, it would increase its number of climate policies from 14 to 17, out of a maximum of 25.[32*]

Other political processes are potentially at least as important, though not plumbed by the statistical analysis: broad coalitions between environmentalists and businesses, the emergence of exceptional executive or legislative leadership, or a climate-related focusing event such as severe storms or drought. In short, changes in their internal processes could produce significant changes in state policies, toward either stronger policies, or, as in New Jersey or Ohio recently, toward weaker policies. At the same time, states cannot move very far on their own, assuming there is no major change in their structural conditions. To bring the low-policy states up to the standards of California or New York would also require federal action, such as that which the EPA has undertaken recently on motor vehicles and power plants.

FEDERAL OPPORTUNITIES

However, federal climate policy still faces major obstacles, including slow national growth in jobs and wages, large and well-mobilized fossil fuel industries, and a combination of the separation of powers, divided government control, and intense party polarization on climate change. These factors make major climate legislation extremely difficult, while executive action is subject to potential veto by a Supreme Court that currently has a five-to-four conservative majority. Yet past experience shows that these obstacles sometimes can be overcome, and hence that the next decade of federal climate policy making is actually quite open. With fortunate timing, increased problem awareness and political commitment could come together in a number of ways to create policy windows for major reforms. Moreover, climate policy advocates do not need to wait for a lucky coincidence of the problem stream and the political stream, but rather can work to enhance both problem perception and political commitment.

Awareness of human-caused climate change has grown in recent years, as evidence has mounted and the scientific consensus on the severity of its impacts has continued to grow. Skepticism has declined in the U.S. media since 2004 and in public opinion since 2009–11.[33] For example, after rising from 17% in 2006 to 34% in 2011, the proportion who deny that the earth is warming at all dropped to 25% in 2014.[34] Public concern about the issue has also risen after hitting a trough in the aftermath of the recession. After declining from 41% in 2007 to 25% in 2011, the share who worry "a great deal" about global warming or climate change rose slightly to 32% in 2015.[35]

Public awareness of the climate change problem and a sense of urgency about it could rise further due to the increasing effects of natural disasters. Every decade since 1980 has seen more frequent billion-dollar climate-related disasters than the previous decade,[36] and this trend is likely to continue, if for no other reason than the increase in the value of assets that are in harm's way. When disasters occur, a variety of actors could use them to publicize and lend credence to the government and to expert studies that already are concluding that climate change is causing damage to the U.S. economy and public health.[37] Particularly likely to affect the public agenda would be hurricanes striking population centers and droughts or floods harming agriculture, because of the economic losses and in some cases the dramatic imagery and loss of life involved.

Even where adaptation spending and policies are major responses to these events, as in New York City, this can also raise awareness of climate change and its costs. Simply getting the issue onto the agenda may improve the prospects for mitigation policies even though mitigation can do nothing to reduce short- and medium-term adaptation costs. Andrew Cuomo's 2013 State of the State address, which promoted the strengthening of RGGI and renewable energy policies as responses to Hurricane Sandy, is a case in point. Furthermore, oil supply disruptions or rising energy prices, which might be driven by conflict in oil-producing regions or strong economic growth in developing countries, could once again increase attention on energy issues. They also would raise interest in climate policies that increase energy efficiency and diversify energy supply.

Furthermore, some political constraints on climate policy have weakened recently. So far, the Supreme Court has given a series of green lights to administrative regulation of greenhouse gas emissions, which potentially can evade the need for new congressional action, although the Court also has set limits on the EPA's authority and the agency's greenhouse gas rule making still occurs in the shadow of pending litigation. Moreover, in early 2015, Republicans showed some small signs of distancing themselves from climate change denial, as 15 Republican senators voted for a symbolic resolution accepting the existence of human-caused climate change.[38] Nevertheless, given the large remaining partisan differences on the issue, if a Republican is elected president in 2016, federal climate policy would likely suffer setbacks, and perhaps a major blow. The president could weaken or delay the EPA's regulation of greenhouse gases and fuel economy rules, and a Republican Congress might even amend the Clean Air Act to end the EPA's authority over greenhouse gases.[39]

However, barring such developments, political commitment to address climate change through mitigation policies could rise further over the next decade. States with relatively strong climate policies, especially California, are demonstrating that such policies are economically, administratively, and politically feasible. The diffusion of climate policies from the leaders to many other states, including through emissions trading in RGGI, also underscores the political feasibility of these policies, and leading states often have pressed for federal action as well. In previous periods, these factors have been strongly associated with the diffusion of environmental and other policies from states to the federal government.[40]

In looking at the next ten years at the federal level, there are three overarching strategic options available to climate policy advocates. One is to pursue executive action under a Democratic or moderate Republican president, despite what probably will be a hostile Congress. A second is to try to help the Democratic Party to gain unified control of Congress and the presidency and then press for ambitious legislation. The third option is to press congressional Republicans to move back toward the center on climate policy and to accept compromise legislation, regardless of who occupies the White House.

At this writing, the first option may appear to be the most likely to succeed, but progress under it would be limited by congressional hostility and the limited ambit for further regulatory innovation provided by the Clean Air Act, energy efficiency legislation, and federal court rulings. The second option is relatively unlikely, for several reasons. Republicans have a long-term structural advantage in House of Representatives elections, due to the inefficient concentration of Democratic voters in large urban areas and Republican control of most state legislatures, which decide on congressional district boundaries.[41] Moreover, Senate filibuster rules make it very unlikely that one party will get effective control of that chamber, which requires 60 seats, and those rules have proved very resilient.

Therefore, the most promising option for advocates, especially in the long term, may be the third one, in which they would press congressional Republicans to return to center-right positions on climate change along with other issues in which major corporations have a stake. This could make bipartisan compromise possible again. Business interests already began to take control of the Republican Party back from Tea Party groups in the 2014 midterm elections. That process might be accelerated if the party nominates a relatively moderate candidate for president, who wins the 2016 general election, or if it nominates an extremist who loses it.

In any case, much hinges on the positions that domestic business and key international actors will take. Corporations have been learning to cope with emissions trading in the European Union, the RGGI states, and now California, among other places, and with greenhouse gas regulation by the U.S. federal government. These jurisdictions have a chance to show the economic feasibility of a lower-carbon economy, and their economic sectors that benefit from climate policy continue to grow. As it gets used to such policies and learns to adapt to them, business has become divided between those who oppose and those who seek to influence and adjust to climate policies, which many of them see as inevitable. Companies which perceive that such policies would give them advantages over competitors, or which expect that they might need to comply with different rules in different states, might shift to favor a federal emissions trading system or carbon tax that would preempt state programs. If passage of such a national program begins to seem like a real possibility, some businesses will shift toward support in an attempt to influence the provisions of the legislation, as they began to do in the mid-2000s.[42]

There are other reasons why large corporations might shift more decisively in favor of national carbon pricing. The recent trend toward disclosing greenhouse gas emissions and including climate change impacts in corporate planning already is enhancing business executives' awareness of climate change issues.[43] In addition, increasing property damage due to climate-related disasters probably will motivate the insurance and financial sectors, as well as businesses expecting to pay rising insurance premiums, to take more interest in climate change mitigation policies.

On the international stage, the December 2015 agreement in Paris, involving voluntary commitments by 195 countries, fell far short of what is needed to prevent two degrees of warming. But the agreement creates opportunities for the multilateral ratcheting up of targets at meetings to be held every five years, and hence for international accords and domestic climate policy to reinforce each other throughout the world. Nonetheless, as was the case before the Paris agreement, much depends on domestic policy development by the largest emitters. Three jurisdictions – the United States, the European Union, and China – account for 48% of global greenhouse gas emissions and 53% of carbon dioxide emissions.[44] In coming years, developments within each of these three jurisdictions could reinforce each other, as they recently have begun to. The EPA's announcement of power plant rules in 2014 sparked debate in China about its domestic climate policy, which was followed by its agreement with the United States, in which China committed to stop its emissions growth around the year 2030.[45] China might be spurred to more ambitious targets and further action by smog events, other environmental disasters, and environmental protests, as well as by feedback from its growing renewable energy sector.[46] If both the U.S. and China continue to tighten their climate policies, their actions could reinforce each other and spur further development of the European Union's policies. A trilateral deal among these actors could be a good basis for strengthening the current global agreement when it is reviewed in 2020, which in turn would help to promote domestic policies in the United States.

Building the necessarily broad advocacy coalitions and countering the economic cost arguments of opponents would not be easy in the United States. Here there is a strong preference for inexpensive energy and a resistance to increasing those costs, especially through policies that can be seen as new taxes.[47] Given stagnant household incomes and slow rates of job creation, the public has given increased priority to economic growth compared to environmental goals since 2009.[48] But public opinion is in many ways supportive of climate policy, with majorities saying that they support the EPA's recent power plant rules, even if they mean higher utility bills, as well as fuel efficiency standards for cars and greenhouse gas regulations on industry.[49] More generally, studies of the public's willingness to pay for climate policy, described in Chapter 4, suggest that politicians have some room to maneuver in calling for climate policies that require financial sacrifices – if the sacrifices are small ones, at least at first.

An adroit use of policy framing also could help to overcome the obstacle presented by cost arguments, especially if the Republican Party moves toward the center on this issue and make less use of extreme estimates of economic costs. Advocates could ease such a shift by framing climate policy in terms of net economic benefits, emphasizing co-benefits for air pollution, health, and job creation, designing policy proposals to shift costs to later years, and reducing the visibility of costs. They can also propose using any additional revenues generated by emissions trading or carbon taxes for popular, broad, public purposes that are linked to climate change mitigation, such as promoting renewable energy, energy efficiency, air quality, and job creation, and by targeting some funds to low-income households. RGGI and California both do this with the revenue from their auctions of emissions allowances.[50]

If some combination of these favorable developments unfold, the United States' structural advantages for climate policy making will become more apparent. The country has high and growing incomes and education levels, with a business sector that possesses large amounts of capital and has transformed itself through technological change relatively rapidly in the past. The U.S. also has a very large renewable energy potential, and the persistent problem of ground-level ozone will continue to be a spur to any climate policies that would reduce common air pollution through energy efficiency or fuel switching. Its pluralist interest group system includes strong environmental organizations, and it can be a conduit for new ideas and an arena for the formation of new coalitions. Environmental organizations and a diverse set of businesses already joined forces to work for a cap-and-trade system in 2009, and they could reactivate their alliance under the right political circumstances. And not least, federalism gives the states, and particularly California, important roles in innovating and testing climate policies.

The adoption of stronger climate policies would depend on a confluence of factors, perhaps coordinated by advocates or events, but probably also depending on some fortunate timing. But this does not mean that advocates need only to wait for policy windows to emerge. Even during the implementation phase of existing policies, environmental mobilization could make a difference, such as the millions of supportive comments on the EPA's draft power plant rules. More fundamentally, the opening and closing of policy windows is highly unpredictable, so advocates need to be ready to act, with policy proposals, strategies, and prospective coalition partners on hand. Some policy windows are very brief, like the one for congressional action that opened in the first part of 2009. At the same time, the window of opportunity for climate policy that began with Obama's inauguration did not close completely with the defeat of climate legislation in Congress; the battles have shifted to EPA rule making and to the courts.

Advocates can do many things to increase their chances of succeeding now, or when the next large policy window happens to open. They can increase their main power resources – organized members, money, and relationships with officials and with other groups. They can deepen their organizational networks, linking national organizations with grassroots groups through joint campaigns. They can form, or test the waters for, broad alliances with other environmental, business, health, labor, and religious organizations. They can hone their arguments about co-benefits and attack the opposition by targeting its weaknesses, such as the poor reputation of the oil and coal industries. They can support candidates who are favorable to climate policy and then try to work with whoever wins the election.

Notes

Chapter 1

An asterisk (*) after an endnote number in the text denotes that the endnote contains substantive material.

1. UNFCCC, "Status of the Doha Amendment," at unfccc.int, accessed December 21, 2015; I exclude countries with less than one million population.
2. Compston and Bailey 2008; Kelly 2008; Künkel, Jacob, and Busch 2006; Bernauer 2013, 433–36.
3. Talberg and Swoboda 2013; Szarka 2007, chapter 6; Walz 2007.
4. UNFCCC data, excluding land use and forestry changes. Of the leading countries, Germany's reductions were driven partly by the deindustrialization of the former East Germany and the United Kingdom's by its switch from coal to natural gas, but even in those cases, explicit climate policies were responsible for at least half of the reductions (Eichhammer 2001, 39; Kelly 2008, 160, 203; Karapin 2012).
5. A national climate policy index (Künkel, Jacob, and Busch 2006, 12) explains 44% of the variance in 1990–2011 emissions trends in industrialized democracies (UNFCCC data, excluding LULUCF; author's analysis).
6. Besides those mentioned in the text, other laggards, with rising emissions from 1990 to 2012, include New Zealand, Spain, Portugal, Japan, Ireland, Greece, and Austria (UNFCCC data).
7. Brewer and Pease 2008; Harrison and Sundstrom 2010, 280; Layzer 2012, 270–307; Hempel 2006; Compston and Bailey 2013, 24; EBRD 2011, 61.
8. However, at the Copenhagen conference in 2009, President Obama committed the United States to a 17% reduction over 2005–20, which is effectively a 3% reduction over 1990–2020. U.S. officials stated in 2012 that the target was not contingent on congressional legislation (Bianco et al. 2013, 2). Under Obama, the United States has asserted a leadership role in international negotiations (Bang and Schreurs 2011).
9. UNFCCC data, excluding land use and forestry changes. However, this increase can be attributed to a rapid (25%) rate of population growth. Per capita emissions actually fell by 12% over this period (EPA and Census data; where data sources are given, calculations are by the author).
10. Rabe 2009a, 37.

11. Byrne et al. 2007; Farber 2011.
12. Rabe 2004; Chatrchyan and Doughman 2008; Posner 2010; Selin and VanDeveer 2011; CCES 2011.
13. Rabe 2004; 2015.
14. Rabe 2008.
15. ICLEI 2009; Krause 2011, 47. However, local action plans and official commitments probably have had only modest effects on the adoption of policies to reduce emissions, since political will, funding, legal jurisdiction, and administrative capacities are often lacking (Krause 2011, 56–57; Tang et al. 2010, 49–50). Local programs are uneven, are often merely symbolic, are mostly limited to government operations, and include little monitoring of implementation (Krause 2015, 87, 91–94, nn. 12–13, 21). States, especially relatively large ones, have additional advantages over municipalities for the purposes of this study. They are large enough to have an impact on national emissions and national climate policy making, and they have greater economic diversity, which makes them more representative of the United States as a whole.
16. Galarraga, González-Eguino, and Markandya 2009, 8–12. In many countries, regional and local governments also play large roles in broader sustainable development programs, which often include climate policies (Bruyninckx, Happaerts, and van den Brande 2012; Happaerts, van den Brande, and Bruyninckx 2011).
17. Selin and VanDeveer 2005; 2009.
18. Selin and VanDeveer 2012; Betsill and Rabe 2009.
19. Carlson 2009; Schreurs and Tiberghien 2007.
20. Vogel 1995, 259–62.
21. For example, Busch, Jörgens, and Tews 2005; Knill, Shikano, and Tosun 2014.
22. Rabe 2009a, 45–46.
23. Harrison 2014.
24. Goulder, Jacobsen, and van Benthem 2012.
25. Rabe 2008, 124; Endrud 2008, 280–84; Rabe 2011, 504–15.
26. Selin and VanDeveer 2012.
27. Hanemann 2007, 3.
28. Hanemann and Busch 2011, 127.
29. Data from World Bank development indicators database.
30. World Bank data, at data.worldbank.org, accessed December 21, 2014.
31. WRI-CAIT data.
32. Clark 2011, 306; Rosen 2009, 180–83.
33. Raymond 2010, 105.
34. Carlson 2009, 1134.
35. Harrison and Sundstrom 2007; Paterson 1996; Schreurs 2002; Elliott 2013; Selin and VanDeveer 2012; Keller 2010.
36. Kingdon 2003.
37. Sabatier and Wieble 2007.
38. Cf. Hellwig 2014.
39. Tilly 1984, 81–84.

Chapter 2

1. E.g., Pelling 2011; Mazmanian, Jurewitz, and Nelson 2013.
2. E.g., CNRA 2009.
3. USEPA 2015a, ES-5–7, ES-13–16.

4. Cf. Krause 2015, 89, 91 & n. 20.
5. Nationally, the siting of coal-fired power plants is a borderline example of implicit climate policy, since the long-term effects of local environmental opposition and siting policies on the fuel mix in electricity generation often may have been weak (Montgomery 1993). Thus, the coal-fired share of national electricity generation fell only marginally, from 53% in 1965 to 52% in 1995. However, in New York State, these forces clearly limited coal burning (see Chapter 8); coal's share of electricity generation there fell from 46% to 17% in that period (USEIA data).
6. However, government promotion of nuclear energy was largely unsuccessful after the mid-1970s, due to local opposition, rising costs, lagging demand, and rising concerns about safety and waste disposal. Hence, it will be treated in the case study chapters much more briefly than renewable energy. Nuclear power as a share of national electricity generation reached about 20% in the early 1990s, after plants ordered in the 1970s were finally completed, and it has plateaued since then (USEIA data).
7. Cavanagh 2009, 81.
8. USEIA data; Wing and Eckaus 2007, 5271–72.
9. Cf. Smith 2013.
10. Most typologies of environmental policy focus on this dimension, contrasting command-and-control regulations with market incentives (e.g., Richards 2000, 284–85; Kelly 2008, 51–53).
11. For example,, in USDOS 2014, 119–26.
12. Cf. Rabe 2010a.
13. UNFCCC data.
14. Policies for reducing carbon dioxide emissions rely on technologies, some of which are proven effective while others have not been or are in early stages of development (Zehner 2012). Hence, I do not include programs to promote biofuels (which in the United States mainly means corn ethanol), hydrogen vehicles, or carbon capture and storage for coal-fired plants. I do include electric vehicle policies if the electricity mix has a relatively low carbon intensity, as in California (NRC 2010).
15. Data from Karapin 2012, 52.
16. Timney 2004, 17–19; Vietor 1984.
17. Data from WRI-CAIT, excluding land use and forestry changes. This rank is among countries with at least one million population and hence excludes Luxembourg; Australia is ranked first, at 23.9 tons/person, and Canada is close behind the United States at 20.1 tons/person.
18. New Zealand and Switzerland have national systems at this writing (cf. Talberg and Swoboda 2013, 11–14), following the repeal of Australia's system in 2014; South Korea also initiated a system in 2015.
19. CCES 2013a, 7–8.
20. Data from USEPA 2015a, ES-7.
21 White House 2015.
22. Counts of state policy adoption in this section are from the Center for Climate and Energy Solutions (www.c2es.org), as of June 2015.
23. Cromwell 2010.
24. Rabe 2010a, 596–99; Raymond 2010.
25. Data from Hanemann and Busch 2011, 136; CCES 2013b, 9–10.
26. Kazazis 2012, 1210.
27. Rabe 2013, 9.
28. ASAP 2014.

29. USDOS 2014, 135; Ungar et al. 2014, 3.
30. USDOS 2010, 45–46; USEPA 2011.
31. Timney 2004, 27; USDOE 2014a; 2014b.
32. May and Koski 2007, 51.
33. Timney 2004, 28.
34. Stein 1980, 22–23.
35. Matisoff and Edwards 2014, app. F.
36. CCES data; Carley 2011, 280.
37. Molina 2014, vi. The estimate is for 2009–12.
38. Nemet and Kammen 2007, 750.
39. ELP 2010.
40. Wiser, Bollinger, and Barbose 2007, 79.
41. IEA 2014, 5–6.
42. Data from USEIA 2015, Tables 1.6.B., 1.14.B., 1.18.B., 1.20.B.
43. Gipe 1995; Martinot, Wiser, and Hamrin 2005, 3–4.
44. Carley and Miller 2012, 740.
45. Stern and Wobus 2008, 16, 20; Bird, Kreycik, and Friedman 2008, 10.
46. Menz and Vachon 2006.
47. Nivola 1999, 111, n. 10.
48. American Petroleum Institute data; EFTE 2011, 29–30; Nivola 2010, 166.
49. Cassidy 2014, 5.
50. USEPA 2013, 2.
51. NHTSA 2011, 3–4; Graetz 2011, 57.
52. ACEA 2011, 20.
53. Under the impetus of higher fuel prices, it then rose 15% from 2005 to 2011 (NHTSA 2011, 4).
54. Nivola 2010, 161, 169; UBA 2007, 24.
55. Brown et al. 1995.
56. Perl and Dunn 2007, 13.
57. EDF 2007.
58. Previously, Congress had passed the federal Energy Independence and Security Act in 2007, which gradually raised the CAFE standard from 25 to 35 mpg for the combined new vehicle fleet during 2011–20 (Yacobucci 2008, 10; EISA 2007, Sec. 102).
59. USEPA 2012a, 4; CCES 2014; EU 2014.
60. USEPA 2010, 25328; USEPA 2012b, 62630 n. 10.
61. Nichols 2009a; Voorhees 2009; Motavalli 2012; Eilperin 2011a; 2011b.
62. Hanemann and Busch 2011, 132.
63. Felker 2009.
64. Davenport 2014a; 2014b.
65. Posner 2010; Rabe 2011.
66. Derthick 2010.
67. Byrne et al. 2007.
68. Aulisi et al. 2007; Carlson 2009; Rabe 2011.
69. Rabe 2011, 504–15. In addition, the legal doctrine of the dormant commerce clause prevents states from adopting regulations that discriminate against economic activity in other states (Domanski 2008, 51–52).
70. Cf. Probst and Szambelan 2009.

71. CCES 2015a, 1.
72. Eilperin 2007.
73. Nichols 2009b, 191.
74. USEPA 2010, 25328. Litigants raised the preemption argument, and they temporarily succeeded when the EPA justified its December 2007 rejection of California's waiver request by arguing that scheduled increases in CAFE standards would be sufficient (EDF 2007). But in the end, the carmakers agreed to drop their legal claims of preemption as part of their 2009 agreement with the Obama administration.
75. CCES 2015b.
76. ASAP 2014.
77. The map and the following discussion refer to an adjusted version of the February 2014 CCES climate policy index. I have adjusted the index by dropping two items not related to climate change mitigation, by giving carbon capture-and-storage policies an equal weight with the other individual energy sector policies, and by giving equal weights to four main areas: the electricity, buildings, and transportation sectors and the group of cross-sectoral measures. The resulting adjusted number of potential policies ranges from 0 to 25.
78. Rabe 2004, 20, 30.
79. Author's analysis of data from the CARB greenhouse gas inventory and CARB 2014b, 24 (including sinks and using AR4 GWP values); Hanemann and Busch 2011, 127.
80. CARB 2008, 17, 21–22.
81. CARB 2011a.
82. CARB data; AB 1532; SB 535; Lazo 2014.
83. NYSERDA 2002, pp. 1–42 to 1–44.
84. Cahill 2008, 446.
85. Aulisi et al. 2007, 28–35.
86. Raymond 2010, 103–5; U.S. EPA data.
87. The initial states were Connecticut, Delaware, Maine, Maryland, Massachusetts, New Hampshire, New Jersey, New York, Rhode Island, and Vermont.
88. Duane 2010, 733; data from NYSERDA 2014, S-2.
89. Raymond 2010, 108.
90. Raymond 2010, 109.
91. Huber 2013, 87 & n. 133.
92. RGGI 2013a; RGGI data.
93. RGGI 2013a.
94. Love 2010.
95. RGGI 2013b.
96. Hanemann 2008, 123; Nadel and Goldstein 1996, 164, 171.
97. Hanemann and Busch 2011, 142.
98. CCES 2015b.
99. Downs et al. 2013, 125.
100. CEC 2007, 27.
101. CEC 2013a, 34.
102. Hanemann and Busch 2011, 143.
103. USDOE 2010, 4, 6.
104. Farrell and Hanemann 2009, 94.

105. Brownstein 2009, 69; Sciortino et al. 2011, 19–20; CPUC n.d..
106. Mazmanian, Jurewitz, and Nelson 2008, 413.
107. Farrell and Hanemann 2009, 89.
108. The federal Energy Policy and Conservation Act (1975) and Energy Conservation and Production Act (1976) created the State Energy Conservation Program, which provided $170 million in grants through 1979, and the Weatherization Assistance Program, which provided $290 million for low-income residences (Stein 1980, 12–15).
109. Stein 1980, 22–23.
110. Decotis and Ferguson 2012, 881, 897 n. 19; Decotis 2009, 40.
111. Decotis 2009, 40.
112. Miller 2009, 3.
113. Decotis 2009, 43.
114. Salkin 2014, 735–36.
115. NYSCAC, p. 3–4; WRI-CAIT data for 1990 and 2008.
116. OTA 1992, 123; Decotis and Ferguson 2012, 882.
117. The American Society of Heating, Refrigerating and Air-Conditioning Engineers.
118. Data from the Online Code Environment and Advocacy Network, at energycodesocean.org, accessed November 14, 2013.
119. DEC 2004.
120. NYSERDA 2010, ES-1.
121. Stall 1977; Weil 2009, 127.
122. Righter 1996, 204–6.
123. Taylor 2008, 2847; Cox, Blumstein, and Gilbert 1991, 347; Righter 1996, 204–6.
124. Gipe 1995, 30.
125. Righter 1996, 204–6.
126. Cox, Blumstein, and Gilbert 1991, 356.
127. Swisher and Porter 2006, 187; Harris and Navarro 1999, 16.
128. Taylor 2008, 2849–50.
129. CEC 2013b.
130. DSIRE data; Hall and Taplin 2010, 70.
131. Hanemann and Busch 2011, 131; Farrell and Hanemann 2009, 97; CEC 2014a.
132. Hall and Taplin 2010, 70.
133. USEIA data.
134. USEIA 2013, 355.
135. Swisher and Porter 2006, 187–88; Sharma 2002, 158–59.
136. Lambright, Pralle, and Boscarino 2006, 26, 38.
137. AWEA and NYSERDA data.
138. Rubenstein 2012, 60, 88.
139. NYT 1997.
140. Salkin 2014, 756–59.
141. CARB 2009, 12. Although the point is somewhat controversial, the balance of evidence suggests that electric vehicles in California have life cycle emissions lower than conventional gasoline vehicles (Wilson 2013; Hawkins et al. 2012); for the critical view, see Lomborg 2013; Petersen 2011.
142. Perl and Dunn 2007, 12–13; O'Dell 2012.
143. Lutsey and Sperling 2008, 678; Maples 2013.

144. CARB 2011b data.
145. Maples 2013; Hakim 2004.
146. Yeh and Sperling 2010; Mazmanian, Jurewitz, and Nelson 2008, 406.
147. CEC 2014b.
148. Pettit 2012; 2013.
149. At the same time, many local governments in California have become active in climate policy. In 2011, 27% of cities and counties (where 50% of the state's residents live) were members of the U.S. Conference of Mayors Climate Protection Agreement or the Sierra Club's Cool Counties program. About three-quarters of local governments that responded to a 2010 survey were planning to or had already done emissions inventories and climate action plans, although the effectiveness of these activities is uncertain (Bedsworth and Hanak 2013, 665–67, 670).
150. Darakjian 2009, 387–90.
151. Baker 2003; Hakim 2007a.
152. NYSCAC 2010, D-1.
153. Lambright, Pralle, and Boscarino 2006, 33–37.
154. NYSDOT 2014.
155. Over the 1990–2020 period, California's projected population increase is 36.6%, while New York State's is 9.5% (data from the California Department of Finance and the Cornell Program on Applied Demographics). Hence, California's 2020 greenhouse gas target requires a 26% decrease in per capita emissions, while New York's target requires an 18% decrease in that metric.
156. Raymond 2014.
157. Hanemann and Busch 2011, 136–37; NYSERDA 2014, S-2.
158. ACEEE data.
159. Righter 1996, 204; Swisher and Porter 2006, 187–88; Gipe 1995, 32.
160. California's baseline is taken from the share sold by investor-owned utilities (CPUC data). New York's target is 30% by 2015, but this includes 1% from voluntary consumer choice programs and an anticipated 21% from large hydroelectric sources that were already in place in 2004, leaving 8% to come from new renewable sources (DSIRE 2014).
161. Carley and Miller 2012, 740.
162. NYSERDA 2015, 1.
163. CPUC data.
164. Melewski 2002, 176–78; Aulisi et al. 2007, 27–28; Raymond 2010, 105, 107.
165. Trinkle 1995, 176–78; Aulisi et al. 2007, 33.
166. Barringer 2012; Wines 2013.
167. The legislative failure was due in part to opposition from the oil industry and Republicans; it has led Inslee's administration to pursue executive action to directly regulate greenhouse gas emissions by industry, oil and natural gas producers, power generators, and landfill operators (Bernton 2015; Banse 2015).
168. Nadel and Goldstein 1996, 164–65; Aulisi et al. 2007, 29–30; Nadel 2002, 162.
169. Rosenfeld 2009, 71–72; Hirsh 1999, 149–52.
170. Randolph 2014, 128–35.
171. ZEV 2014.
172. Rabe 2009a, 40; Lutsey and Sperling 2008, 677.

173. Data are from CARB, NYSERDA, and U.S. EPA. I have used data from these government agencies rather than WRI-CAIT data because the latter misses some sources, relies on the EPA's generic assumptions about all states (which presumably are less accurate than state officials' methods), and is less up to date than agency data. The data used here exclude changes in sinks (forestry and other land use) and include electricity imports from out of state. I adjusted CARB data for 1991–99 to make them comparable to the 2000–12 data, which use more recent global warming potential values; the adjustment used the ratio of the values for the 2000–04 years, which are present in both time series. The 1990 baseline for California is CARB's recalculated value (from CARB 2014b, 24), adjusted to exclude sinks.
174. The Organisation for Cooperation and Development estimates that the 2005–09 decline in U.S. carbon dioxide emissions actually was greater when calculated on a consumption basis (16%) than on a production basis (10%), because imports dropped sharply during the recession. Data from the Global Carbon Project shows that consumption and production based emissions declined to the same extent in the United States from the 2007 peak, with both down 12% during 2007–12 (cf. Global Carbon Project 2014, 10; also Boitier 2012, 8; Davis and Caldeira 2010). Another source of inaccuracy in emissions inventories is the undercounting of fugitive methane emissions from natural gas production. However, it is unlikely that this problem has biased official emissions trends downward in recent years. Trend data on this source of greenhouse gas emissions is not available, but it probably did not increase greatly due to the rise of hydraulic fracturing. This is because the dominant sources of uncounted emissions are probably leaks in the production, distribution, and use of natural gas, rather than fracking (Brandt et al. 2014, 734, 735).
175. In particular, using 2012 as the endpoint would be somewhat misleading, since in that year the U.S. emissions decline matched California's for the first time, due to severe drought in California, which reduced hydroelectric power generation, and the unexpected closing of the San Onofre nuclear power plant (CARB 2014a).
176. Data sources: USDOS 2014, 119–24 (generally); Lowenberger 2012, 5–6 and USDOS 2014, 121 (energy efficiency standards, averaged); NRC 2002, 20, 111 and USDOS 2010, 69 (CAFE); Schneider, Dutzig, and Sargent 2013, 25 (wind power; author's calculations); USEPA 2015b, 1365 (power plant standards, average of rate-based and mass-based estimates). In this table, I attribute half the emissions reductions due to wind power to the federal Production Tax Credit and the other half to state policies. The CAFE 1978–2004 program estimate from the NRC study is 367 MT, but this does not include the effects of oil price increases after 1978 (Nivola 2010, 161–62). To roughly account for those, I have reduced the NRC estimate by one-half.
177. Estimates in the 2008 column are by the author from data in Sudarshan 2011 (natural gas efficiency), CEC 2009, 241 (electricity efficiency), and from CARB 2008 and USEIA (renewable energy). The 2020 column includes those estimates and additional policy effect estimates from CalEPA 2013, table 2 and CARB 2014b, based on items in CARB 2008, 17. Included in these tables and totals are only those policy measures that are projected to produce reductions of at least 1% of the 1990 baseline emissions by 2020 (62 MT), excluding land use changes and policies affecting them. Contrary to the EPA, I follow Cassidy 2014 and CARB 2012 in estimating the federal Renewable Fuel Standard to have produced at most

negligible net emissions reductions, because of fossil fuel inputs and land use changes involved in corn ethanol production.

178. As in the U.S. analysis, the cutoff for inclusion was that projected emissions reductions in 2020 must equal at least 1% of 1990 emissions (4.3 MT). Also, only policy measures that were in the process of implementation in 2013 were included (CalEPA 2013, table 2). The totals are conservative estimates for these reasons and because they do not include the effects of California appliance standards that have been adopted by the federal government.
179. Byrne et al. 2007, 4565–66.
180. Lutsey and Sperling 2008, 676.

Chapter 3

1. Cf. Hay 2009; March and Olsen 1984, 735–37.
2. Exceptions on climate policy include Harrison and Sundstrom 2007; Falkner 2008.
3. Kingdon 2003; Zahariadis 2007.
4. Sabatier 1988; Sabatier and Weible 2007.
5. For example, Collier and Collier 1991.
6. Künkel, Jacob, and Busch 2006, 18–19; Harrison and Sundstrom 2007, 9–10; Lantis 2006; Busby and Ochs 2004, 51; Desai 2002, 357–63.
7. Dolsak 2001, 424.
8. Scruggs 2003, 185–88; Jahn 2008, 9, 23.
9. Busby and Ochs 2004, 50–51; Skjaerseth, Bang, and Schreurs 2013, 73–75.
10. Harrison 2010, 68.
11. Kraft 2002, 45–48; Elliott 2013, 2–3.
12. Scruggs 2003, 167–68; Klyza and Sousa 2013, chapter 1; Engel 2015, 32.
13. Chatrchyan and Doughman 2008, 250.
14. Carlson 2009; Derthick 2010.
15. Schreurs 2002, 242–43; Künkel, Jacob, and Busch 2006.
16. The concept of neocorporatism was developed to explain economic policy making and performance (e.g., Lehmbruch and Schmitter 1982), and studies of environmental politics often have borrowed measures that are based on labor corporatism. A valuable exception is Scruggs (2003, 219–28), which assesses the degree of environmental corporatism in 14 countries.
17. Jahn 2008.
18. Scruggs 1999, 30–31; 2003, 13–14, chapter 5; Crepaz 1995, 391–92, 395–400; Neumayer 2003, 205; Jahn 1998, 125; Ricken 1997, 205–10, 238.
19. Layzer 2012.
20. Schreurs 2002, 242–44; Harrison and Sundstrom 2007, 9; Busby and Ochs 2004, 52–53; Scruggs 2003, 175–76.
21. Scruggs 2003, 27 n. 46.
22. Jahn 1998, 124; Neumayer 2003, 218.
23. Scruggs 1999, 27–28; Armingeon 2008, 10; Poloni-Staudinger 2008, 422–23.
24. Scruggs 1999, 18.
25. Electoral system characteristics were taken from Norris 2004, 87; green parties were identified largely from Neumayer 2003, 211; electoral results are from Wikipedia data, through 2013. Comparative studies of the causes of green party

electoral success are surprisingly lacking, but Bolleyer (2013, 88–91) finds that proportional electoral rules have promoted the long-term survival of green and other new minor parties in 17 democracies.

26. Norris 2004, 44.
27. That is, in countries with at least some degree of electoral proportionality, such as that afforded by runoff elections in France.
28. Jänicke 1992, 50; Künkel, Jacob, and Busch 2006, 17–18; Weale 1992; McBeath and Rosenberg 2006, 120–22.
29. Lester and Lombard 1990, 309; Thomson 2014; Ringquist 1993a; Berry and Berry 2007, 236; Haskell and Price 1973; May and Koski 2007, 55–56; Rabe 2009b, 71–72.
30. Rabe 2009b, 71–72; Rabe and Mundo 2007, 272–73.
31. Mueller 1998, 471, 474.
32. Paterson 1996, 81–82; Schreurs 2002, 256; Elliott 2013, 5; Kraft 2002, 50–51.
33. Selin and VanDeveer 2012, 345.
34. Ringquist 1993a; 1993b; Matisoff 2008, 538–40; Derthick 2010; Thomson 2014, 38–39; Fowler and Breen 2013, 86.
35. Paterson 1996, 83; Berry and Berry 2007, 234–37; Ringquist 1993b; cf. Scruggs 2003, 56–65.
36. Scruggs 2003, 60–61; Jahn 2008.
37. Zahran et al. 2007, 42–43; Berry and Berry 2007, 234; Jänicke 2008, 137; McBeath and Rosenberg 2006, 118–19.
38. Lester and Lombard 1990; Walker 1969, 885.
39. Inglehart 1977; Scruggs 2003, 105.
40. Falkner 2008, chapter 4.
41. Meckling 2011, 69–72.
42. Künkel, Jacob, and Busch 2006; Falkner 2008, 97–99; Chatrchyan and Doughman 2008, 253; Brewer 2015, 62–63.
43. Ringquist 1993a; 1993b; Byrne et al. 2007, 4566; EBRD 2011, 64; Meckling 2011, 66–69.
44. Paterson 1996, 81; Leggett 2000.
45. Dunlap and McCright 2011; Boykoff and Boykoff 2004.
46. Paterson 1996, 77–82; Dolsak 2001, 424; Griffin 1985, 215–16.
47. Paterson 1996, 79; Zahran et al. 2007, 47; Dolsak 2001, 426; Matisoff 2008, 538–40, 543; Sekar and Sohngen 2014, 9.
48. Busby and Ochs 2004, 42–43.
49. For example, Zahran et al. 2007, 47; Paterson 1996, 78; Künkel, Jacob, and Busch 2006, 22; Matisoff 2008, 538–40; May and Koski 2007, 54–56; Byrne et al. 2007; Kosloff, Trexler, and Nelson 2004.
50. Matisoff 2008, 538–40, 543.
51. Lowry 1992; Lester and Lombard 1990.
52. Matisoff 2008, 538–40, 543.
53. Paterson 1996, 85.
54. Rabe 2005, 22; Rabe 2009b, 70–71; Byrne et al. 2007, 4566.
55. Engel 2009, 440.
56. Harrison and Sundstrom 2007; Paterson 1996; Schreurs 2002; Elliott 2013; Keller 2010.

57. This work draws on the agenda-setting literature (e.g., Birkland 1997; Kingdon 2003).
58. Jänicke 2008, 92.
59. Rabe 2008, 107; Rabe and Mundo 2007, 280; Rosen 2009, 173.
60. Oshitani 2006, chapter 9; Harrison and Sundstrom 2007, 8.
61. Chatrchyan and Doughman 2008, 253; Rabe and Mundo 2007, 273–74; Rabe 2008, 107; Rosen 2008, 12–14.
62. Rabe and Mundo 2007, 273–74; Engel 2009, 441.
63. Provost 2010.
64. Thomson 2014, 33.
65. Harrison and Sundstrom 2007, 8; Lester and Lombard 1990, 309; Teske 2004, 176; Deitchman 2014.
66. Calvert 1989, 170.
67. Rabe 2006, 6; cf. Huang 2007, 5574.
68. Roman and Carson 2009, 45, 49; Brewer 2012.
69. Klyza and Sousa 2013.
70. Coley and Hess 2012, 582.
71. RT 2014; Foster 2014.
72. Gifford and Comeau 2011.
73. Layzer 2012, 9.
74. Vezirgiannidou 2013, 605–6.
75. Nisbet 2009, 18–19.
76. Koehn 2008; Urpelainen 2009, 85–86; Rabe 2006, 6–7; 2011, 501–4; Byrne et al. 2007, 4566; Matisoff 2008, 538–40, 543.
77. Rabe 2004, 50–51.
78. Skjaerseth, Bang, and Schreurs 2013.
79. Huberty and Zysman 2014, xiii.
80. Urpelainen 2009, 86.
81. Jänicke 2005, 134.
82. Vogel 1995, 260–61.
83. Randolph 2014, 134–38; Pew Charitable Trusts 2009.
84. Engel 2009, 443.
85. Selin and VanDeveer 2007, 12–15.
86. Urpelainen 2009.
87. Cf. Michaelowa 1998; Harrison and Sundstrom 2010, 265–68.
88. Künkel, Jacob, and Busch 2006; Tjernshaugen and Lee 2007.
89. Bomberg 2012, 412–20; Kelemen and Vogel 2010, 438–44.
90. Ringquist 1993a, 330; Lee 1983, 168–70.
91. Jacques, Dunlap, and Freeman 2008; Schreurs 2002; Layzer 2012, 282–86; Selin and VanDeveer 2012, 346; Brewer 2015, 79–80.
92. Falkner 2008, 98–100.
93. Brewer 2015, 55–61.
94. Mildenberger 2013, 3, 25–30; Dipeso 2005, 101–2; Layzer 2007, 116–20.
95. Harrison and Sundstrom 2010, 261–65; Selin and VanDeveer 2012, 345; EBRD 2011, 64.
96. Thomson 2014; Chatrchyan and Doughman 2008, 252.
97. Brewer and Pease 2008; Brulle, Carmichael, and Jenkins 2012; Chong 2015, 121–26.

98. Kraft 2002, 51; Harrison and Sundstrom 2007; Thomson 2014; Johnson, Brace, and Arceneaux 2005.
99. Chong 2015, 113–15, 127; Scruggs and Benegal 2012; Brulle, Carmichael, and Jenkins 2012.
100. Bartosiewicz and Miley 2013, 8; Layzer 2011.
101. Bartosiewicz and Miley 2013, 43.
102. The latter is also a weakness in most of the work that uses a combination of structures and processes to explain policies (e.g., Harrison and Sundstrom 2007; Compston and Bailey 2008; Falkner 2008).
103. Kingdon 2003; for applications to climate policy making, see Pralle 2009; Farley et al. 2007; Karapin 2014; to other environmental policy making, see Dunn 2006; Jänicke 2008; Layzer 2012.
104. Kingdon 2003, 15–20, 165–67, 203.
105. Kingdon 2003, 202.
106. For example, Collier and Collier 1991.
107. For example, Baumgartner and Jones 1993; Repetto 2006.
108. Kingdon 2003, 197–200; Zahariadis 2007, 74.
109. The climate policy area is one where a large number of policy solutions were developed in the 1990s and were tried out in a number of states (see Chapter 2). Many of them met basic requirements of technical feasibility, fit in with dominant values, were workable in budgetary terms, and were widely communicated to policy communities across the United States. Therefore, weaknesses in the policy stream are not likely to have constrained policy change at times when the problem and political streams converged, at least not in states like California and New York, which due to their size and geographical proximity to other leading states had good access to information about available policy proposals.
110. Zahariadis 2007, 70.
111. Kingdon 2003, 197–99; Birkland 1997, 3.
112. Kingdon 2003, 90–115, 197–98.
113. Birkland 1998, 54.
114. Kingdon 2003, 198–99, 204–5.
115. Kingdon 2003, 19.
116. Kingdon 2003, 198.
117. True, Jones, and Baumgartner 2007, 161–62.
118. Zahariadis 2007, 74; Kingdon 2003, 205.
119. Kingdon 2003, 204–5.
120. Although they sometimes are seen as competitors and have different emphases, the two theories have much in common and few real points of disagreement. Both hold that the usual inertia in policy making can be overcome when sharp changes in perceived problems and political alignments create openings for political actors to press successfully for major policy changes.
121. Advocacy coalition theory also emphasizes belief systems and long-term changes in ideas, which are not analyzed here.
122. Sabatier 1988; Sabatier and Weible 2007.
123. Jänicke 2005, 138; Jost and Jacob 2004; Watanabe 2009.
124. Falkner 2008, 97–99; Bartosiewicz and Miley 2013; Skocpol 2013; Brewer 2015, 56–60; Bryner 2008, 325.

125. Of course, the size, resources, and effectiveness of these coalitions are also affected by social structures and political institutions, as are the other political processes included in this chapter.
126. Sabatier and Weible 2007, 202.
127. On critical junctures theory, see Collier and Collier 1991; Cappocia and Kelemen 2007; Soifer 2012; on path dependence in environmental policy, see Vogel 1986; in climate policy, see Laird and Stefes 2009; Knox-Hayes 2012. Daniel Mazmanian suggested to me that path dependence could help explain the California case.
128. Collier and Collier 1991, 31–37. Similarly, the theory of punctuated equilibrium holds that policy making alternates between long periods of stasis, when it is confined to policy specialists, and occasional bursts of major policy change – when pressure builds, issues get the attention of political generalists, and the scope of conflict increases (Baumgartner and Jones 1993; True, Jones, and Baumgartner 2007).
129. In addition, some critical junctures directly affected only some states, such as the acid rain problem for Northeastern states in the 1980s and 1990s, the Atlantic hurricane seasons for Gulf Coast states in 2004–05, or Hurricanes Irene and Sandy for Northeastern states in 2011–12.
130. Huberty and Zysman 2014; Selin and VanDeveer 2007, 12–15.
131. However, this factor is actually only quasi-structural, since it is subject to a fairly large degree of change in the short or medium term.
132. Tilly 1984, 81–84.
133. These indicators are imperfect. News articles at least partly reflect the agenda of policy makers, while congressional activity and presidential speeches are also public actions and may not reflect the more privately held priorities of policy makers. However, these indicators have the advantage of being consistent measures that are available across a long time period that includes multiple policy-making episodes.
134. Cf. George and Bennett 2005, chapter 10.

Chapter 4

1. Layzer 2012; Brewer 2012; Skocpol 2013, 60–69.
2. Jacques, Dunlap, and Freeman 2008; McCright and Dunlap 2011, 156.
3. Skocpol 2013, 58.
4. Chong 2015, 121–26; McCright and Dunlap 2011, 176; Brulle, Carmichael, and Jenkins 2012; Swift 2014.
5. Engel 2015, 43–46; Rabe 2015, 60.
6. Schlozman, Verba, and Brady 2013, chapters 10–14.
7. Data from the Center for Responsive Politics, at www.opensecrets.org, accessed on September 18, 2014; fossil fuel industries include the categories coal mining, oil and gas, and natural gas pipelines.
8. Brewer 2015, 79–80.
9. Layzer 2007, 98–99.
10. Dunlap and McCright 2011, 148–51; Goldenberg 2013, 2.
11. Jacques, Riley, and Freeman 2008.
12. Boykoff and Boykoff 2004, 129; Boykoff 2007, 474.

13. Revkin 2003.
14. Bush 2001.
15. Brulle 2015, 149–51, 161–63.
16. McCright and Dunlap 2011, 156; Evans and Steven 2007, 9; Weart 2013, 29, 41–42; Layzer 2007, 95; Bryner 2008, 321–26.
17. Paterson 1996, 77–82; Griffin 1985, 215–16.
18. European Environment Agency data for 2009; USEIA data for 2010.
19. Vezirgiannidou 2013, 597–98; Bang 2010; Graetz 2011, chapter 10.
20. USEIA data; cf. Greene 2010, 1615–17. Here I define consumption simply as imports plus domestic production.
21. Vezirgiannidou 2013.
22. However, that surge was indirectly and partially due to the country's increased import dependence in the early 2000s, which increased medium-term interest in reducing oil consumption; reforms resulted partially from a combination of this interest and the political shift toward the Democrats in 2006 and 2008.
23. Kelemen and Vogel 2010.
24. For example, Layzer 2012.
25. Hence, the 1970 Clean Air Act and its 1977 and 1990 amendments, which date from an era of bipartisan consensus on the need to regulate environmental hazards, remain the basis for most of the climate policies that the Obama administration has been able to adopt.
26. Harrison and Sundstrom 2007, 6–7; Selin and VanDeveer 2012, 345; EBRD 2011, 64; Dolsak 2001, 423.
27. Data from www.worldvaluessurvey.org, accessed September 22, 2014; the relatively strong climate policy countries were identified from Künkel, Jacob, and Busch 2006.
28. Evans and Steven 2007, 18, 20.
29. With so much polling data available, I took steps to avoid making a biased selection. For the analyses in this section, I used www.pollingreport.com (accessed on September 22, 2014) and AEI 2013 to identify all the time series of national polls where the same question was asked over at least a five-year period and through at least 2010. All those trends are summarized here, in averages over the given periods unless otherwise specified; where data from specific polls or ranges of polls are cited in this section, they are from the above two sources.
30. Rabe 2015, 76.
31. Gallup data for 1997–2014. Similarly, CBS News during 2009–14 found that an average of 73% thought climate change is happening now or will be in the future. In addition, an average of 58% say that most scientists believe global warming is occurring, and an average of 74% say they understand the issue well (Gallup data for 1997–2013 and 1992–2013, respectively).
32. Gallup data for 2001–14.
33. Gallup data for 2003–14.
34. CNN/ORC data for 2007–13.
35. Pew data for 2006–13.
36. Pew data for 2006–11.
37. Gallup data for 1989–2013.
38. Pew data for 2007–13.
39. Chong 2015, 119.

40. ABC News data, 2010.
41. Borick 2010, 40.
42. ABC News, Pew, and Quinnipiac data.
43. Gallup data for 2001–12.
44. Pew data for 2005–12.
45. Data for 2008–09 from Borick 2010, 44–45.
46. Data for 2004–08 from Borick 2010, 52.
47. Borick and Rabe 2012, 5.
48. Borick 2010, 49–51, 53.
49. Johnson and Nemet 2010, 1.
50. Gallup data for 2011–13.
51. Gallup data for 2001–12.
52. Gallup data for 2013.
53. Borick and Rabe 2012, 3.
54. Gallup data for 2008–13.
55. AEI 2013, 74–75.
56. Data for 2008–09 from Borick 2010, 53; Pew data for 2009–10.
57. Borick and Rabe 2012, 2; Rabe 2010b, 152.
58. Gerston and Christensen 2012, 106, 103; Janiskee and Masugi 2008, 83–84.
59. Morehouse and Jewell 2003, 174–75.
60. Hamm and Moncrief 2008, 155.
61. Data from "List of California Ballot Propositions" and "List of New York Ballot Measures," accessed at ballotpedia.org on September 24, 2014.
62. Siaroff 1999, 198; Scruggs 2003, 219–28.
63. Gerston and Christensen 2012, 53–59; Janiskee and Masugi 2008, 52.
64. Kersh 2006, 96; Schneier and Murtaugh 2001, 248.
65. Thomas and Hrebenar 1999, 13.
66. Gerston and Christensen 2012, 56.
67. Rosen 2009, 170.
68. Gerston and Christensen 2012, 58.
69. Biber 2013, 404, 413–14.
70. Wikipedia data.
71. Peterson and Rose 2006, 628.
72. USEIA data.
73. On the other hand, California has very little coal production, and only 20% of its electricity generation is from coal; this helps to make its electricity generation more diversified than the rest of the United States (Rosen 2009, 157; Duane 2010, 730).
74. USEIA and U.S. Census data.
75. Cf. Paterson 1996, 81.
76. USEIA data.
77. Updated data made available by Richard Fording, accessed May 6, 2013 at rcfording.wordpress.com/state-ideology-data; cf. Berry et al. 1998.
78. Data from Mazur and Welch 1999, 390–91.
79. The question for the first point was whether "stricter environmental laws and regulations are worth the cost." Californians favored this view by 58% to 37%, a margin of 21 percentage points, in 1998; a 1996 national survey by the Pew

Research Center showed that the U.S. public favored it by 63% to 30%, a margin of 33 percentage points (Baldassare 1998, 15). In 2001, respondents opposed Bush's reversal on greenhouse gas regulation by a 56% to 34% margin in California and by a nearly identical 54% to 34% margin nationally (data from a *Los Angeles Times* April 2001 poll, via Polling the Nations).

80. Rosen 2009, 151–52, 179.
81. U.S. Census data.
82. Cf. May and Koski 2007; Rabe 2009b, 71–72.
83. Byrne et al. 2007, 4566; Chatrchyan and Doughman 2008, 250.
84. U.S. Census data.
85. Farrell and Hanemann 2009, 87.
86. National Association of Manufacturers data, accessed at nam.org on January 7, 2013.
87. U.S. Bureau of Economic Analysis data.
88. Thomson 2014, 9.
89. Data from CARB 2011c, 14; USEPA 2014, ES-22.
90. Data from U.S. Bureau of Economic Analysis, averaged over 1997–2000; I selected highly polluting industries by adapting the categories used in Ringquist 1993a, 337 n. 4.
91. Data from Hovey and Hovey 2001, 52; U.S. Bureau of Economic Analysis.
92. Vasi 2009, 325.
93. But cf. Matisoff 2008, 539.
94. Data from Lopez et al. 2012.
95. USEIA data.

Chapter 5

1. The CCES climate policy index used here shows only limited change over the brief period that it has been assembled; the correlation between the 2008 and 2014 weighted indices is 0.88. Moreover, the changes over time are often due to corrections to the index made by CCES staff, rather than to changes in policies.
2. Carley and Miller 2012, 740; Downs et al. 2013, 22–23.
3. The index, developed by this organization when it was called the Pew Center on Global Climate Change, has been published at irregular intervals since 2008; I used the latest available version at the time I performed the analysis, because it was probably the most accurate as well as the most up to date.
4. While the index includes the District of Columbia, I omitted that jurisdiction from this analysis because it lacks the degree of autonomous self-government that the states enjoy.
5. There is no other index or dataset on state climate policies as comprehensive as the CCES index. The second most comprehensive index, found in the annual energy efficiency scorecards published by the American Council for an Energy-Efficient Economy (ACEEE), lacks renewable energy and cross-sectoral climate policies. It also has other disadvantages for my purposes, including somewhat arbitrary weights given to different policies and, for many measures, volatile scores depending on short-term changes in government spending.

6. Specifically, the February 2014 CCES index included these items used to construct my dependent variable: five cross-sectoral policies (emissions targets, emissions reporting, climate action plans, participation in regional initiatives, and caps for electricity emissions); eight electricity sector policies (public benefits funds, renewable portfolio standards, net metering, mandatory green pricing programs, rate decoupling, renewable energy certificate tracking systems, energy efficiency standards, and carbon capture-and-storage policies); six transportation policies (biofuels mandates and incentives, low-carbon fuel standards, truck idling policies, plug-in electric vehicle policies, vehicle miles traveled policies, and zero-emission vehicles programs); and six buildings policies (residential and commercial codes, appliance efficiency standards, state building standards, and the on-bill and PACE methods for financing energy efficiency improvements and renewable energy installations).
7. Cf. Thomson 2014, 7–20; Rabe 2008, 110–15.
8. Sufficient data to calculate median policy adoption dates is available for seven of the CCES index items (Humes 2009, 261; Matisoff and Edwards 2014, App. F; EDF 2007; CCES data); for these policies, the median date of adoption for states that had adopted them ranged from 1999 to 2008, with an overall median value of 2006.
9. These are residential building codes, commercial building codes, and appliance efficiency standards.
10. In 2011, electricity generation accounted for 33% of U.S. emissions, transportation for 27%, and the residential and commercial sectors (largely due to energy consumption by buildings) for 34% (U.S. EPA data). These figures double count electricity consumption in the residential and commercial sectors; similarly, the electricity and buildings policies in the CCES index overlap. The figures do not include emissions by industry (28% of the total) or agriculture (9%), which largely have not been subject to state climate policies and are not treated in the CCES index.
11. After carefully considering the list of policies, I decided that with the exception of what were initially five carbon capture-and-storage items, the mitigation policies are of roughly equal importance in state-level climate policy, so reweighting particular policies would not have improved the index.
12. Skew is –0.03 and kurtosis is –0.77.
13. Cf. Tucker 1982.
14. Time would be needed for a variety of mechanisms to unfold, including awareness and understanding of the structural conditions by elected officials, interest groups, and the public.
15. Morehouse and Jewell 2003, 174–75.
16. This was the best available measure of how the concentration of authority varies across states, since other dimensions used in cross-national research do not vary much in this context. Each state has a separation of powers between executive, legislative, and judicial branches, and each is a unitary political system with significant autonomy within the U.S. federal system. Other indices of governors' powers were less suitable for tapping this structural factor, because they included items that measured short-term changes in power, such as partisan control of the legislature (Beyle and Ferguson 2008, 212–13; Beyle 1995, 29).
17. Thomas and Hrebenar 1999, 13.
18. While using an ordinal variable is not ideal for linear regression, it is unlikely to have substantive consequences (King 1986, 666).

19. Dowling and Lem 2009, 3.
20. Although this variable is the absolute number of candidates, it is not a proxy for state population, with which it correlates only weakly (R = 0.22).
21. Adjusting this variable for population size is arguably a better measure of the state administration's ability to intervene in society. However, when such a variable was tested in the structural and combined models, it was not significant and had a negative coefficient, apparently because largely rural, low-population states scored high on it.
22. National Conference of State Legislatures, "Full and Part-Time Legislatures," at www.ncsl.org, accessed January 7, 2013.
23. Updated data for citizen ideology and government ideology are made available by Richard Fording at rcfording.wordpress.com/state-ideology-data, accessed May 6, 2013.
24. Berry et al. 1998, 331, 334.
25. Cf. Berry et al. 2007, 114–15.
26. Since the proportion of people with advanced degrees is in both components of the index, it gives double weight to this segment of the population. That is appropriate because of the latter's role in both the theoretical mechanisms described here.
27. From USEIA and U.S. Bureau of Economic Analysis data.
28. The theoretical literature expects problem severity to drive climate policy. Reverse causation, in which climate policy would reduce air pollution, is not plausible in this dataset, since air pollution is measured in the 1990s and the climate policy index reflects policies adopted mainly since 2000. In any case, the coefficients for the air pollution variable in the models in this chapter are positive, and its significance levels would not have been altered if a two-tailed test had been used.
29. Cf. Matisoff and Edwards 2014.
30. For each year, the county-level non-attainment data were weighted by county population as a share of state population, and the population shares living in non-attainment areas for each of the six pollutants were then averaged.
31. The two factors could not be used simultaneously, because partisan control of state government positions (as measured by the Ranney index) is very highly correlated with government ideology (R = 0.88). In any case, this party control variable produced very similar results when substituted for government ideology in the process and combined models described in this chapter.
32. Berry et al. 2010.
33. This is the "NOMINATE" version, which Berry et al. (2010, 121) conclude is marginally more valid than their original version based on unadjusted interest group ratings. Updated data provided by Richard Fording at the source given in note 23.
34. Berry et al. 1998, 333; 2010, 120.
35. Cf. Lowry 1998.
36. I assessed organizational engagement in the climate change issue by reading archived versions of organization websites starting in 1996 (using the Wayback Machine at archive.org/web). I defined having a "major focus" as conducting campaigns on, or devoting a large portion of the website to, the issue. Three of these seven organizations, comprising 56% of the total membership, are also among those found to have been most involved in congressional testimony on

climate change during 1980–2010 and among those most central to the networks of environmental organizations active on the issue (Brulle 2015, 160, 163).

37. It also can be considered an intervening variable, since it may be the result of early climate policies as well as a driver of later policies. Therefore, if it were a significant variable in the multiple regressions, a possible interpretation would be that at least part of the association is due to reverse causation.
38. Pew Charitable Trusts 2009, 52.
39. The inclusion of both is not redundant; their Pearson's R correlation is only –0.51.
40. From U.S. Bureau of Labor Statistics data.
41. From U.S. Bureau of Economic Analysis data.
42. These were fossil fuel production, air pollution, and environmental organization membership.
43. Scatterplots showed mild, linear heteroskedasticity in the combined model, but the Breusch-Pagan and Koenker tests were negative (p = .12 and .19, respectively).
44. Lewis-Beck 1980, 33–34; Schroeder, Sjoquist, and Stephan 1986, 44–48. For those who nevertheless prefer two-tailed tests, it is worth noting that all the significant variables in the models described in this chapter's process, structural, and combined models would be significant at the .05 level if two-tailed tests were used, except for education in the combined model (p = .062, two-tailed).
45. Out of 120 bivariate relationships among the independent variables, only 4 had a Pearson's R value above 0.60.
46. Lewis-Beck 1980, 58–63.
47. The coefficient for heavy industry, surprisingly, was positive, though not significant (p = .25) in a two-tailed test, which would have been appropriate to use, given the unexpected sign. A possible explanation for this result is that heavy industry share contributes to environmental degradation, which leads to stronger mobilization for environmental policy. Although this model already controls for the effects of criteria air pollution, heavy industry produces other environmental problems, such as toxic air and water pollution and degradation of wildlife habitats, that are not captured in the air pollution measure and that could promote environmental mobilization. This interpretation is supported by the fact that the coefficient for the heavy industry variable was reduced by three-quarters and became far from significant (p = .69, two-tailed) in the combined model, which includes environmental organization membership.
48. For air pollution, fossil fuel production, and environmental organization membership, the coefficients refer to the log-transformed variables used in the regression equations; hence, their impacts on climate policies depend on percentage changes in the original, untransformed variables.
49. More accurately, since this is a cross-sectional analysis, the following numbers refer to the difference in the independent variable that causally explains a one-policy difference in the climate policy index across states.
50. The latter could be due either to 5.3 percentage points more with bachelor's degrees, or 2.6 percentage points more with advanced degrees.
51. Typical states were selected from those ranked between 24th and 27th out of 50 states; the high-value state in each row is the state ranked 5th for the variable.
52. For government ideology, these were citizen ideology, education, and fossil fuel production; for environmental organization membership, I used citizen ideology,

education, air pollution, and heavy industry, the latter as a proxy for environmental problems not captured by the air pollution variable (see note 47).

53. The cutoffs used were 4/n for Cook's Distance, +/– 2.0 for the residuals, and $2^*(k/n)^{(1/2)}$ for dfits (Belsey, Kuh, and Welsch 1980; Fox 1991, 33).
54. Downs et al. 2013, x.
55. Hayes and Cai 2007, 712.
56. Alternative indicators of economic composition (the GDP share of manufacturing industries or of highly polluting industries) were also insignificant when substituted for the heavy industry variable in the combined model, as was a measure of renewable energy potential based on National Renewable Energy Laboratory data (log-transformed).
57. U.S. Census data.
58. EPA data.
59. While this index recently has been sensitive to national trends, it still shows much stability in rank ordering the states; the correlation between the 2000–04 and 2005–09 scores was 0.92.
60. USEIA data for 2000–13.
61. Citizen ideology and education are somewhat more difficult to investigate there, since their possible mechanisms are diffuse. They were treated in Chapter 4, where it was shown that California and New York are relatively high but not exceptionally high on these variables. Chapters 7 through 9 will include them mainly by proxy, through the analysis of the election of liberal government officials (which is conditioned by a generally liberal citizen ideology) and the roles of environmental organizations (which benefit from the relatively educated populations of those states).
62. Harrison and Sundstrom 2007; Bang 2003; Lantis 2006; Dolsak 2001; Schreurs 2002.
63. Lester and Lombard 1990; May and Koski 2007; Byrne et al. 2007.
64. Paterson 1996; Matisoff 2008; Schreurs 2002.
65. Dolsak 2001, 423; Paterson 1996; but cf. Matisoff 2008.
66. For an exception, see Zahran et al. 2007.
67. Florida 2002; cf. Portney 2013, 310–11.
68. Matisoff 2008.
69. Paterson 1996; Dolsak 2001.
70. Oshitani 2006; Harrison and Sundstrom 2007; Chatrchyan and Doughman 2008; Thomson 2014; Tjernshaugen and Lee 2007; Bang 2003; Schreurs 2002; Knox-Hayes 2012.
71. Mitchell, Mertig, and Dunlap 1991, 221; Center for Responsive Politics data.
72. From the Extreme Weather Sourcebook, available at the Center for Science and Technology Policy Research, University of Colorado (www.sip.ucar.edu/source book/, accessed February 23, 2015).
73. Education and income had a bivariate correlation of 0.80. With both variables in the structural model, the coefficient of income became negative; the instability of the coefficient is a sign of multicollinearity. The two variables also had VIF values between 3.7 and 4.0 when entered together. Separately, I regressed income and education on all the other structural variables in the model (cf. Lewis-Beck 1980, 60–63); in this procedure, 76% of the variance in income and 73% of the variance in education was explained.

74. The index, when substituted for education, was significant in the structural model, but with a smaller standardized regression coefficient than education. In the combined model, the income-education index was not significant (p = .16).
75. Their bivariate correlation was 0.72. Legislative professionalism was positively associated with climate policy in bivariate correlations and in the structural model excluding administrative capacity, but failed a one-tailed significance test in that model. With both variables in the structural model, the coefficient on legislative professionalism became negative; the instability of the coefficient is a sign of multicollinearity. Legislative professionalism also had a VIF value of 3.2 when both variables were included. When these two variables were regressed individually on all the other structural variables, 68% of the variance in legislative professionalism and 57% of the variance in administrative capacity was explained.
76. This governmental capacity index, like the administrative capacity variable, was significant only in the combined model; in both models, the index had smaller standardized coefficients and weaker significance levels than administrative capacity.

Chapter 6

1. Budget data from California ebudget (www.ebudget.ca.gov) and the CEC.
2. Mazmanian, Jurewitz, and Nelson 2008, 412.
3. Rabe 2009a, 50.
4. Roe 1984.
5. Farrell and Hanemann 2009, 91–92; NRC 2006, 70–73.
6. It granted 110 waivers through February 2014; EPA data accessed at www.epa.gov/otaq/cafr.htm on March 19, 2014.
7. Farrell and Hanemann 2009, 91–92.
8. To obtain the data for this figure, I counted articles through keyword searches in the Proquest archive of the *Los Angeles Times*, using the historical article images through 1984 and the "text" database beginning in 1985. Keywords were "smog" and "air pollution"; "articles" were limited to articles, editorial articles, editorial cartoons, front page, letters, reviews, photo standalones, and tables of contents that mention the search terms. The two Proquest databases had substantially different coverage for the years in which they overlapped (1985–90). Hence, in order to produce a time series that was comparable over time, I adjusted the counts in the historical database downward by the ratio of the two databases' coverage amounts during the period of overlap. It is unlikely that alternative measures of the public agenda would produce substantially different results; for a discussion of this issue, see notes 3–4 in Chapter 9.
9. For data sources and methods, see the previous footnote. Keywords for energy problems were "energy crisis," "oil crisis," "gas crisis," "energy shortage," "oil shortage," "gas shortage," "rising energy price," "rising oil price," "rising gas price," and "electricity crisis."
10. Unfortunately, data on the political agenda is not readily available for California before 1995.
11. This may have been due to a mismatch between the problem and the available, widely accepted policy solutions. In the mid-1950s, automobile exhaust was first

identified as the main cause of smog, but this was still disputed by the car industry until about 1958. Up to the mid-1950s, citizen protests and regulatory measures had focused on industrial sources, and there was not yet a regulatory apparatus for regulating automobile exhaust.

12. Dewey 2000, 39.
13. Ground-level ozone and its precursors, including NO_x, declined in the Los Angeles area after 1970 (Pollack 2013).
14. Krier and Ursin 1977, 52–53, 71; Dewey 2000, 39–42, 57, 105; Jacobs and Kelly 2008, 236; South Coast Air Quality Management District data.
15. Jacobs and Kelly 2008, 35–36, 49–50, 53, 63, 65.
16. On sudden grievances as a cause of protest activity, see Walsh 1981.
17. Dewey 2000, 84; Ainsworth 1947.
18. Jacobs and Kelly 2008, 54; Krier and Ursin 1977, 56, 71, 75–76.
19. Elkind 2011, 52–54, 59; Fawcett 1990, 92.
20. Gonzalez 2002, 223; Krier and Ursin 1977, 57–59.
21. Gonzalez 2002, 224–25; Dewey 2000, 43.
22. Dewey 2000, 53–54; Krier and Ursin 1977, 62–70; Jacobs and Kelly 2008, 117.
23. Fawcett 1990, 83; Dewey 2000, 60, 90–92.
24. Dewey 2000, 93–95; Sellers 2012, 223–25.
25. Jacobs and Kelly 2008, 193–97, 216, 220; Dewey 2000, 98; Gonzalez 2002, 229–30.
26. Gonzalez 2002, 227; Dewey 2000, 61, 63.
27. Krier and Ursin 1977, 89, 137–38; Gonzalez 2002, 224–27, 232; Fawcett 1990, 83.
28. Krier and Ursin 1977, 138–41; Dewey 2000, 65.
29. South Coast Air Quality Management District data; Fawcett 1990, 192–93.
30. Dewey 2000, 73; Fawcett 1990, 89.
31. Dewey 2000, 68–71, 73, 107; Krier and Ursin 1977, 138–41, 166.
32. Dunlap 1991, 289; Dewey 2000; Jones 1975; Gonzales 2005, 9.
33. Gottlieb 1993, 126; Penna and Geels 2011, 10; Jones 1975, 142, 153.
34. Krier and Ursin 1977, 175, 178.
35. Krier and Ursin 1977, 178–79, 184–89.
36. Krier and Ursin 1977, 178–79, 184–89, 191, 210; Dewey 2000, 77–78.
37. Denning 2011, 106–7.
38. Krier and Ursin 1977, 178; Jacobs and Kelly 2008, 238.
39. Dewey 2000, 106.
40. Krier and Ursin 1977, 181–83.
41. Jacobs and Kelly 2008, 176–78; Dewey 2000, 76–77.
42. Gottlieb 1993, chapters. 3–4.
43. Rudolph and Ridley 1986, 123–24; Schrepfer 1983, 165–66.
44. Zierold 1984, 19; Birkland 1997, 76–77.
45. Mitchell, Mertig, and Dunlap 1991, 221.
46. Getze 1969; Bergholz 1969; LAT 1969; Elwell 1969; Moore 1970; Keating 1970; Gonzalez 2005, 82.
47. Schrepfer 1983, 168, 172–75; Berry 1984, 61–64.
48. Zierold 1984, 51.
49. Zierold 1984, 53–54.
50. Roe 1984, 23–26, 28.

51. San Francisco Chronicle 1973.
52. Warren 1982, 19.
53. Pryor 1973.
54. Black and Flarend 2010, 94.
55. van Est 1999, 32; Pryor 1973; Warren 1982, 21–22, 25.
56. Rudolph and Ridley 1986, 123–25.
57. Schrepfer 1983, 179–84.
58. Zierold 1984, 96.
59. Warren 1982, 19–21.
60. Zierold 1984, 97; San Francisco Chronicle 1973.
61. McCarthy 1996, 80; Zierold 1984, 96.
62. Zierold 1984, 96.
63. Warren 1982, 21.
64. Alquist and Warren 1974a.
65. Warren 1973.
66. Alquist and Warren 1974b.
67. Warren 1982, 25.
68. Warren 1982, 25.
69. Warren 1982, 22, 25–26.
70. Gillam 1975; 1976.
71. Fairbanks 1976; LAT 1976a; Lampe 1978.
72. Rosenfeld 2009, 69–70.
73. LAT 1976b; 1977.
74. LAT 1978; CEC 1978.
75. Rosenfeld 2009, 64.
76. Lifset 2014b.
77. Roe 1984.
78. Verleger 1979, 463, 467.
79. Simon 2007, 70.
80. Roe 1984, 127–28; Redburn and Blakeslee 1979.
81. Roe 1984, 127–28.
82. Wellock 1998, 178–79.
83. Fairbanks 1974; Reich 1974a; Boyarsky 1974.
84. LAT 1974; Goff 1974.
85. Reich 1974b; Endicott 1974.
86. Zierold 1984, 147; Bollens and Williams 1978, 165–68.
87. van Est 1999, 35, 43; Righter 1994, 147.
88. Zierold 1984, 102.
89. Joppke 1993, 67.
90. Roe 1984, 104–7.
91. van Est 1999, 34; Wellock 1998, 182, 186–89; Joppke 1993, 68; Hill 1978a; 1978b; 1978c.
92. The share of petroleum in California's electricity generation fuel mix dropped 35 percentage points, while natural gas rose 19, nuclear power 17, geothermal 7, biomass 5, and wind power 2 percentage points; coal consumption rose only 1 percentage point in this period (USEIA data).
93. Rosenfeld 2009, 45–46, 69; Hanemann 2007, 6–7; Fairbanks 1976b.

94. van Est 1999, 34.
95. Hume 1977; Stall 1977; Skelton 1978a; Weil 2009, 127.
96. Skelton 1978b; Harris 1978.
97. Righter 1994, 148; Harris and Navarro 1999, 14.
98. Rood 1980.
99. Gipe 1995, 32.
100. Righter 1994, 147.
101. Roe 1984, 85–86, 112–15, 118, 127.
102. Righter 1996, 204.
103. Roe 1984, 120–22; van Est 1999, 48.
104. Roe 1984, 159; Harris and Navarro 1999, 12–13.
105. Gipe 1995, 32.
106. Hirsh 1999, 181–82.
107. Roe 1984, 195–96.
108. Roe 1984, 23–26, 28, 215.
109. Wilson 1985, 147; cf. Rosenfeld 2009, 71.
110. Farrell and Hanemann 2009, 92.
111. Brownstein 2009, 69; Cavanagh 2009, 89 n. 14.
112. Rosenfeld 2009, 70.
113. Harris and Navarro 1999, 13–14.
114. Righter 1996, 203–6; Cox, Blumstein, and Gilbert 1991, 356.
115. USEIA data.
116. Sovacool 2008, 150–54.
117. van Est 1999, 55.
118. Swisher and Porter 2006, 187; Harris and Navarro 1999, 16; Weil 2009, 137.
119. After this period, California's wind power capacity grew only 2.6% per year, from 1,121 Megawatts in 1985 to 1,616 Megawatts in 1999; however, at that time it still made up two-thirds of the U.S. total (National Renewable Energy Laboratory data).
120. Sovacool 2008, 154–55; 2009, 4502.
121. Wellock 1998, 178–79; McFadden 2013, 65; Reeves 1975.
122. Dewey 2000, 97.
123. U.S. Census 1970, p. 112.
124. Schrepfer 1983, 164, 172.
125. Berry et al.'s measure of state government ideology averages the legislative and executive branches. The liberalness of the California state legislature is evident in the large swings in the overall measure when Republicans were elected to the governorship. The state legislature evidently was about 15 to 20 percentage points more liberal than the U.S. average for state officials during the 1960s and 1970s (author's analysis).

Chapter 7

1. Search results are from the National Newspapers Core database; keywords were "global warming" and "climate change." Results are the total number of items each year, not filtered by document type; however, 93% of the items for 1985–2013 were classified as news, commentary, articles, features, or editorials.

2. The *Los Angeles Times* published over 1500 articles on energy problems, and only 350 on climate change, in 2001; the former was slightly higher than the peak during the oil embargo in 1974. Attention on energy issues remained relatively high through 2005, with more than 200 articles published that year, compared to only 65 per year during 1982–99.
3. The CEC issued several reports and held a symposium in 1989–91 (CEC 2005b).
4. Speeches for this analysis were taken from the California Assembly's *Daily Journal*. The length of attention to energy or climate problems and policies was measured by counting words in full sentences on those topics, divided by the total words in the speech. The climate change category includes implicit climate policies such as renewable energy or energy efficiency policies during the post-1997 period, with such mentions also classified as energy policy.
5. I searched executive orders at the archive at www.gov.ca.gov, which goes back to 1999, on March 4, 2014. The executive orders after 2004 concerned greenhouse gas emissions targets (2005), biofuel targets and the role of the California Environmental Protection Agency in climate policy (both in 2006), the low-carbon vehicle fuel standard (2007), programs to reduce vehicle miles traveled and a revision of the renewable portfolio standard (both in 2008), RPS targets (2009), directions to the Strategic Growth Council (2010), and the zero-emission vehicle program and energy consumption requirements for state buildings (both in 2012).
6. Bogoshian and Alex 2009, 346; Hanemann 2007, 13.
7. Sweeney 2002, 152–53, 158, 170–74.
8. Rosen 2009, 166.
9. Only 5% and 11%, respectively, blamed the latter (*Los Angeles Times* data via Polling the Nations).
10. Vogel 2002.
11. LAT 2002a.
12. Mazmanian, Jurewitz, and Nelson 2008, 408.
13. Hanemann 2007, 9. It stopped accepting new registrations in 2010 and encouraged its members to join the Climate Registry, which has a North American scope.
14. Farrell and Hanemann 2009, 93; Polakovic 2001.
15. Foster 2001.
16. King 2001.
17. Warren 2001.
18. Barabak 2001.
19. Paine 2005, 21.
20. Paine 2005, 24; Thomson 2014, xxiii.
21. Lucas 2002.
22. LAT 2001; San Francisco Chronicle 2002; Zuckerman 2002.
23. Franco et al. 2008, S14; Hanemann 2007, 14.
24. Field et al. 1999, 2.
25. Franco et al. 2008, S9.
26. Bustillo 2002; Bustillo and Ingram 2002; LAT 2002c; San Francisco Chronicle 2002.
27. Polakovic 2002.
28. Bradley 2001.
29. Pralle 2006, 400.

30. Paine 2005, 22–24.
31. Paine 2005, 21; Brownstein 2009, 70.
32. Carlson 2013, 79 & n. 62.
33. In 2014, this was 12 cents per ton of CO_2 emissions, intended to provide about $50 million that year (CARB 2014c).
34. Carlson 2013, 64–66, 80–81, 83.
35. Perl and Dunn 2007, 14.
36. Paine 2005, 24–25.
37. Senate Rules Committee 2002.
38. Paine 2005, 24.
39. Paine 2005, 31–32.
40. Vasi 2011, 103; Assembly Committee on Utilities 2002.
41. Senate Rules Committee 2002.
42. Bustillo and Vogel 2002.
43. Bustillo 2002; Paine 2005, 29, 32.
44. Paine 2005, 28, 32–34.
45. Schulte 2006, 48–49.
46. Paine 2005, 35–36; Assembly Bill 1493 (July 22, 2002), chapter 200, section 43018.5d.
47. Rosen 2009, 170, 184.
48. LAT 2002a; Simon 2002. In 2006, the 2010 target year was restored, and in 2011 the municipal utilities were included while the target was again increased, to 33% by 2020.
49. Farrell and Hanemann 2009, 94.
50. Rosen 2009, 165–66.
51. Baldassare 2000, 7; 2001, 1; 2002, 2.
52. Baldassare 2002, 15–16.
53. Rosen 2009, 169.
54. Polakovic and Bustillo 2002.
55. Ingram 2002b; LAT 2002d.
56. Paine 2005, 35.
57. Farrell and Hanemann 2009, 92, 94.
58. CEC 2003a.
59. Farrell and Hanemann 2009, 94.
60. CEC 2003b, 2003c.
61. Hanemann 2007, 12.
62. Drought data from United States Drought Monitor at droughtmonitor.unl.edu. For states of emergency in this period, see Dirmann 2001; Bailey 2001; LAT 2002b; Martin 2003; Jones and Morain 2003. Ironically, California droughts became more severe, prolonged, and widespread beginning in 2007, after the passage of most of the state's major climate legislation.
63. Author's analysis of *Los Angeles Times* articles for 1985–2014.
64. Schoen 2007.
65. Ingram 2002a; Bustillo and Ingram 2002; Polakovic 2002.
66. LAT 2006; Lifsher 2006a; Moore 2006.
67. Hayhoe 2004; UCS 2004.
68. Franco et al. 2008, S14; Hanemann 2007, 13.

69. Hayhoe et al. 2004, 12423; Hanemann 2007, 14.
70. Franco et al. 2008, S14.
71. Franco et al. 2008, S18; UCS 2004a; 2004b.
72. Franco et al. 2008, S8; CalEPA 2006.
73. Franco et al. 2008, S17.
74. Mazmanian, Jurewitz, and Nelson 2008, 408.
75. Schoch and Wilson 2006.
76. Brulle, Carmichael, and Jenkins 2012, 174.
77. Rosen 2009, 175–76.
78. I performed the content analysis on the portions of the final bills and executive orders that justified the policy actions. I counted words for all justifications in the categories given in the table; when the content pertained to more than one category, it was counted in both, such as when global warming threats to air pollution or energy security were described.
79. Another Republican, Tom McClintock, got 13% of the vote.
80. Hanemann 2007, 13.
81. Gerston and Christensen 2004, 31, 86–87.
82. Schwarzenegger 2003.
83. Weiss and Bustillo 2003; Farrell and Hanemann 2009, 95; Hanemann 2007, 13.
84. Bustillo 2003.
85. Gerston and Christiansen 2004, 93–94.
86. Morain 2005.
87. Hanemann and Busch 2011, 131.
88. Hanemann 2007, 15.
89. Hanemann 2007, 15–16.
90. Mazmanian, Jurewitz, and Nelson 2008, 407–8.
91. Hioureas and Cain 2011, 38.
92. Baldassare et al. 2007, 25.
93. Hanemann 2007, 15–17; Hall and Taplin 2010, 75; Hanemann and Busch 2011, 15–17; Rosen 2009, 171.
94. Schwarzenegger 2005.
95. Hanemann 2008, 119; CalEPA 2006, 17, 54.
96. Hall and Taplin 2010, 75.
97. Farrell and Hanemann 2009, 102.
98. Russo 2007.
99. Meckling 2011, 56; Hanemann 2008, 119–20.
100. Hanemann 2007, 20–21; CalEPA 2006, 39–84.
101. Farrell and Hahnemann 2009, 102–3; Hanemann 2007, 21.
102. Duane 2010, 737 n. 90.
103. Russo 2006.
104. Farrell and Hanemann 2009, 91, 102–3; Hanemann 2007, 21.
105. The issue is complex. Six months after his re-election, Schwarzenegger fired CARB Chair Robert Sawyer in June 2007, apparently for refusing to slow down the implementation of AB 32 (Rosenberg 2007; Dugan 2007). He chose Mary Nichols, who had oil industry ties, to replace Sawyer. But Nichols also had a strong background in air pollution control in the Brown and Clinton administrations, and she proved to be effective at implementing AB 32 (Carlson 2013, 67).

In other ways, Schwarzenegger showed a clear commitment to climate policy in his second term. He kept funding for nearly all the CARB and California Environmental Protection Agency jobs related to climate change during a 2008 budget crisis. He also adopted some measures by executive order, such as the low-carbon fuel standard, and pursued emissions trading agreements with other states in the Western Climate Initiative (Rosen 2009, 186; Duane 2010, 736 n. 87; Hanemann 2008, 126).

106. Carlson 2013, 76.
107. Hanemann 2008, 127; Humes 2009, 162–63; Darakjian 2009, 379 & n. 36.
108. Nationally, a 54% to 34% majority opposed Bush's policy reversal, basically identical to the 56% to 34% margin in California (*Los Angeles Times* April 2001 poll data via Polling the Nations). See also the section on political culture in California and New York, in Chapter 4.
109. Rosen 2009, 153, 163, 178.
110. Baldassare 2005, 8; 2006, 7; 2007, 8.
111. Baldassare 2005, 8–9; 2006, 7–9; Baldassare et al. 2007, 8–9.
112. Mazmanian, Jurewitz, and Nelson 2008, 408–9. However, two studies published the year after passage criticized these studies for underestimating the costs (e.g., Stavins, Jaffe, and Schatzki 2007).
113. Mazmanian, Jurewitz, and Nelson 2008, 407.
114. Farrell and Hanemann 2009, 102, 103; Rosen 2009, 170.
115. Assembly Bill 32 (September 27, 2006), chapter 488, section 38501c.
116. Mazmanian, Nelson, and Jurewitz 2013, 425.
117. Lifsher 2006b.
118. Schwarzenegger 2012, 548–50.
119. Brownstein 2009, 75.
120. Meckling 2011, 141, 145–46.
121. Hall and Taplin 2010, 70–71; Brownstein 2009, 75.
122. Senate Rules Committee 2006.
123. The American Federation of State, County, and Municipal Employees.
124. Hall and Taplin 2010, 71–72.
125. Hanemann 2007, 23.
126. Senate Rules Committee 2006.
127. Assembly Bill 32 (September 27, 2006), chapter 488, section 38570.
128. Hanemann 2008, 125.
129. Knox-Hayes 2012.
130. Hanemann 2008, 126; Nichols 2009b, 200.
131. Hall and Taplin 2010, 72; Ballotpedia 2014.
132. Buffa et al. 2008a; Buffa 2008b, 58, 62.
133. Biber 2013, 404–5.
134. Biber 2013, 412–16 & n. 58; Ballotpedia 2015.
135. Bartosiewicz and Miley 2013, 84–85.
136. Roosevelt 2010.
137. Biber 2013, 413–18.
138. Sollen 1998, 64; Birkland 1997, 101–2; Zierold 1984, 19; Smith 2002, 72, 76.
139. Field Research Corporation 2010, 4.
140. Randolph 2014, 132.

141. Gonzalez 2005, 59–64.
142. Ten states had adopted RPSs before California, while at least four states had adopted emissions reduction targets before California (Rabe 2006, 4; CCES data).

Chapter 8

1. For example, Raymond 2010; 2014; Rabe 2013; Funk 2009; Huber 2013.
2. Air pollution policy has occurred mainly in other periods, with the adoption of major policies on urban air pollution in the late 1960s and on acid rain in the 1980s (Duane 2010; Melewski 2002). However, these did not affect fossil fuel consumption except indirectly in ways noted in this chapter.
3. In a separate development, New York City in 2007 initiated the PlaNYC planning process, including a greenhouse emissions target of a 30% reduction over 2005–30.
4. Data from the *Albany Times Union* are from the newspaper's online archive and consist of all articles; data from the *New York Times* are from Gale/Infotrac and include all records (for more details, see notes 3–4 in Chapter 9). Keywords were "global warming" and "climate change."

   The 2001 rise was less pronounced and the drop-off in attention after 2007 was more rapid and larger in the Albany newspaper than in the *New York Times*. However, both sources probably are about equally valid indicators of the public agenda in New York State: the *Times Union* because it is not colored by the need to serve as a national newspaper and the *Times* because of its much larger circulation in the state.
5. Counts of Assembly, Senate, and joint bills from the New York State Assembly website (assembly.state.ny.us/leg), accessed on November 4, 2014 using full text searches for "global warming." Searches for "climate change" were not useful because of technical limitations of the search engine. There is no searchable database of legislative bills for New York State covering the years before 1999, so I did not measure bill introductions related to energy problems.
6. Analysis of State of the State speeches from news sources; where full transcripts or prepared texts were not available (for 1970–73, 1980, 1982, and 1987–2010), I analyzed excerpts or news reports. The length of attention to energy or climate problems and policies was measured by counting words in full sentences or clauses, divided by the total words in the speech, excerpts, or report. The climate change problems/policies category includes implicit climate policies such as renewable energy and energy efficiency policies during the post-1997 period, with such mentions also classified as energy problems/policies. Besides the increases mentioned in the text, energy issues got onto the governor's agenda in the late 1980s, due to the state government's attempt to take over the Long Island power company, and in 1996, due to restructuring of the utility sector.
7. In addition, the 1979 energy crisis, combined with Carey's leadership on the issue, led to the adoption of an energy master plan in 1980.
8. Lifset 2014a, 287.
9. Dewey 2000, 130–33; Lifset 2014a, 287.
10. Lifset 2006, 29–31.
11. Lifset 2014a, 289–90; Forsyth 1983, 61.
12. Newman 2011; Rudolph and Ridley 1986, 122–23.

13. Gottlieb 1993, chapters 3–4; Sellers 2012.
14. Dewey 2000, 130–33, 156.
15. Lurkis 1982, 90, 92, 159.
16. Talbot 1972, 133–37.
17. Marist College 2013.
18. Lifset 2006; Forsyth 1983, 61–74.
19. Jones 1963; Newman 2011; Lurkis 1982, 107.
20. Levine 1984, n. 3.
21. Birkland 2004; Pratt 2001, 2; Cordes 2001, 39 n. 60.
22. Lifset 2014a, 294.
23. Few comparable polls are available. New York ranked second among all states in support for environmental protection spending in 1973–96 (Mazur and Welch 1999, 391). Similarly, in *New York Times* polls, a 63% to 31% majority (+32 points) in the state favored high environmental standards regardless of cost, compared with a 57% to 36% majority (+21 points) nationally in 2002 (data via Polling the Nations and www.americans-world.org).
24. Pratt 2001, 3.
25. Lifset 2014a, 291.
26. Cordes 2001, 39–40.
27. NYT 1975a.
28. Meislin 1978.
29. Cordes 2001, nn. 61–62.
30. Lifset 2014a, 294; Montgomery 1993, 71.
31. During 1965–95, the coal share in the electricity generation fuel mix dropped by 29 percentage points and petroleum's fell by 12 points, while the shares of natural gas rose by 22, nuclear power by 19, and biomass by 3 percentage points. The rise in nuclear power was not enough to meet electricity demand increases like those of the 1960s, when consumption increased 6% per year. Higher energy prices and conservation policies reduced that growth rate to 1.2% per year during the 1970s; even after fossil fuel prices fell sharply in the early 1980s, electricity consumption grew only 1.7% during the 1980s and 0.2% per year during the 1990s (data from USEIA 2013, 361).
32. Lifset 2008, 163.
33. However, as in California, the shift was painfully slow. As late as 1979, most utilities, including Con Edison, were still dragging their feet. They were doing little to implement the 1977 law requiring low-cost energy audits, and many of them opposed the conservation focus of the energy master plan being drawn up at that time (Kihss 1979). Their energy efficiency efforts did not begin in earnest until 1984, after continued construction delays for new power plants made the latter's ultimate completion uncertain (Decotis 2009, 40).
34. Data from USEIA 2013, 361; Swidler 2002, 216.
35. Shwom 2011, 712; Lifset 2008, 159.
36. Lifset 2014a, 285; Lurkis 1982, 56–58, 61–75. Also, in July 1977, a Consolidated Edison blackout affected the city for about 24 hours, and this time the loss of power led to widespread vandalism, looting, and an estimated $30 million in property losses (Lurkis 1982, 75).
37. Lifset 2014b.

38. Lifset 2008, 160–1; data from USEIA 2013, 361.
39. U.S. Bureau of Labor Statistics data.
40. Rubenstein 2012, 55–56.
41. Reeves 1974.
42. Ronan 1974; Bird 1974. Carey's positions against air pollution were also criticized because of a potential conflict of interest, since his brother, who was also his campaign bankroller, owned companies that imported low-sulfur oil from Libya (Reeves 1974).
43. Vecsey 1974.
44. Democrats who sponsored major energy legislation here included Daniel Haley and Seymour Posner; in the State Senate, Republicans James McFarland and Joseph Pisani were sponsors.
45. Carey 1978.
46. NYT 1975b.
47. NYLS 1975, 347–48.
48. Peterson 1976.
49. NYLS 1976, 321–22.
50. Rubenstein 2012, 56.
51. Carey 1977; 1980.
52. Carey 1980.
53. Horsley 1978; Shaman 1979.
54. Data from USEIA 2013, 361.
55. Decotis and Ferguson 2012, 881.
56. Larocca 1981.
57. Stein 1980, 22–23.
58. Weart 2013.
59. Polakovic 2001.
60. Perez-Pena 2001; Cahill 2008, 444.
61. Revkin 1996a.
62. Lambright, Pralle, and Boscarino 2006, 32.
63. Revkin 1996a; 1997.
64. Revkin 1996b.
65. Dao 1996; Hernandez 1996.
66. Lambright, Pralle, and Boscarino 2006, 33–37.
67. Revkin 1999.
68. Hicks 2002; Perez-Pena and McGeehan 2002.
69. Pataki 2002; Lambright, Pralle, and Boscarino 2006, 41; Rubenstein 2012, 85.
70. CCAP 2003, p. es-1.
71. Lambright, Pralle, and Boscarino 2006, 40; Cahill 2008, 445.
72. By 2005, Pataki was openly running for the 2008 Republican nomination for president (Cooper 2005; Reuters 2005).
73. NYT 2002; Hakim 2005.
74. Huber 2013, 95, 100–101; Rabe 2013, 20–22.
75. Rubinstein 2012, 43–44.
76. Decotis and Ferguson 2012, 876.
77. Coal-consuming utilities also opposed RGGI (Meckling 2011, 143–44).
78. Rubenstein 2012, 59, 63; Evers 2003; Hughes 2006.
79. Cahill 2008, 442–44.

80. Rubenstein 2012, 87, 67; CCAP 2003, 2.
81. Lambright, Pralle, and Boscarino 2006, 29; Huber 2013, 86–87.
82. The membership of these two organizations increased from 36,000 in 1972 to over 300,000 in 1990 (Mitchell, Mertig, and Dunlap 1991, 221).
83. Perez-Pena 2002.
84. Schulz 2008, 5–6. They contributed to the declining share of coal in the state's electricity generation, which dropped from 18% in 2000 to 7% in 2011 (data from USEIA 2013, 361).
85. Lambright, Pralle, and Boscarino 2006, 31.
86. Decotis and Ferguson 2012, 873–74, 883; Lambright, Pralle, and Boscarino 2006, 31, 39; NYSERDA 2012, 5, 8.
87. Salkin 2014; Raymond 2014, 31.
88. A factor specific to New York, which was too late to have much effect on policy making in this period, was a 600-page NYSERDA report that detailed projected climate change impacts in New York State and made recommendations for adaptation measures; it was issued in November 2011 (NYSERDA 2011).
89. U.S. Bureau of Labor Statistics data.
90. NYSERDA 2009; Cuomo 2010; 2013.
91. Paterson 2009.
92. Cuomo 2013, 213–21.
93. From the sources described in note 6.
94. Hakim 2006.
95. Confessore 2007.
96. Hakim 2007b; Salkin 2014, 723.
97. Chan 2009.
98. Salkin 2014, 738–41.
99. NYSCAC 2010, App. C.
100. Navarro 2010.
101. Cuomo 2010, 6, 14.
102. Barringer 2013; Raymond 2014, 31.
103. Salkin 2014, 751–62.
104. Dewey 2000, 20, 116.
105. Decotis and Ferguson 2012, 882–83.
106. New York State was above average in higher education from the 1960s to the 1990s, as well; 8.9% of its over-25 population had four years or more of college in 1960, compared with 7.7% of the U.S. population (U.S. Census 1970, 112).
107. In 1997–2000, New York State produced an annual average of only 25 BTUs per dollar of state economic product, compared with a national average of 6200 BTUs per dollar of GDP. Although natural gas production in the state nearly tripled from 2000 to 2006, it then declined (USEIA data); in 2014, Governor Cuomo announced a ban on natural gas fracking in the state.
108. Navarro 2012.
109. Carey, whose father and brother were in the oil business, voted against key water pollution control amendments in the U.S. House of Representatives, downplayed the dangers of polychlorinated biphenyls (PCBs) by offering to drink a glass of them (Rosen and Slocum 1998, 88), and called for switching power plants from oil to coal in 1980.

Chapter 9

1. Künkel, Jacob, and Busch 2006, 16.
2. Lifset 2014b.
3. The source is the *New York Times* via Proquest; "articles" here includes all records of articles, editorials, front-page headlines, letters to the editor, and a small number of miscellaneous items. To detect energy problems, I used the keywords "energy crisis," "oil crisis," "gas crisis," "energy shortage," "oil shortage," "gas shortage," "rising energy price," "rising oil price," and "rising gas price." Adding additional print or broadcast media to the *New York Times* would improve the accuracy of this index only marginally, since there is very high correlation in attention paid to issues by different news sources. For this point concerning print media, see Baumgartner and Jones 1993, 257–59; concerning broadcast media, see the next note.
4. For the newspaper data and methods, see the previous note; keywords were "global warming" or "climate change." Using a narrower definition of media attention, such as front-page stories in the *New York Times*, produces no differences in the peaks noted here; the correlation between all records and front-page stories is 0.93. The *New York Times* data in Figure 9.1 also correlates highly with television data provided by Robert Brulle, drawn from the Vanderbilt Television News Archive (R = 0.84). It correlates even more strongly with other indices of public attention, such as the global warming/climate change issue index based on the *Los Angeles Times* and used in Figure 7.1, in Chapter 7 (R = 0.98).
5. Weart 2013.
6. This point is supported by data provided by Max Boykoff on five leading U.S. newspapers over the 2000–15 period. However, Brulle's measures of broadcast television news coverage of climate change (e.g., the number of ABC, CBS, NBC, and FOX stories) show a sharper post-2009 decline in public attention; see note 4.
7. I identified drivers of attention by using keywords to filter the articles on global warming and climate change, as follows: "science" or "scientific" combined with "study" or "report" (for science events); "sea level," "drought," "flood," "wildfire," "hurricane," or "storm" (for climate events); the name of the President (and major party presidential candidates in election years), "Congress," or "Senate" (for political events), and the "UN," "United Nations," "Germany," "France," "Britain," "Japan," or "European Union" (for international events). The data source was Proquest for 1985–2011 and Infotrac for 2012–14.
8. Gallup data for 1977–2012. For information about data sources for public opinion polls cited in this chapter, see note 29 in Chapter 4.
9. ABC News data for 2001–12.
10. Gallup data for 1997–2013.
11. Pew data for 2006–11; NBC/*Wall Street Journal* data for 1999–2014.
12. Gallup data for 1984–2014.
13. The data source is Proquest Congressional. Keywords for the climate change issue were "global warming" or "climate change" in all fields excluding full text; for energy policy innovations, the keywords were "energy conservation," "energy efficiency," or "renewable energy."
14. I calculated four correlations, of each institutional actor's attention on energy policy and on explicit climate policy with all *New York Times* articles and with

front-page articles only; for Congress, correlations ranged from 0.88 to 0.94, while for the president, they ranged from 0.65 to 0.85.

15. Presidential State of the Union messages are well suited for measuring the president's agenda, since they are major speeches in which he addresses Congress as well as the public.
16. Speeches for this analysis were taken from the American Presidency Project at the University of California at Santa Barbara (www.presidency.ucsb.edu). I measured the length of attention to energy or climate problems and policies by counting words in full sentences or clauses, divided by the total words in the speech. The January addresses by outgoing presidents (e.g., Reagan's in 1989) are not included. Since 1981, a president in his first year in office has given a quasi-state of the union address in February, which is used here instead. In 1977, Carter did not give such an address, so his major television address on energy policy in April was used for that year, with the average length of his three State of the Union addresses serving as the denominator.
17. In this speech, Bush did not mention climate change, but he advocated for renewable energy and reduced gasoline consumption (which are at least implicit climate policies) in order to reduce dependence on oil imports.
18. Pew data for 2007–13.
19. *Newsweek* data, August 2007; CBS News data, November 2007.
20. Brulle, Carmichael, and Jenkins 2012; Brewer 2012.
21. Gallup data for 1989–2013.
22. Weart 2013, 67; Borick 2010, 28, 34, 35.
23. Pew data for 2006–13; CNN/ORC data for 2007–13; the average is for both series combined.
24. Smith 2002, 24–25.
25. Lifset 2014a.
26. Smith 2002, 24–25.
27. USEIA data.
28. Simon 2007, 68.
29. Ford 1975; Timney 2004, 22.
30. Katz 1984, 60–63; Graetz 2011, 55–56; de Marchi 1981, 489–90.
31. They also created the Strategic Petroleum Reserve and the Energy Research Development Agency, while maintaining oil and natural gas price controls.
32. Katz 1984, 70.
33. Graetz 2011, 55–56.
34. EPCA, Public Law 94–163, sections 325, 327; Shwom 2011, 715; Katz 1984, 60–61. The laws provided federal grants for energy conservation and the creation of state energy offices, and they gave state governments incentives to draw up energy plans that would aim to reduce energy consumption by at least 5% (Stein 1980).
35. Timney 2004, 22.
36. Laird 2001, 90.
37. Kraft 1981, 40.
38. Kraft 2004, 173.
39. Carter 1977.
40. Katz 1984, 87–88; Timney 2004, 22–23.

41. Energy Conservation Policy Act, Titles II–IV; Schipper et al. 1979; Katz 1984, 102–3, 111.
42. Rosenbaum 1981, 65; Nemet and Kammen 2007, 750.
43. Graetz 2011, 120; 1978 Energy Tax Act, section 301.
44. Cochrane 1981, 585–87; Katz 1984, 102–3.
45. Kraft 1981, 42, 51.
46. Carter 1977; Kraft 1981, 41; Graetz 2011, 139–41.
47. Kraft 1981, 39–40, 50; Katz 1984, 117.
48. Real coal prices fell 55% during 1979–89, and real natural gas prices dropped 30% during 1983–92 (USEIA data).
49. Timney 2004, 24–26.
50. Kraft and Axelrod 1984, 319–20.
51. Reagan 1981.
52. Miller 1995, 719.
53. Kraft and Axelrod 1984, 319; Kraft 2004, 115.
54. Nemet and Kammen 2007, 750; Kraft 2004, 174.
55. Gipe 1991, 758; Timney 2004, 25.
56. Timney 2004, 25.
57. Nadel 2002, 162–65; Shwom 2011, 716–18.
58. Weart 2013, 39, 42, 51.
59. IPCC 1990.
60. Brewer and Pease 2008, 85–86.
61. USEIA data.
62. Eikeland 1993b, 25–26; for example, NYT 1990; Hart 1990; Hershey 1990.
63. Bush 1988; Dionne 1988.
64. Bryner and Duffy 2012, 48; Weart 2013, 47.
65. Brewer and Pease 2008, 86; Eikeland 1993a, 3–4.
66. Timney 2004, 27.
67. Montgomery 1993, 75–77.
68. Weiner 2008, 172–76.
69. UN 1992; Brewer and Pease 2008, 86.
70. Stevens 1991.
71. The act also created a Director of Climate Protection within the Department of Energy and required the department to produce a biennial "least-cost" strategy for increasing energy efficiency by 30% over 1988–2010 and renewable energy by 75% over 1988–2005. In this part of the law, one goal was to stabilize and eventually reduce greenhouse gas emissions (sections 1602–3).
72. Timney 2004, 27–28; Yacobucci 2008, 9.
73. Shwom 2011, 719. By 1992, mandatory appliance standards were regularly negotiated between manufacturers, leading states (especially California, New York, and Massachusetts), the American Council for an Energy-Efficient Economy (a trade group), and the NRDC, which presented their results to the Department of Energy for implementation (Shwom 2011, 719).
74. Section 1212.
75. Eikeland 1993b, 63.
76. Eikeland 1993a, 6; Nemet and Kammen 2007, 750.
77. Joskow 2001, 14; Timney 2004, 28.

78. Eikeland 1993a, 5–6; Joskow 2001, 13–14.
79. Represented by the American Council for an Energy-Efficient Economy (founded in 1980) and the Alliance to Save Energy (founded in 1977).
80. Eikeland 1993b, 65–68; 1993a, 10–12.
81. Yacobucci 2008, 2; NAS 1992.
82. Brewer and Pease 2008, 86–87.
83. These were his chief of staff, John Sununu, and officials in the Treasury Department, Council of Economic Advisors, and Office of Management and Budget.
84. Eikeland 1993c, 989.
85. Eikeland 1993b, 56.
86. Simon 2007, 77.
87. Royden 2010, 419; CBO 1993, III–6.
88. Erlandson 1994, 175, 182.
89. Erlandson 1994, 177–80.
90. Christensen 2003, 3.
91. Royden 2010, 420 n. 15.
92. Erlandson 1994, 174.
93. Royden 2010, 420 n. 16.
94. Royden 2010, 418, 432; Erlandson 1994, 184.
95. Although the Rio conference had occurred in June 1992, only eight months before Clinton's BTU tax proposal in February 1993, public attention on global warming had dropped sharply by the time Clinton took office in January; since the Framework Convention had been negotiated and ratified under the Bush administration, Clinton had little opportunity to use it to draw attention to global warming.
96. Cavanaugh, Hahn, and Stavins 2001, 29.
97. Royden 2010, 426.
98. Weart 2013, 61–62.
99. The increase began sharply in June and attention fell off rapidly after December. In 1997, over half of the *New York Times* articles mentioned the United Nations or Kyoto, though 20% mentioned climate science and 17% climate phenomena. This suggests that the international negotiations were drawing attention to the global warming problem as well as to the politics surrounding it.
100. Congressional Republicans' average score from the League of Conservation Voters dropped from about 40% to 15% during the 1990s, while the average score for Democrats increased from about 65% to 80% (from data in Skocpol 2013, 67).
101. Royden 2010, 425–31, 441, 449–50.
102. Royden 2010, 453, 462.
103. Joskow 2001, 53; Yaccobucci and Bamberger 2007, 5–6.
104. Royden 2010, 416, 420 n. 18, 449; Joskow 2001, 17, 53.
105. Simon 2007, 77–78.
106. USDOS 1997, 72, 77, 83–84, 93–94; 2002, 53–55. The program was initiated in 1992.
107. USEPA 2001, 23.
108. USEPA 1991, 11; 1996, 9908–9; Yacobucci 2008, 7–8, 10.
109. Layzer 2012, 293; Gallup data for 1989–2013.

110. Revkin 2001; Bryner and Duffy 2012, 49; Meckling 2011, 134–35.
111. Layzer 2012, 292–93.
112. Chong 2015, 126; McCright and Dunlap 2011, 176; but cf. Skocpol 2013, 67.
113. Stout 2001.
114. Christensen 2003, 5; Layzer 2012, 293–94.
115. Bryner and Duffy 2012, 49 n. 1.
116. Engel 2015, 37–38.
117. Vasi 2011, 103–4.
118. Brewer and Pease 2008, 87.
119. Layzer 2012, 299.
120. Kraft 2004, 179; Timney 2004, 30.
121. Layzer 2012, 293.
122. Kraft 2004, 180.
123. Engel 2015, 37; Grunwald and Eilperin 2005; NBC News 2005.
124. USEIA data.
125. Hulse 2005; Grist 2005.
126. Aulisi et al. 2007, 29–30.
127. The credit was expanded to include municipal solid waste combustion, landfill gas, open-loop biomass, and several other sources, at lower rates.
128. Simon 2007, 80; Rahm 2006, 9–10.
129. The EPA has also begun to tighten methane standards in landfills (rules proposed in 2014), to regulate methane emissions from the oil and natural gas industry (rules proposed in 2015), and to regulate greenhouse gas emissions from aircraft (endangerment finding proposed in 2015).
130. Broder 2007.
131. Herszenhorn 2008.
132. These were Hurricanes Rita, Wilma, and Dennis.
133. Evans and Steven 2007, 9–11.
134. Business Week 2006; Weart 2013, 63–64.
135. Weart 2013, 48; Boykoff 2007, 474.
136. Time, 2005.
137. Farley et al. 2007, 348.
138. Weart 2013, 65.
139. Boykoff 2007, 474.
140. From the author's analysis of organizations' websites, described in Chapter 5.
141. Bartosiewicz and Miley 2013, 29-30; cf. also EGA 2012, 12.
142. Weart 2013, 66.
143. USEIA data.
144. Weart 2013, 64, 66.
145. Weart 2013, 68–69; Boykoff 2013, 798; Nisbet 2011, 54. More recent data evidently is not available.
146. The 2008 data are from three polls with open-ended questions: CBS News, March 2008; *Los Angeles Times*, May 2008; and ABC News, May 2008. The further trend is from Gallup data.
147. Gallup data for 1984–2014.
148. Andrews and Barringer 2007.
149. Meckling 2011, 147–57.

150. Rowlands 2010, 187.
151. However, not all aspects of administration policy tended to reduce greenhouse gas emissions. Obama pursued an "all-of-the-above energy policy" that included offshore oil drilling and the continued leasing of federal lands in the Powder River Basin for coal mining (Cockerham 2015).
152. Bryner and Duffy 2012, 59.
153. McCright and Dunlap 2011, 176; Skocpol 2013, 67; Swift 2014.
154. Pew Charitable Trusts 2009, 40.
155. Rabe 2010c, 279.
156. Wallach 2012, 8 n. 29; Probst and Szambelan 2009, 7.
157. Bartosiewicz and Miley 2013; Mildenberger 2013.
158. Skocpol 2013, 20.
159. Skocpol 2013.
160. Rabe 2010a, 603; Rabe 2010c, 279.
161. Bartosiewicz and Miley 2013, 6 n. 4, 31, 35 & n. 32, 48–49.
162. Bartosiewicz and Miley 2013, 8–9, 14.
163. Bartosiewicz and Miley 2013, 8, 49-50.
164. Lorinc 2010.
165. Bartosiewicz and Miley 2013, 63–65; Rabe 2010c, 279.
166. Lizza 2013.
167. Bartosiewicz and Miley 2013, 66–67.
168. Nivola 2010, 162.
169. Data from McCubbin 2013, 51–52. This effectively raised the fuel economy standard for cars and light trucks to 35.5 mpg. CAFE standards already were scheduled to be raised to 35 mpg during 2011–20, according to the 2007 law signed by Bush, which allowed the Transportation Secretary to raise standards further through 2030 (Yacobucci 2008, 10; EISA 2007, section 102). The deal with the carmakers tightened the standards by moving up the compliance date to 2016 and edging the target up to 35.5 mpg (Nivola 2010, 158).
170. Burtraw 2011, 5.
171. Landler 2014.
172. White House 2011; USEPA 2012a, 4.
173. Bryner and Duffy 2012, 135; Friedman 2011.
174. McCubbin 2013, 57.
175. However, broadcast television coverage of the issue declined rapidly.
176. McCubbin 2014, 54–57.
177. The first case was *Coalition for Responsible Regulation v. Environmental Protection Agency*; the 2014 decision was in *Utility Air Regulatory Group v. EPA*.
178. McCubbin 2014, 57, 67–68.
179. Geman 2013.
180. Scruggs 2003, 101, 110.
181. USEIA data for crude oil and products.
182. Dunlap and McCright 2011; McCright and Dunlap 2011.
183. Weart 2013, 47; McCright and Dunlap 2000; McCright and Dunlap 2003.
184. Dunlap and McCright 2011, 155.
185. Pew 2012, 4, 6, 20–21.
186. Jon Huntsman, a minor candidate, was an exception.
187. Spross, Koronowski, and Germain 2013; Koronowski and Germain 2014; Klieg-man 2014.

Chapter 10

1. Several socioeconomic-structural variables also were not statistically significant in Chapter 5: high income, a lack of heavy industry, polluting industries, or manufacturing industries, and high renewable energy potential.
2. Cf. Klyza and Sousa 2013.
3. California and New York rank 1st and 3rd, respectively, among all states in the total number of state government employees.
4. The United States is about average among industrialized democracies in higher education attainment (OECD data).
5. Wikle 1995, 42; Milbrath 1984, 76.
6. This book's research design has limitations for addressing this question, since its selection of cases is biased toward successful policy-making episodes.
7. After a burst of major reforms, a period of consolidation rather than further major reform is likely to follow, for several reasons. Vetted policy proposals may not be available, and businesses require time to absorb anticipated economic costs or to see economic uncertainties reduced. Moreover, the implementation of recently adopted policies occupies administrative agencies and sometimes requires political will, and a leading state may wait for other states or the federal government to catch up and to see how effective the new measures will be.
8. Kaplan 2014.
9. McGreevey 2003.
10. Pew Research Center data.
11. Kingdon 2003, 86–89, 201–03, 205.
12. Cf. Zahariadis 2007, 81–82.
13. For example, the severity of the 1970s energy crises did not dictate the nature of the solutions that were enacted in California and New York. In the absence of environmental opposition to nuclear and coal power, those might have become the main long-term options institutionalized in the 1970s, rather than energy efficiency and renewable energy.
14. Cf. Farley et al. 2007; Gifford and Comeau 2011; Layzer 2012; Vezirgiannidou 2013; Nisbet 2009.
15. Kingdon 2003, 166.
16. Ballotpedia 2014. Other major decisions were made since 2010, but these mainly have implemented the 2005–08 reforms: Proposition 23 was defeated (2010); and the low-carbon fuel standard rules (2010) and emissions trading system (2013, 2015) were put into effect. Despite the political commitment of the current Brown administration and a largely favorable problem context (the salience of climate change in the news and major drought in California), major new initiatives have not been forthcoming in this period so far. However, Brown's January 2015 inaugural address, in which he called for a 50% RPS by 2030 and a halving of emissions in transportation and the building sector, might mean that the period of consolidation has ended and a new round of major climate policy making in California is starting. Brown has followed up with an executive order ordering CARB to implement a 40% cut in emissions over 1990–2030 (Executive Order B-30-15).
17. Kingdon 2003, 203.
18. Others reach similar findings, e.g., Liu, Lindquist, and Vedlitz 2011, 414–15.

19. Klyza and Sousa 2013, 265–69, 296–303. The rising importance of executive authority in federal climate policy making also makes it difficult to characterize the political stream in the policy windows theory. If the executive strongly favors and Congress opposes the adoption of a major policy, it is not clear whether political commitment is adequate for policy reform.
20. For example, Gillis 2014.
21. In addition, the much older Public Service Commission was involved in renewable energy and energy efficiency until 1996. The New York Department of State gained responsibility for building codes after 1996, and the Department of Environmental Conservation implements RGGI.
22. The other federal agencies important for climate policy, the Transportation Department (for fuel economy) and the EPA (greenhouse gas regulation), date to 1967 and 1970, respectively, before the period of this book's federal case study.
23. Carlson 2013.
24. Roe 1984, 47.
25. Maw 2006.
26. Randolph 2014, 129–35; Biber 2013, 439–40.
27. Knox-Hayes 2012, 554–57; Huber 2013, 100–101; Raymond 2014.
28. McGreevey and Mason 2015.
29. Gerdes 2013.
30. Greenblatt and Long 2012.
31. Koronowski 2013.
32. These results are from the combined model. Since the model is cross-sectional, not longitudinal, it only explains differences between states and cannot be used directly to predict the results of interventions. Nevertheless, it is reasonable to assume that the same factors which explain differences between states would also operate over time within any given state.
33. Skocpol 2013, 78; Pew data for 2006–13; CNN/ORC data for 2007–13. For information about data sources for public opinion polls in this and the following note, see note 29 in Chapter 4.
34. Pew data for 2006–14.
35. Gallup data for 1989–2015 in Saad 2015.
36. Based on inflation-adjusted data from NOAA's Billion Dollar Disaster program, accessed at www.ncdc.noaa.gov/billions on January 6, 2015.
37. Melillo, Richmond, and Yohe 2014.
38. Schor 2015.
39. Plumer 2015.
40. Aulisi et al. 2007, 12.
41. Chen and Rodden 2013.
42. Mildenberger 2013.
43. Hoffman 2007.
44. WRI-CAIT data for 2012, excluding land use and forestry changes.
45. Davenport 2014c; Landler 2014.
46. Xu 2014; Schoen 2013, 6.
47. Borick 2010, 49–51; Rabe 2010a.
48. Gallup's time series, which goes back to 1984, shows that the public gave economic growth a higher priority over environmental protection for the first time in

2009–10 and 2012–13; although these priorities were reversed again in 2014, the public still gave a historically low priority to environmental protection that year (Swift 2014).

49. Harder 2014; Borick 2010, 44.
50. Raymond 2014; CCES 2013b, 8–9; Taylor 2014.

References

ACEA (Association des Constructeurs Européens d'Automobiles). 2011. *EU Economic Report, July 2011*. Brussels, ACEA.

AEI (American Enterprise Institute). 2013. *Polls on the Environment, Energy, Global Warming, and Nuclear Power*. Washington, DC: AEI, April.

Ainsworth, Ed. 1947. "Crowd Heckles Supervisors in Smog Protest," *Los Angeles Times*, February 19.

Alquist, Albert and Charles Warren. 1974a. "Minimum Energy Efficiency Standards for Appliances," paper attached to a May 4, 1974 letter to members of the California State Senate, Warren-Alquist Act correspondence file, 129, at www.energy.ca.gov, accessed March 6, 2014.

1974b. Letter to members of the California State Senate, May 7, 1974, Warren-Alquist Act correspondence file, 125–26, at www.energy.ca.gov, accessed March 6, 2014.

Andrews, Edmund and Felicity Barringer. 2007. "Bush Seeks Vast, Mandatory Increase in Alternative Fuels and Greater Vehicle Efficiency," *New York Times*, January 24.

Armingeon, Klaus. 2008. "Encompassing Institutions and CO_2 Emissions," paper for the ECPR Joint Sessions, Rennes, France, April 11–16.

ASAP (Appliance Standards Awareness Project). 2014. "National Standards," at www.appliance-standards.org/national, accessed September 1, 2014.

Assembly Committee on Utilities (California). 2002. "SB 1078 Bill Analysis," August 15, 2002, at leginfo.ca.gov, accessed February 25, 2015.

Aulisi, Andrew et al. 2007. *Climate Policy in the State Laboratory*. Washington, DC: World Resources Institute.

Bailey, Eric. 2001. "Water War Pits Farms Against Fish," *Los Angeles Times*, May 6.

Baker, Al. 2003. "Pataki Opposing Increase in Taxes to Close Deficits," *New York Times*, January 9.

Baldassare, Mark. 1998. *PPIC Statewide Survey: Californians and Their Government*. San Francisco: Public Policy Institute of California, October.

2000. *PPIC Statewide Survey: Special Survey on Californians and the Environment*. San Francisco: Public Policy Institute of California, June.

2001. *PPIC Statewide Survey: Californians and Their Government*. San Francisco: Public Policy Institute of California, July.

2002. *PPIC Statewide Survey: Special Survey on Californians and the Environment*. San Francisco: Public Policy Institute of California, June.

2005. *PPIC Statewide Survey: Special Survey on the Environment*. San Francisco: Public Policy Institute of California, July.

2006. *PPIC Statewide Survey: Special Survey on the Environment*. San Francisco: Public Policy Institute of California, July.

Baldassare, Mark et al. 2007. *PPIC Statewide Survey: Californians and the Environment*. San Francisco: Public Policy Institute of California, July.

Ballotpedia. 2014. "California Proposition 39, Income Tax Increase for Multistate Businesses (2012)," last modified April 13, at ballotpedia.org, accessed May 5, 2014.

2015. "California Proposition 23, the Suspension of AB 32 (2010)," last modified January 1, at ballotpedia.org, accessed February 25, 2015.

Bang, Guri. 2003. *Sources of Influence in Climate Change Policymaking*, Dr. Polit. thesis submitted to the Department of Political Science, University of Oslo, June.

2010. "Energy Security and Climate Change Concerns," *Energy Policy* 38:1645–53.

Bang, Guri and Miranda Schreurs. 2011. "A Green New Deal," in Rüdiger Wurzel and James Connelly, eds., *The European Union as a Leader in International Climate Change Politics*. London: Routledge, 235–51.

Banse, Tom. 2015. "Washington State Plowing Ahead With Greenhouse Gas Limits," *NW News Network*, November 18, at nwnewsnetwork.org, accessed December 11, 2015.

Barabak, Mark. 2001. "Bush Criticized as Fear for Environment Grows," *Los Angeles Times*, April 30.

Barringer, Felicity. 2012. "California and Quebec Near an Agreement for Trading of Carbon Permits," *New York Times*, May 10.

2013. "States' Group Calls for 45% Cut in Amount of Carbon Emissions Allowed," *New York Times*, February 7.

Bartosiewicz, Petra and Marissa Miley. 2013. "The Too Polite Revolution," paper prepared for a symposium on The Politics of America's Fight Against Global Warming, Harvard University, February 14.

Baumgartner, Frank and Bryan D. Jones. 1993. *Agendas and Instabilities in American Politics*. Chicago: University of Chicago Press.

Bedsworth, Louise and Ellen Hanak. 2013. "Climate Policy at the Local Level," *Global Environmental Change* 23:664–77.

Belsley, David, Edwin Kuh, and Roy Welsch. 1980. *Regression Diagnostics*. New York: John Wiley and Sons.

Bergholz, Richard. 1969. "Fight Opens for Vote to Outlaw Smog Producers," *Los Angeles Times*, August 1.

Bernauer, Thomas. 2013. "Climate Change Politics," *Annual Review of Political Science* 16:421–48.

Bernton, Hal. 2015. "Inslee a Student of California's Carbon-Cap Lessons," *Seattle Times*, February 5.

Berry, Frances and William Berry. 2007. "Innovation and Diffusion Models in Policy Research," in Paul Sabatier, ed., *Theories of the Policy Process*, 2nd edn. Boulder, CO: Westview Press, 223–60.

Berry, Phillip. 1984. "Sierra Club Leader, 1960s–1980s," oral history interview conducted by Ann Lage, Bancroft Library, University of California at Berkeley, for the Sierra Club History Series.

Berry, William et al. 1998. "Measuring Citizen and Government Ideology in the American States," *American Journal of Political Science* 42:327–48.

2007. "The Measurement and Stability of State Citizen Ideology," *State Politics and Policy Quarterly* 7:111–32.

2010. "Measuring Citizen and Government Ideology in the U.S. States," *State Politics and Policy Quarterly* 10:117–35.

Betsill, Michele and Barry Rabe. 2009. "Climate Change and Multilevel Governance," in Daniel Mazmanian and Michael Kraft, eds., *Toward Sustainable Communities*. Cambridge, MA: MIT Press, 201–26.

Beyle, Thad. 1995. "Enhancing Executive Leadership in the States," *State and Local Government Review* 27 (Winter):18–35.

Beyle, Thad and Margaret Ferguson. 2008. "Governors and the Executive Branch," in Virginia Gray and Russell Hanson, eds., *Politics in the American States*, 9th edn. Washington, DC: CQ Press, 192–228.

Bianco, Nicholas et al. 2013. *Can The U.S. Get There From Here?* Washington, DC: World Resources Institute, February.

Biber, Eric. 2013. "Cultivating a Green Political Landscape," *Vanderbilt Law Review* 66:399–462.

Bird, David. 1974. "Environment Poses Thorny Problems," *New York Times*, October 25.

Bird, Lori, Claire Kreycik, and Barry Friedman. 2008. *Green Power Marketing in the United States*, 11th edn., National Renewable Energy Laboratory Technical Report, October.

Birkland, Thomas. 1997. *After Disaster*. Washington, DC: Georgetown University Press.

1998. "Focusing Events, Mobilization, and Agenda Setting," *Journal of Public Policy* 18:53–74.

2004. "Environmental Successes and Continued Challenges in the Hudson Valley," *Albany Law Environmental Outlook Journal* 8:187–211.

Black, Brian and Richard Flarend. 2010. *Alternative Energy*. Santa Barbara, CA: Greenwood Press.

Bogoshian, Matt and Ken Alex. 2009. "The Essential Role of State Enforcement in the Brave New World of Greenhouse Gas Emission Limits," *UCLA Journal of Environmental Law and Policy* 27:337–52.

Boitier, Baptiste. 2012. "CO_2 Emissions Production-Based Accounting vs. Consumption," paper presented at the Final WIOD Conference, Groningen, Netherlands, April 24–26.

Bollens, John and G. Robert Williams. 1978. *Jerry Brown: In a Plain Brown Wrapper*. Pacific Palisades, CA: Palisades Publishers.

Bolleyer, Nicole. 2013. *New Parties in Old Party Systems*. New York: Oxford University Press.

Bomberg, Elizabeth. 2012. "Mind the (Mobilization) Gap," *Review of Policy Research*, 29:408–30.

Borick, Christopher. 2010. "American Public Opinion and Climate Change," in Barry Rabe, ed., *Greenhouse Governance*. Washington, DC: Brookings Institution, 24–57.

Borick, Christopher and Barry Rabe. 2012. "Public Views on Climate Policy Options," Brookings Institution. June 11.
Boyarsky, Bill. 1974. "Brown and Flournoy," *Los Angeles Times*, November 3.
Boykoff, Maxwell. 2007. "Flogging a Dead Norm?" *Area* 39:470–81.
2013. "Public Enemy No. 1?" *American Behavioral Scientist* 57:796–817.
Boykoff, Maxwell and Jules Boykoff. 2004. "Balance as Bias," *Global Environmental Change* 14:125–36.
Bradley, William. 2001. "See Two Governors Run From Blame," *Los Angeles Times*, August 19.
Brandt, A. R. et al. 2014. "Methane Leaks from North American Natural Gas Systems," *Science* 343 (14 February):733–35.
Brewer, Paul. 2012. "Polarisation in the USA," *European Political Science* 11:7–17.
Brewer, Paul and Andrew Pease. 2008. "Federal Climate Politics in the United States," in Hugh Compston and Ian Bailey, eds., *Turning Down the Heat*. New York: Palgrave Macmillan, 85–103.
Brewer, Thomas. 2015. *The United States in a Warming World*. New York: Cambridge University Press.
Broder, John. 2007. "Industry Flexes Muscle, Weaker Energy Bill Passes," *New York Times*, December 14.
Brown, Mark et al. 1995. "Technological Innovation through Environmental Policy," *Public Productivity and Management Review* 19:77–93.
Brownstein, Ronald. 2009. "The California Experiment," *The Atlantic* 304 (October), 66–76.
Brulle, Robert. 2015. "The US National Climate Change Movement," in Yael Wolinsky-Nahmias, ed., *Changing Climate Politics*. Washington, DC: CQ Press, 146–70.
Brulle, Robert, Jason Carmichael, and J. Craig Jenkins. 2012. "Shifting Public Opinion on Climate Change," *Climatic Change* 114:169–88.
Bruyninckx, Hans, Sander Happaerts, and Karoline van den Brande. 2012. *Sustainable Development and Subnational Governments*. New York: Palgrave Macmillan.
Bryner, Gary. 2008. "Failure and Opportunity: Environmental Groups in US Climate Change Policy," *Environmental Politics* 17:319–36.
Bryner, Gary and Robert Duffy. 2012. *Integrating Climate, Energy, and Air Pollution Policies*. Cambridge, MA: MIT Press.
Buffa, Andrea et al. 2008a. "California's Global Warming Solutions Act of 2006," Center for Labor Research and Education, University of California at Berkeley, August.
Buffa, Andrea. 2008b. "California Labor Unions Support Global Warming Solutions," *California Progress Report*, August 13.
Burtraw, Dallas. 2011. "US Climate Change Policy Efforts," Center for European Studies, Brussels, Policy Brief No. 255, September.
Busby, John and Alexander Ochs. 2004. "From Mars and Venus Down to Earth," in David Michel, ed., *Climate Policy for the 21st Century*. Washington, DC: Johns Hopkins University Press, 35–76.
Busch, Per-Olof, Helge Jörgens, and Kerstin Tews. 2005. "The Global Diffusion of Regulatory Instruments," *Annals of the American Academy of Political and Social Sciences* 598:146–67.

Bush, George H. W. 1988. "From Afar, Both Candidates Are Environmentalists ...," *New York Times*, September 24.

Business Week. 2006. "Global Warmng," *Business Week* (December 18), 102.

Bustillo, Miguel. 2002. "Exhaust Legislation May Hit a Red Light," *Los Angeles Times*, May 21.

2003. "Schwarzenegger Becomes the GOP's Green Candidate," *Los Angeles Times*, September 21.

Bustillo, Miguel and Carl Ingram. 2002. "Auto-Emission Bill OKd, Sent to Davis," *Los Angeles Times*, July 2.

Bustillo, Miguel and Nancy Vogel. 2002. "Renewable-Energy Bill Passed by Assembly," *Los Angeles Times*, August 28.

Byrne, John et al. 2007. "American Policy Conflict in the Greenhouse," *Energy Policy* 35, 4555–73.

Cahill, John. 2008. "Environmental Law in New York State," *Pace Environmental Law Review* 25:441–49.

CalEPA (California Environmental Protection Agency). 2006. "Climate Action Team Report to Governor Schwarzenegger and the Legislature," March.

2013. "State Agency Greenhouse Gas Reduction Report Card," January.

Calvert, Jerry. 1989. "Party Politics and Environmental Policy," in James Lester, ed., *Environmental Politics and Policy*. Durham, NC: Duke University Press, 158–78.

Cappocia, Giovanni and R. Daniel Kelemen. 2007. "The Study of Critical Junctures," *World Politics* 59:341–69.

CARB (California Air Resources Board). 2008. "Climate Change: Proposed Scoping Plan," October, at www.arb.ca.gov, accessed April 29, 2013.

2009. "2050 Greenhouse Gas Emissions Analysis: Staff Modeling in Support of the Zero Emission Vehicle Regulation," at www.arb.ca.gov, accessed February 25, 2015.

2011a. "ARB Emissions Trading Program Overview," revised January 21, 2011b, at www.arb.ca.gov, accessed February 7, 2013.

2011b. "California Announces Groundbreaking Advanced Clean Car Rules," press release, November 16, at www.arb.ca.gov, accessed September 1, 2014.

2011c."California Greenhouse Gas Emissions Inventory: 2000–2009," December, at www.arb.ca.gov, accessed January 25, 2013.

2012. "Table 6. Carbon Intensity Lookup Table for Gasoline and Fuels that Substitute for Gasoline," updated December 2012, at www.arb.ca.gov, accessed June 30, 2014.

2014a. "California Greenhouse Gas Emissions for 2000 to 2012: – Trends of Emissions and Other Indicators," May 13, at www.arb.ca.gov, accessed June 21, 2014.

2014b. "First Update to the Climate Change Scoping Plan," May, at www.arb.ca.gov, accessed June 27, 2014.

2014c. "AB 32 Cost of Implementation Fee Regulation," at www.arb.ca.gov, accessed December 15, 2014.

Carey, Hugh. 1977. "Text of Carey's 'State of State' Message Proposing Income Tax Cut and Bond Issue," *New York Times*, January 6.

1978. "Special Message: Energy," February 8, *Public papers of Hugh L. Carey, Fifty-First Governor of the State of New York*. Albany: State of New York.

1980. "Excerpts From Governor Carey's State of the State Message to the Legislature," *New York Times*, January 10.

Carley, Sanya. 2011. "The Era of State Energy Policy Innovation," *Review of Policy Research* 28:265–94.
Carley, Sanya and Chris Miller. 2012. "Regulatory Stringency and Policy Drivers," *Policy Studies Journal* 40:730–56.
Carlson, Ann. 2009. "Iterative Federalism and Climate Change," *Northwestern University Law Review* 103:1097–1161.
2013. "Regulatory Capacity and State Environmental Leadership," *Fordham Environmental Law Review* 24 (Spring):63–86.
Carter, Jimmy. 1977. "Address to the Nation on Energy," televised speech, April 18, at millercenter.org, accessed February 26, 2015.
Cassidy, Emily. 2014. *Ethanol's Broken Promise*. Washington, DC: Environmental Working Group, May.
Cavanagh, Ralph. 2009. "Graphs, Words and Deeds," *Innovations* 4:81–89.
Cavanaugh, Sheila, Robert Hahn, and Robert Stavins. 2001. *National Environmental Policy During the Clinton Years*. Washington, DC: Resources for the Future.
CBO (Congressional Budget Office). 1993. *An Analysis of the President's February Budgetary Proposals*. Washington, DC: CBO, March.
CCAP (Center for Clean Air Policy). 2003. *Recommendations to Governor Pataki for Reducing New York State Greenhouse Gas Emissions*. Washington, DC: CCAP, April.
CCES (Center for Climate and Energy Solutions). 2011. "Climate Change 101: State Action," January, at www.c2es.org, accessed July 28, 2012.
2013a. "Options and Considerations for a Federal Carbon Tax," February 28, at www.c2es.org, accessed February 26, 2015.
2013b. "California Cap-and-Trade Program Summary," at www.c2es.org, accessed March 11, 2014.
2014. "Federal Vehicle Standards," at www.c2es.org, accessed May 14.
2015a. "Q & A: EPA Regulation of Greenhouse Gas Emissions from Existing Power Plants," at www.c2es.org, accessed July 14, 2015.
2015b. "Appliance Efficiency Standards in Detail," at www.c2es.org, accessed June 25, 2015.
CEC (California Energy Commission). 1978. "Regulations Establishing Energy Conservation Standards for New Residential and New Nonresidential Buildings, as Amended," July 26.
2003a. *Final 2003 Energy Action Plan I*, at www.energy.ca.gov, accessed February 26, 2015.
2003b. *Reducing California's Petroleum Dependence*, August.
2003c. *2003 Integrated Energy Policy Report*, December.
2005. *State of California's Actions to Address Global Climate Change*, at www.climatechange.ca.gov, accessed February 24, 2015.
2007. *2007 Integrated Energy Policy Report*, CEC-100–2007-008-CMF, at www.climatechange.ca.gov, accessed February 26, 2015.
2009. *California Energy Demand 2010–2020: Adopted Forecast*, December, at www.climatechange.ca.gov, accessed June 27, 2014.
2013a. *2013 Integrated Energy Policy Report*. CEC-100–2013-001-CMF, at www.climatechange.ca.gov, accessed May 16, 2014.
2013b. "History of California's Renewable Energy Programs," at www.energy.ca.gov, accessed May 19, 2014.

2014a. "About Go Solar California," at www.gosolarcalifornia.org, accessed February 23, 2015.

2014b. "Low Carbon Fuel Standard," at www.energy.ca.gov, accessed May 20, 2014.

Chan, Sewell. 2009. "Paterson Warns of 'Historic Economic Challenge'," *New York Times*, January 7.

Chatrchyan, Allison and Pamela Doughman. 2008. "Climate Policy in the USA," in Hugh Compston and Ian Bailey, eds., *Turning Down the Heat*. New York: Palgrave Macmillan, 241–62.

Chen, Jowei and Jonathan Rodden. 2013. "Unintentional Gerrymandering," *Quarterly Journal of Political Science* 8:239–69.

Chong, Dennis. 2015. "Explaining Public Conflict and Consensus on the Climate," in Yael Wolinsky-Nahmias, ed., *Changing Climate Politics*. Washington, DC: CQ Press, 110–45.

Christensen, Atle. 2003. "Convergence or Divergence?" FNI Report 6/2003, Fridtjof Nansen Institute, Lysaker, Norway.

Clark, Ian. 2011. "Lessons Learned and Suggestions for Improving Regulatory Cooperation between California and the EU," in David Vogel and Johan Swinnen, eds., *Transatlantic Regulatory Cooperation*. Northampton, MA: Edward Elgar, 295–316.

CNRA (California Natural Resources Agency). 2009. *2009 California Climate Adaptation Strategy*, at www.climatechange.ca.gov, accessed May 8, 2014.

Cochrane, James. 1981. "Carter Energy Policy and the Ninety-Fifth Congress," in Craufurd Goodwin, ed., *Energy Policy in Perspective*. Washington, DC: Brookings Institution, 547–600.

Coley, Jonathan and David Hess. 2012. "Green Energy Laws and Republican Legislators in the United States," *Energy Policy* 48:576–83.

Collier, Ruth and David Collier. 1991. *Shaping the Political Arena*. Princeton, NJ: Princeton University Press.

Compston, Hugh and Ian Bailey, eds. 2008. *Turning Down the Heat*. New York: Palgrave Macmillan.

Compston, Hugh and Ian Bailey. 2013. "Comparing Climate Policies," paper for presentation at the Political Studies Association Conference, Cardiff, Wales, March 25–27.

Confessore, Nicholas. 2007. "Spitzer Unveils Initiatives That He Says Would Have State Consuming Less Energy in 2015," *New York Times*, April 20.

Cooper, Michael. 2005. "The Shadow of His Predecessor Dominates the Pataki Legacy," *New York Times*, July 29.

Cordes, Jennifer. 2001. "Article X," *Albany Law Environmental Outlook Journal* 6 (Fall):37–46.

Cox, Alan, Carl Blumstein, and Richard Gilbert. 1991. "Wind Power in California," in Gilbert, ed., *Regulatory Choices: A Perspective on Developments in Energy Policy*. Berkeley, CA: University of California Press, 347–74.

CPUC (California Public Utilities Commission). n.d. "California's Decoupling Policy," at www.noe21.org, accessed February 26, 2015.

Crepaz, Markus. 1995. "Explaining National Variations of Air Pollution Levels," *Environmental Politics* 4:391–414.

Cromwell, Aaron. 2010. *Greenhouse Gas Reduction Strategies in State Climate Action Plans*. Master's thesis, Hubert H. Humphrey Institute of Public Affairs, University of Minnesota, May 11.

Cuomo, Andrew. 2010. *The New NY Agenda: A Cleaner, Greener NY*, at www.nyalt.org, accessed February 26, 2015.

2013. *NY Rising: 2013 State of the State*, at www.governor.ny.us, accessed March 11, 2014.

Dao, James. 1996. "Pataki Leads Rare Coalition For Bond Act," *New York Times*, October 22.

Darakjian, John. 2009. "SB 375," *Journal of Environmental Law* 27:371–411.

Davenport, Coral. 2014a. "A Dozen States File Suit Against New Coal Rules," *New York Times*, August 1.

2014b. "Key Details of E.P.A. Carbon Emissions Proposal," *New York Times*, June 2.

2014c. "Governments Await Obama's Move on Carbon to Gauge U.S. Climate Efforts," *New York Times*, May 26.

Davis, Steven and Ken Caldeira. 2010. "Consumption-Based Accounting of CO_2 Emissions," *PNAS: Proceedings of the National Academy of Sciences* 107:5687–92.

de Marchi, Neal. 1981. "The Ford Administration," in Craufurd Goodwin, ed., *Energy Policy in Perspective*. Washington, DC: Brookings Institution, 475–546.

DEC (Department of Environmental Conservation, New York State). 2004. "Green Building Tax Credit (GBTC) – Request for Comment on Green Refrigerants," at www.dec.ny.gov, accessed November 14, 2013.

Decotis, Paul. 2009. "Coming of Age in New York," *The Bridge: Linking Engineering and Society* 39 (Summer):37–43.

Decotis, Paul and Sean Ferguson. 2012. "The Politics of Energy in New York State," in Gerald Benjamin, ed., *The Oxford Handbook of New York State Government and Politics*. New York: Oxford University Press, 865–97.

Deitchman, Benjamin. 2014. *Why U.S. States Became Leaders in Climate and Energy Policy*. Ph.D. dissertation, School of Public Policy, Georgia Institute of Technology, August.

Denning, Robert. 2011. *The Creative Society*. Ph.D. dissertation, Ohio State University.

Derthick, Martha. 2010. "Compensatory Federalism," in Barry Rabe, ed., *Greenhouse Governance*. Washington, DC: Brookings Institution, 58–72.

Desai, Uday. 2002. "Institutional Profiles and Policy Performance," *Environmental Politics and Policy in Industrialized Countries*. Cambridge, MA: MIT Press, 357–81.

Dewey, Scott. 2000. *Don't Breathe the Air*. College Station, TX: Texas A&M University Press.

Dionne, E.J. 1988. "Bush Is Elected by a 6-5 Margin with Solid G.O.P. Base in South," *New York Times*, November 9.

Dipeso, Jim. 2005. "Business Makes the Case for Climate Change Policy," *Environmental Quality Management* 15:99–105.

Dirmann, Tina. 2001. "County Backs Ranchers' Declaration of Disaster," *Los Angeles Times*, January 24.

Dolsak, Nivas. 2001. "Mitigating Global Climate Change," *Policy Studies Journal* 29:414–36.

Domanski, Jeffrey, ed. 2008. *States and Climate Change*. Princeton, NJ: Policy Research Institute for the Region, Princeton University.

Dowling, Conor and Steve Lem. 2009. "Explaining Major and Third Party Candidate Entry in U.S. Gubernatorial Elections, 1980–2005," *State Politics and Policy Quarterly* 9:1–29.

Downs, Annie et al. 2013. *The 2013 State Energy Efficiency Scorecard*. Washington, DC: American Council for an Energy-Efficient Economy, November.

DSIRE (Database of State Incentives for Renewables and Efficiency). 2014. "New York Incentives/Policies for Renewables & Efficiency," at www.dsireusa.org, accessed May 20, 2014.

Duane, Timothy. 2010. "Greening the Grid," *Vermont Law Review* 34:711–80.

Dugan, Judy. 2007. "Clueless Mary Nichols," August 20, at www.consumerwatchdog.org, accessed May 2, 2014.

Dunlap, Riley. 1991. "Trends in Public Opinion Toward Environmental Issues," *Society and Natural Resources* 4:285–312.

Dunlap, Riley and Aaron McCright. 2011. "Organized Climate Change Denial," in John Dryzek, Richard Norgaard, and David Schlosberg, eds., *The Oxford Handbook of Climate Change and Society*. New York: Oxford University Press, 144–60.

Dunn, James. 2006. "Automobile Fuel Efficiency Policy," in Robert Repetto, ed., *Punctuated Equilibrium and the Dynamics of U.S. Environmental Policy*. New Haven, CT: Yale University Press, 197–231.

EBRD (European Bank for Reconstruction and Development). 2011. *Special Report on Climate Change: The Low Carbon Transition*. London: EBRD.

EDF (Environmental Defense Fund). 2007. "EPA Denies States the Right to Fight Global Warming," press release, December 19, at www.edf.org, accessed February 26, 2015.

EFTE (European Federation for Transport and Environment). 2011. *Fuelling Oil Demand*. Brussels: EFTE, April.

EGA (Environmental Grantmakers Association). 2012. *Tracking the Field, Volume 3*. New York: EGA.

Eichhammer, Wolfgang, et al. 2001. "Greenhouse Gas Reductions in Germany and the UK," Report prepared for the 6th Conference of the Parties (COP 6), Bonn, 16–27 July.

Eikeland, Per. 1993a. "The Shaping of US Energy Policy," *International Challenges* 13 (4):3–16.

1993b. "US Energy Policy in the Greenhouse," EED Report 1993/1, Fridtjof Nansen Institute, Lysaker, Norway.

1993c. "US Energy Policy at a Crossroads?" *Energy Policy* 21:987–99.

Eilperin, Juliet. 2007. "EPA Chief Denies California Limit on Auto Emissions," *Washington Post*, December 20.

2011a. "New Vehicle Rules to Curb Greenhouse Gas Emissions Spark Debate," *Washington Post*, July 3.

2011b. "Automakers, Obama Administration Agree on Fuel Efficiency Standards Through 2025," *Washington Post*, July 27.

Elkind, Sarah. 2011. *How Local Politics Shape Federal Policy*. Chapel Hill, NC: University of North Carolina Press.

Elliott, E. Donald. 2013. "Why the United States Does Not Have a Renewable Energy Policy," *Environmental Law Reporter* 43 (2):10095–10101.

ELP (Electric Light and Power). 2010. "Senate Introduces Stand Alone Renewable Energy Standard Bill," at www.elp.com, accessed August 28, 2014.

Elwell, Charles, 1969. "Claremont Professors Leap Into Smog Battle," *Los Angeles Times*, September 21.

Endicott, William. 1974. "Geothermal Energy Work Stalled by 2 Oil Firms," *Los Angeles Times*, October 5.

Endrud, Nathan. 2008. "State Renewable Portfolio Standards," *Harvard Journal on Legislation* 45:259–86.

Engel, Kristin. 2009. "Whither Subnational Climate Change Initiatives in the Wake of Federal Climate Legislation?" *Publius* 39:432–54

2015. "The Limits of National Climate Policy Making and the Role of the Courts," in Yael Wolinsky-Nahmias, ed., *Changing Climate Politics*. Washington, DC: CQ Press, 32–54.

Erlandson, Dawn. 1994. "The BTU Tax Experience," *Pace Environmental Law Review* 12:173–84.

EU (European Union). 2014. "Road Transport: Reducing CO_2 Emissions from Vehicles," at ec.europa.eu, accessed December 15, 2014.

Evans, Alex and David Steven. 2007. "Climate Change: The State of the Debate," Center on International Cooperation, October.

Evers, John. 2003. "Comments of the Business Council of New York State, Inc.," New York State Public Service Commission Case 03-E-0188, September 12, at www.dps.ny.us, accessed February 25, 2015.

Fairbanks, Robert. 1974. "Flournoy Offers 7-Point Plan to Reform PUC," *Los Angeles Times*, July 23.

1976. "Energy Panel Adopts Refrigerator Rules," *Los Angeles Times*, October 15.

Falkner, Robert. 2008. *Business Power and Conflict in International Environmental Politics*. New York: Palgrave Macmillan.

Farber, Daniel. 2011. "Legal Guidelines for Cooperation between the EU and American State Governments," in David Vogel and Johan Swinnen, eds., *Transatlantic Regulatory Cooperation*. Northampton, MA: Edward Elgar, 3–19.

Farley, et al. 2007. "Opening the Policy Window for Ecological Economics," *Ecological Economics* 63:344–54.

Farrell, Alexander and Michael Hanemann. 2009. "Field Notes on the Political Economy of California Climate Policy," in Henrik Selin and Stacey VanDeveer, eds., *Changing Climates in North American Politics*. Cambridge, MA: MIT Press, 87–109.

Fawcett, Jeffry. 1990. *The Political Economy of Smog in Southern California*. New York: Garland.

Felker, Edward. 2009. "EPA Issues CO2 'Endangerment Finding'," *Washington Times*, December 7.

Field, Christopher et al. 1999. *Confronting Climate Change in California*. Cambridge, MA: Union of Concerned Scientists.

Field Research Corporation. 2010. "State Ballot Propositions," press release # 2356, September 26, at www.field.com, accessed February 26, 2015.

Florida, Richard. 2002. *The Rise of the Creative Class*. New York: Basic Books.

Ford, Gerald. 1975. "Address Before a Joint Session of the Congress Reporting on the State of the Union," January 15, at www.presidency.ucsb.edu, accessed January 7, 2014.

Forsyth, Alfred. 1983. "The Sierra Club in New York and New Mexico, 1965–1978," interview conducted by Jeffrey Ingram, Bancroft Library, University of California at Berkeley, for the Sierra Club Oral History Project.

Foster, Douglas. 2001. "Jerry Brown: An Energy Gap, Not a Crisis," *Los Angeles Times*, May 27.

Foster, Joanna. 2014. "Gov. Pence Lets Indiana Energy Efficiency Program Die," *Thinkprogress*, March 31, at thinkprogress.org, accessed December 11, 2015.

Fowler, Luke and Joseph Breen. 2013. "The Impact of Political Factors on States' Adoption of Renewable Portfolio Standards," *Electricity Journal* 26 (March):79–94.

Fox, John. 1991. *Regression Diagnostics*. Newbury Park, CA: Sage Publications.

Franco, Guido et al. 2008. "Linking Climate Change Science with Policy in California," *Climatic Change* 87 (Supplement 1), S7–S20.

Friedman, Thomas. 2011. "This Is a Big Deal," *New York Times*, December 3.

Funk, William. 2009. "Constitutional Implications of Regional CO_2 Cap-and-Trade Programs," *Journal of Environmental Law* 27:353–69.

Galarraga, Ibon, Mikel González-Eguino, and Anil Markandya. 2009. "The Role of Regions in Climate Change Policy," Working paper 2009–04, Basque Centre for Climate Change, September.

Geman, Ben. 2013. "White House 'Delighted' by UK Coal Move," *The Hill*, November 20, at www.thehill.com, accessed November 20, 2013.

George, Alexander and Andrew Bennett. 2005. *Case Studies and Theory Development in the Social Sciences*. Cambridge, MA: MIT Press.

Gerdes, Justin. 2013. "How Will California Slash Greenhouse Gas Emissions 80% by 2050?" *Forbes*, February 27.

Gerston, Larry and Terry Christensen. 2004. *California Politics and Government*, 8th edn. Boston: Wadsworth.

2012. *California Politics and Government*, 11th edn. Boston: Wadsworth.

Getze, George. 1969. "Mothers Charge Laxity in Smog-Control Fight," *Los Angeles Times*, July 17.

Gifford, Robert and Louise Comeau. 2011. "Message Framing Influences Perceived Climate Change Competence, Engagement, and Behavioral Intentions," *Global Environmental Change* 21:1301–07.

Gillam, Jerry. 1975. "Energy Panel Falters, Says Assemblyman," *Los Angeles Times*, December 20.

1976. "Energy Panel Chief Defends Record," *Los Angeles Times*, March 16.

Gillis, Justin. 2014. "Panel's Warning on Climate Risk," *New York Times*, March 31.

Gipe, Paul. 1991. "Wind Energy Comes of Age," *Energy Policy* 19:756–67.

1995. *Wind Energy Comes of Age*. New York: John Wiley.

Global Carbon Project. 2014. "Global Carbon Budget 2014," September 21, at www.globalcarbonproject.org, accessed September 21, 2014.

Goff, Tom. 1974. "Brown Labels PUC a 'Private Utilities' Group," *Los Angeles Times*, May 1.

Goldenberg, Suzanne. 2013. "Secret Funding Helped Build Vast Network of Climate Denial Think Tanks," *Grist*, December 15, at grist.org, accessed September 18, 2014.

Gonzalez, George. 2002. "Urban Growth and the Politics of Air Pollution," *Polity* 35:213–36.

2005. *The Politics of Air Pollution*. Albany, NY: SUNY Press.

Gottlieb, Robert. 1993. *Forcing the Spring*. Washington, DC: Island Press.

Goulder, Lawrence, Mark Jacobsen, and Arthur van Benthem. 2012. "Unintended Consequences from Nested State and Federal Regulations," *Journal of Environmental Economics and Management* 63:187–207.

Graetz, Michael. 2011. *The End of Energy*. Cambridge, MA: MIT Press.

Greenblatt, Jeffrey and Jane Long. 2012. *California's Energy Future*. Sacramento, CA: California Council on Science and Technology.

Greene, David. 2010. "Measuring Energy Security," *Energy Policy* 38:1614–21.

Griffin, Karen Griffin. 1985. "State Government Conservation Programs," in John Sawhill and Richard Cotton, eds., *Energy Conservation*. Washington, DC: Brookings Institution, 205–36.

Grist. 2005. "Passage of Energy Bill Highlights Lack of United Democratic Opposition," *Grist*, August 9, at grist.org, accessed February 23, 2015.

Grunwald, Michael and Juliet Eilperin. 2005. "Energy Bill Raises Fears About Pollution, Fraud," *Washington Post*, July 30.

Hakim, Danny. 2004. "California Backs Plan for Big Cut in Car Emissions," *New York Times*, September 24.

2005. "Battle Lines Set as New York Acts to Cut Emissions, *New York Times*, November 26.

2006. "10 States Sue E.P.A. on Emissions," *New York Times*, April 28.

2007a. "States Set to Sue the U.S. Over Greenhouse Gases," *New York Times*, October 24.

2007b. "Spitzer Budget Seeks Overhaul of Health Care," *New York Times*, February 1.

Hall, Nina and Ros Taplin. 2010. "Environmental Nonprofit Campaigns and State Competition," *Voluntas* 21: 62–81.

Hamm, Keith and Gary Moncrief. 2008. "Legislative Politics in the States," in Virginia Gray and Russell Hanson, eds., *Politics in the American States*, 9th edn. Washington, DC: CQ Press, 154–91.

Hanemann, Michael. 2007. "How California Came to Pass AB 32, the Global Warming Solutions Act of 2006," Department of Agricultural and Resource Economics, University of California at Berkeley, March, at escholarship.org/uc/item/1vb0j4d6, accessed April 23, 2012.

2008. "California's New Greenhouse Gas Laws," *Review of Environmental Economics and Policy* 2:114–29.

Hanemann, Michael and Chris Busch. 2011. "Climate Change Policy in California," in David Vogel and Johan Swinnen, eds., *Transatlantic Regulatory Cooperation*. Northampton, MA: Edward Elgar, 125–60.

Happaerts, Sander, Karoline van den Brande, and Hans Bruyninckx. 2011. "Subnational Governments in Transnational Networks for Sustainable Development," *International Environmental Agreements* 11:321–39.

Harder, Amy. 2014. "Obama Carbon Rule Backed by Most Americans – WSJ/NBC Poll," *Wall Street Journal*, June 18.

Harris, Ellen Stern. 1978. "Observing Sun Day for Energy," *Los Angeles Times*, April 30.

Harris, Frank and Peter Navarro. 1999. "Policy Options for Promoting Wind Energy Development in California," Graduate School of Management, University of California at Irvine, November.

Harrison, Kathryn. 2010. "The United States as Outlier," in Harrison and Lisa Sundstrom, eds., *Global Commons, Domestic Decisions*. Cambridge, MA: MIT Press, 67–104.

2014. "Federalism and Climate Policy Innovation," *Canadian Public Policy* 39 (2): S95–S108.

Harrison, Kathryn and Lisa Sundstrom. 2007. "Introduction," *Global Environmental Politics* 7 (November):1–18.

2010. "Conclusion," in idem., eds., *Global Commons, Domestic Decisions*. Cambridge, MA: MIT Press, 261–89.

Hart, Gary. 1990. "Ill Equipped for the Crises of the 90's," *New York Times*, August 7.

Haskell, Elizabeth and Victoria Price. 1973. *State Environmental Management*. New York: Praeger Publishers.

Hawkins, Troy, et al. 2012. "Comparative Environmental Life Cycle Assessment of Conventional and Electric Vehicles," *Journal of Industrial Ecology* 17:53–64.

Hay, Colin. 2009. "Political Ontology," Robert Goodin, ed., *The Oxford Handbook of Political Science*. New York: Oxford University Press, chapter 23.

Hayes, Andrew and Li Cai. 2007. "Using Heteroskedasticity-Consistent Standard Error Estimators in OLS Regression," *Behavior Research Methods* 39:709–22.

Hayhoe, Katharine et al. 2004. "Emissions Pathways, Climate Change, and Impacts on California," *PNAS: Proceedings of the National Academy of Sciences* 101:12422–27.

Hellwig, Timothy. 2014. *Globalization and Mass Politics: Retaining the Room to Maneuver*. New York: Cambridge University Press.

Hempel, Lamont. 2006. "Climate Policy on the Installment Plan," in Norman Vig and Michael Kraft, eds., *Environmental Policy*, 6th edn. Washington: CQ Press, 288–310.

Hernandez, Raymond. 1996. "Pataki's Support Tops 50% for First Time, a Poll Says," *New York Times*, September 22.

Hershey, John. 1990. "Energy Policy Trade-Off," *New York Times*, August 18.

Herszenhorn, David. 2008. "After Verbal Fire, Senate Effectively Kills Climate Change Bill," *New York Times*, June 7.

Hicks, Jonathan. 2002. "While Spitzer Shrugs It Off, Others Talk of His Future," *New York Times*, October 28.

Hill, Gladwin. 1978a. "California Board Won't Give Permit for New Nuclear Energy Project," *New York Times*, February 1.

1978b. "California County Votes, 2 to 1, to Bar Nuclear Plant," *New York Times*, March 9.

1978c. "Coast Legislators Apparently Doom Nuclear Plant," *New York Times*, April 18.

Hioureas, Christina and Bruce Cain. 2011. "Transatlantic Environmental Regulation-Making," in David Vogel and Johan Swinnen, eds., *Transatlantic Regulatory Cooperation*. Northampton, MA: Edward Elgar, 20–48.

Hirsh, Richard. 1999. *Power Loss*. Cambridge, MA: MIT Press.

Hoffman, Andrew. 2007. *Carbon Strategies*. Ann Arbor, MI: University of Michigan Press.

Horsley, Carter. 1978. "Home Fuel Equation Changing," *New York Times*, June 11.

Hovey, Kendra and Harold Hovey. 2001. *CQ's State Fact Finder 2001*. Washington, DC: CQ Press.

Huang, Ming-Yuan et al. 2007. "Is the Choice of Renewable Portfolio Standards Random?" *Energy Policy* 35:5571–75.

Huber, Bruce. 2013. "How Did RGGI Do It?" *Ecological Law Quarterly* 40:59–106.

Huberty, Mark and John Zysman. 2014. "Preface," in Zysman and Huberty, eds., *Can Green Sustain Growth?* Stanford, CA: Stanford Business Books, xi–xiv.

Hughes, Gary. 2006. "Council Remains Concerned About Potential Effect of Renewable-Energy Mandate on Costs, Reliability," March 6, at www.bcnys.org, accessed November 17, 2014.

Hulse, Carl. 2005. "Senate Makes Environment the Focus of Energy Bill," *New York Times*, June 17.

Hume, Ellen. 1977. "Governors Urge President to Use Energy Resources," *Los Angeles Times*, July 10.

Humes, Edward. 2009. *Eco Barons*. New York: Ecco.

ICLEI (International Council for Local Environmental Initiatives). 2009. "U.S. Local Sustainability Plans and Climate Action Plans," updated November 30, at www.icleiusa.org, accessed May 30, 2014.

IEA (International Energy Agency). 2014. *IEA Wind 2013 Annual Report*, July, at www.ieawind.org, accessed February 26, 2015.

Inglehart, Ronald. 1977. *The Silent Revolution*. Princeton, NJ: Princeton University Press.

Ingram, Carl. 2002a. "Senate Panel Votes to Cut Auto Emissions," *Los Angeles Times*, April 2.

2002b. "Senate Passes Alternate Exhaust Emissions Bill," *Los Angeles Times*, June 30.

IPCC (Intergovernmental Panel on Climate Change). 1990. *Climate Change*. New York: Cambridge University Press.

Jacobs, Chip and William Kelly. 2008. *Smogtown*. New York: Overlook Press.

Jacques, Peter, Riley Dunlap, and Mark Freeman. 2008. "The Organisation of Denial," *Environmental Politics* 17:349–85.

Jahn, Detlef. 1998. "Environmental Performance and Policy Regimes," *Policy Sciences* 31:107–31.

2008. "The Politics of Climate Change," paper prepared for the ECPR Joint Sessions of Workshops, Rennes, France, April 11–16.

Jänicke, Martin. 1992. "Conditions for Environmental Policy Success," *The Environmentalist* 12:47–58.

2005. "Trend Setters in Environmental Policy," *European Environment* 15:129–42.

2008. *Megatrend Umweltinnovation*. Munich: Oekom Verlag.

Janiskee, Brian and Ken Masugi. 2008. *Democracy in California*, 2nd edn. Lanham, MD: Rowman and Littlefield.

Johnson, Martin, Paul Brace, and Kevin Arceneaux. 2005. "Public Opinion and Dynamic Representation in the American States," *Social Science Quarterly* 86:87–108.

Johnson, Evan and Gregory Nemet. 2010. "Willingness to Pay for Climate Policy," Robert M. La Follette School of Public Affairs, University of Wisconsin at Madison, June.

Jones, Charles O. 1975. *Clean Air: The Policies and Politics of Pollution Control*. Pittsburgh: University of Pittsburgh Press.

Jones, Theodore. 1963. "Lilienthal Fears Atom Plant Here," *New York Times*, April 5.

Jones, Gregg and Dan Morain. 2003. "U.S. Rejected Davis on Aid to Clear Trees," *Los Angeles Times*, October 31.

Joppke, Christian. 1993. *Mobilizing Against Nuclear Energy*. Berkeley, CA: University of California Press.

Joskow, Paul. 2001. "U.S. Energy Policy During the 1990s," NBER Working Paper 8454, July.

Jost, Gesine and Klaus Jacob. 2004. "The Climate Change Policy Network in Germany," *European Environment* 14:1–15.

Kaplan, Thomas. 2014. "Cuomo, Joined by Biden, Details Disaster Aid Plans," *New York Times*, January 7.

Karapin, Roger. 2012. "Explaining Success and Failure in Climate Policies," *Comparative Politics* 45:46–68.

2014. "Wind-Power Development in Germany and the United States," in Andreas Duit, ed., *State and Environment*. Cambridge, MA: MIT Press, 111–45.

Katz, James. 1984. *Congress and National Energy Policy*. New Brunswick, NJ: Transaction Books.

Kazazis, Alexander. 2012. "The Western Climate Initiative," *Brooklyn Journal of International Law* 37:1177–1214.

Keating, Hal. 1970. "Youths to Demonstrate Concern for Ecology, Environmental Pollution," *Los Angeles Times*, April 19.

Kelemen, Daniel and David Vogel. 2010. "Trading Places: The Role of the United States and the European Union in International Environmental Politics," *Comparative Political Studies* 43:427–56.

Keller, Sarina. 2010. "Sources of Difference," *Energy Policy* 38:4741–42.

Kelly, Geoffrey. 2008. *National Policy Choices for an International Problem*. Ph.D. thesis, Faculty of Commerce, University of Wollongong, Australia.

Kersh, Rogan. 2006. "Interest-Group Lobbying in New York State," in Robert Pecorella and Jeffrey Stonecash, eds., *Governing New York State*, 5th edn. Albany, NY: SUNY Press, 93–110.

Kihss, Peter. 1979. "Energy Official Is 'Disappointed' by Utilities' Conservation Drive," *New York Times*, November 18.

King, Gary. 1986. "How Not to Lie with Statistics," *American Journal of Political Science* 30:666–87.

King, Peter. 2001. "Thanks to Bush, Green Interest Soars," *Los Angeles Times*, June 24.

Kingdon, John. 2003. *Agendas, Alternatives, and Public Policies*, 2nd edn. New York: Longman.

Kliegman, Julie. 2014. "Jerry Brown Says 'Virtually No Republican' in Washington Accepts Climate Change Science," *Politifact*, September 25, at www.politifact.com, accessed February 25, 2015.

Klyza, Christopher and David Sousa. 2013. *American Environmental Policy*, updated edn. Cambridge, MA: MIT Press.

Knill, Christoph, Susumu Shikano, and Jale Tosun. 2014. "Explaining Environmental Policy Adoption," in Andreas Duit, ed., *State and Environment*. Cambridge, MA: MIT Press, 53–79.

Knox-Hayes, Janelle. 2012. "Negotiating Climate Legislation," *Regulation and Governance* 6:545–67.

Koehn, Peter. 2008. "Underneath Kyoto," *Global Environmental Politics* 8:53–77.

Koronowski, Ryan. 2013. "California, Oregon, Washington, and British Columbia Agree to Cooperate on Reducing Carbon Pollution," *Thinkprogress*, October 28, at thinkprogress.org, accessed July 5, 2014.

Koronowski, Ryan and Tiffany Germain. 2014. "The Climate Guide to Governors," *Thinkprogress*, July 1, at thinkprogress.org, accessed February 25, 2015.

Kosloff, Laura, Mark Trexler, and Hal Nelson. 2004. "Outcome-Oriented Leadership," *Widener Law Journal* 14:173–204.

Kraft, Michael. 1981. "Congress and National Energy Policy," in Regina Axelrod, ed., *Environment, Energy, Public Policy*. Lexington, MA: D.C. Heath, 37–60.

2002. "Environmental Policy and Politics in the United States," in Uday Desai, ed., *Environmental Politics and Policy in Industrialized Countries*. Cambridge, MA: MIT Press, 29–70.

2004. *Environmental Policy and Politics*, 3rd edn. New York: Pearson.

Kraft, Michael and Regina Axelrod. 1984. "Political Constraints on Development of Alternative Energy Sources," *Policy Studies Journal* 13:319–30.

Krause, Rachel. 2011. "Symbolic or Substantive Policy?" *Environment and Planning C* 29:46–62.

2015. "Climate Policy Innovation in American Cities," in Yael Wolinsky-Nahmias, ed., *Changing Climate Politics*. Washington, DC: CQ Press, 82–108.

Krier, James and Edmund Ursin. 1977. *Pollution and Policy*. University of California Press.

Künkel, Nana, Klaus Jacob, and Per-Olof Busch. 2006. "Climate Policies – (the Feasibility of) a Statistical Analysis of Their Determinants," unpublished manuscript, Environmental Policy Research Center, Free University of Berlin.

Laird, Frank. 2001. *Solar Energy, Technology Policy, and Institutional Values*. New York: Cambridge University Press.

Laird, Frank and Christoph Stefes. 2009. "The Diverging Paths of German and United States Policies for Renewable Energy," *Energy Policy* 37:2619–29.

Lambright, Henry, Sarah Pralle, and Jessica Boscarino. 2006. "Governing Energy Innovation," in Dianne Rahm, ed., *Sustainable Energy and the States*. Jefferson, NC: McFarland and Company, 26–47.

Lampe, David. 1978. "Law Is About to Snuff Pilot Lights in California," *Los Angeles Times*, June 19.

Landler, Mark. 2014. "U.S. and China Reach Climate Accord After Months of Talks," *New York Times*, November 11.

Lantis, Jeffrey. 2006. "The Life and Death of International Treaties," *International Politics* 43:24–52.

Larocca, James. 1981. "Empire State Pioneers in Conserving Energy," *New York Times*, February 23.

Larue, Anna, Noelle Cole, and Peter Turnbull. 2013. "What if This Actually Works?" in Fereidoon Sioshansi, *Energy Efficiency*. Waltham, MA: Elsevier, 275–304.

LAT (Los Angeles Times). 1969. "Youths Push Smog Fight," *Los Angeles Times*, August 21.

1974. "Brown Urges Energy Panel," *Los Angeles Times*, April 16.

1976a. "California Moves to Ban Pilot Lights on Appliances," *Los Angeles Times*, April 2.

1976b. "Energy Panel Approves Strict Conservation Rules for Buildings," *Los Angeles Times*, February 7.
1977. "Energy Panel Bans Electric Coil Heat," *Los Angeles Times*, March 24.
1978. "Marin Judge Overturns Energy Board's Changes," *Los Angeles Times*, June 4.
2001. "Big Cars, Shrinking Glaciers," *Los Angeles Times*, December 29.
2002a. "Fueling a Green Energy Quest," *Los Angeles Times*, March 25.
2002b. "Drought Has Cost Them $2 Million, Ranchers Say," *Los Angeles Times*, October 4.
2002c. "California, Acting Globally," *Los Angeles Times*, May 13.
2002d. "The State Takes the Wheel," *Los Angeles Times*, July 2.
2006. "Be a Cool Leader," *Los Angeles Times*, August 13.
Layzer, Judith. 2007. "Deep Freeze," in Michael Kraft and Sheldon Kamieniecki, eds. *Business and Environmental Policy*. Cambridge, MA: MIT Press, 93–125.
2010. *Open for Business*. Cambridge, MA: MIT Press.
2011. "Cold Front," in Theda Skocpol and Lawrence Jacobs, eds., *Reaching for a New Deal*. New York: Russell Sage Foundation.
2012. *The Environmental Case*, 3rd edn. Washington: CQ Press.
Lazo, Alejandro. 2014. "How Cap-and-Trade Is Working in California," *Wall Street Journal*, September 28.
Lee, Henry. 1983. "National Energy Policy from State and Local Perspectives," in Dorothy Zinberg, ed., *Uncertain Power*. New York: Pergamon Press, 157–81.
Leggett, Jeremy. 2000. *The Carbon War*. London: Penguin.
Lehmbruch, Gerhard and Phillippe Schmitter. 1982. *Patterns of Corporatist Policy-Making*. Beverly Hills, CA: Sage.
Lester, James and Emmett Lombard. 1990. "The Comparative Analysis of State Environmental Policy," *Natural Resources Journal* 30:301–19.
Levine, Gerald. 1984. "The New York State Environmental Quality Review Act of 1975," *Fordham Urban Law Journal* 12:1–72.
Lewis-Beck, Michael. 1980. *Applied Regression*. Newbury Park, CA: Sage Publications.
Lifset, Robert. 2006. "The Environmental Is Political," *Hudson River Valley Review* 22 (2):28–53.
2008. "Energy Conservation in America," in Mogens Rüdiger, ed., *The Culture of Energy*. Newcastle, UK: Cambridge Scholars Publishing, 158–66.
2014a. "Environmentalism and the Electrical Energy Crisis," in Lifset, ed., *American Energy Policy in the 1970s*. Norman, OK: University of Oklahoma Press, 283–302.
2014b. "A New Understanding of the American Energy Crisis of the 1970s," *Historical Social Research* 39 (4):22–42.
Lifsher, Marc. 2006a. "Emission Bill Sets Off Clash in Sacramento," *Los Angeles Times*, August 24.
2006b. "Gov. Calls for Curbs on Emissions," *Los Angeles Times*, April 12.
Liu, Xinsheng, Eric Lindquist, and Arnold Vedlitz. 2011. "Explaining Media and Congressional Attention to Global Climate Change, 1969–2005," *Political Research Quarterly* 64:405–19.
Lizza, Ryan. 2013. "The President and the Pipeline," *New Yorker* (September 16), 38–51.
Lomborg, Bjorn. 2013. "Green Cars Have a Dirty Little Secret," *Wall Street Journal*, March 11.

Lopez et al. 2012. *U.S. Renewable Energy Technical Potentials*. Golden, CO: National Renewable Energy Laboratory, July.

Lorinc, John. 2010. "3 Companies Quit Group Over Moves on Climate," *New York Times*, February 16.

Love, Norma. 2010. "Three States Raid Anti-Pollution Funds," *Associated Press*, December 20.

Lowenberger, Amanda. 2012. *The Efficiency Boom*. Washington, DC: American Council for an Energy-Efficient Economy, March.

Lowry, William. 1992. *Dimensions of Federalism*. Durham: Duke University Press.

1998. "Religion and the Demand for Membership in Environmental Citizen Groups," *Public Choice* 94:223–40.

Lucas, Greg. 2002. "Push to Pass Greenhouse Gas Measure," *San Francisco Chronicle*, January 22.

Lurkis, Alexander. 1982. *The Power Brink*. New York: Icare Press.

Lutsey, Nicholas and Daniel Sperling. 2008. "America's Bottom-Up Climate Change Mitigation Policy," *Energy Policy* 36:673–85.

Maples, John. 2013. "California Greenhouse Gas Emissions Standards for Light-Duty Vehicles," U.S. Energy Information Agency, at www.eia.gov, accessed January 10, 2013.

March, James and Johan Olsen. 1984. "The New Institutionalism," *American Political Science Review* 78:734–49.

Marist College. 2013. "The Scenic Hudson Decision," at library.marist.edu, accessed September 9, 2013.

Martin, Hugo. 2003. "Fire Season Arrives, but Not Funds to Clear Out Dead Trees," *Los Angeles Times*, May 29.

Martinot, Eric, Ryan Wiser, and Jan Hamrin. 2005. "Renewable Energy Policies and Markets in the United States," Center for Resource Solutions, San Francisco, at www.martinot.info, accessed May 31, 2010.

Matisoff, Daniel. 2008. "The Adoption of State Climate Change Policies and Renewable Portfolio Standards" *Review of Policy Research* 25:527–45.

Matisoff, Daniel and Jason Edwards. 2014. "Kindred Spirits or Intergovernmental Competition?" School of Public Policy, Georgia Institute of Technology.

Maw, Greg. 2006. "AB 32 (Nunez): Oppose," Senate Republican Office of Policy, AB 32 bill analyses, California State Archives.

May, Peter and Chris Koski. 2007. "State Environmental Policies," *Review of Policy Research* 24:49–64.

Mazmanian, Daniel, John Jurewitz, and Hal Nelson. 2008. "California's Climate Change Policy," *Journal of Environment and Development* 17:401–23.

2013. "The Paradox of 'Acting Globally While Thinking Locally'," *Journal of Environment and Development* 22:186–206.

Mazmanian, Daniel, Hal Nelson. and John Jurewitz. 2013. "Climate Change Policy," in Ethan Rarick, ed., *Governing California*, 3rd. edn. Berkeley, CA: Institute of Governmental Studies Press, 405–28.

Mazur, Allan and Eric Welch. 1999. "The Geography of American Environmentalism," *Environmental Science and Policy* 2:389–96.

McBeath, Jerry and Jonathan Rosenberg. 2006. *Comparative Environmental Politics*. Dordrecht, Netherlands: Springer.

McCarthy, Leo. 1996. Interview conducted by Carole Hicke, Regional Oral History Office, University of California at Berkeley, for the California State Archives State Government Oral History Program.

McCright, Aaron and Riley Dunlap. 2000. "Challenging Global Warming as a Social Problem," *Social Problems* 47:499–522.

2003. "Defeating Kyoto," *Social Problems* 50:348–73.

2011. "The Politicization of Climate Change and Polarization in the American Public's Views of Global Warming, 2001–2010," *Sociological Quarterly* 52: 155–94.

McCubbin, Patricia. 2013. "Regulation of Greenhouse Gases and Other Air Pollutants in the First Obama Administration and Major Air Issues for the Second Term," *Buffalo Environmental Law Journal* 21:47–68

McFadden, Chuck. 2013. *Trailblazer*. Los Angeles: University of California Press.

McGreevey, James. 2003. "An Energy Plan," *New York Times*, August 24.

McGreevy, Patrick and Melanie Mason. 2015. "Gov. Brown's Speech Evokes Praise, Trepidation among Lawmakers," *Los Angeles Times*, January 5.

Meckling, Jonas. 2011. *Carbon Coalitions*. Cambridge, MA: MIT Press.

Meislin, Richard. 1978. "Carey Seeks to Bar New Power Plants Using Nuclear Fuel," *New York Times*, July 28.

Melewski, Bernard. 2002. "Acid Rain and the Adirondacks," *Albany Law Review* 66:171–205.

Melillo, Jerry, Terese Richmond, and Gary Yohe, eds. 2014. *Climate Change Impacts in the United States*. Washington, DC: U.S. Government Printing Office.

Menz, Fredric and Stephan Vachon. 2006. "The Effectiveness of Different Policy Regimes for Promoting Wind Power," *Energy Policy* 34:1786–96.

Michaelowa, Axel. 1998. "Climate Policy and Interest Groups," *Intereconomics* 33:251–59.

Milbrath, Lester. 1984. *Environmentalists*. Albany, NY: SUNY Press.

Mildenberger, Matto. 2013. "The Politics of Strategic Accommodation," paper prepared for the International Conference on Public Policy, Grenoble, France, June.

Miller, Alan. 1995. "Energy Policy from Nixon to Clinton," *Environmental Law* 25:715–31.

Miller, Alice. 2009. "New York State's Plan for Energy and CO_2 Reductions," New York Department of Public Service, March 17, at www.aeeny.org, accessed November 12, 2013.

Mitchell, Robert, Angela Mertig, and Riley Dunlap. 1991. "Twenty Years of Environmental Mobilization," *Society and Natural Resources* 4:219–34.

Molina, Maggie. 2014. *The Best Value for America's Energy Dollar*. Washington, DC: American Council for an Energy-Efficient Economy, March.

Montgomery, W. David. 1993. "Interdependencies between Energy and Environmental Policies," in Hans Landsberg, ed., *Making National Energy Policy*. Washington, DC: Resources for the Future, 61–94.

Moore, Curtis. 2006. "A Bold Plan to Protect the Planet," *Los Angeles Times*, September 2.

Moore, Scott. 1970. "Student Activists Turn to a Pollution Frontier," *Los Angeles Times*, January 11.

Morain, Dan. 2005. "Senate Rejects Governor's Choice," *Los Angeles Times*, September 2.

Morehouse, Sarah and Malcolm Jewell. 2003. *State Politics, Parties, and Policy*, 2nd edn. Lanham, MD: Rowman and Littlefield.

Motavalli, Jim. 2012. "California's Quiet but Crucial Role in Shaping Fuel Economy Standards," *New York Times*, September 4.

Mulford, David Donald. 1989. Oral history interview conducted by Timothy Fong and Ann Lage, Regional Oral History Office, University of California at Berkeley, for the California State Archives State Government Oral History Program.

Müller, Edda. 1998. "Difference in Climate Change Policy in Germany and the United States From a Political Science Perspective," *Energy and Environment* 9:463–77.

Nadel, Steven. 2002. "Appliance and Equipment Efficiency Standards," *Annual Review of Energy and Environment* 27:159–92.

Nadel, Steven and David Goldstein. 1996. "Appliance and Equipment Efficiency Standards," Proceedings of the ACEEE Summer Study on Energy Efficiency in Buildings, American Council for an Energy-Efficient Economy, 163–72.

NAS (National Academy of Sciences). 1992. *Policy Implications of Greenhouse Warming*. Washington, DC: NAS.

Navarro, Mireya. 2010. "Paterson Wants New York to Slash Greenhouse Gases," *New York Times*, November 9.

2012. "Challenge to New York's Carbon Trading Fails," *New York Times*, June 14.

NBC News. 2005. "Bush Signs $12.3 Billion Energy Bill Into Law," August 8, at www.nbcnews.com, accessed April 14, 2014.

Nemet, Gregory and Daniel Kammen. 2007. "U.S. Energy Research and Development," *Energy Policy* 35:746–55.

Neumayer, Eric. 2003. "Are Left-Wing Party Strength and Corporatism Good for the Environment?" *Ecological Economics* 45:203–20.

Newman, Andy. 2011. "New York's Nuclear Future That Might Have Been," *New York Times*, April 13.

NHTSA (National Highway Transportation Safety Administration). 2011. *Summary of Fuel Economy Performance (Public Version)*. Washington, DC: U.S. Department of Transportation, April 28.

Nichols, Mary. 2009a. Letter to the U.S. Environmental Protection Agency and the U.S. Department of Transportation. May 18, at www.arb.ca.gov, accessed February 25, 2015.

2009b. "California's Climate Change Program," *Journal of Environmental Law* 27:185–212.

Nisbet, Matthew. 2009. "Communicating Climate Change," *Environment: Science and Policy for Sustainable Development*, 51 (2):12–23.

2011. *Climate Shift: Clear Vision for the Next Decade of Public Debate*. Washington, D.C.: American University, School of Communication.

Nivola, Pietro. 1999. "Are Europe's Cities Better?" *Public Interest* No. 137:73–84.

2010. "The Long and Winding Road," in Barry Rabe, ed., *Greenhouse Governance*. Washington, DC: Brookings Institution, 158–80.

Norris, Pippa. 2004. *Electoral Engineering*. New York: Cambridge University Press.

NRC (National Research Council). 2002. *Effectiveness and Impact of Corporate Average Fuel Economy (CAFE) Standards*. Washington, DC: National Academy Press.

2006. *State and Federal Standards for Mobile-Source Emissions*. Washington, DC: National Academies Press.

2010. *Hidden Costs of Energy*. Washington, DC: National Academies Press.
NYLS (New York Legislative Services). 1975. *New York State Legislative Annual 1975*. New York: NYLS.
1976. *New York State Legislative Annual 1976*. New York: NYLS.
NYSCAC (New York State Climate Action Council). 2010. *Climate Action Plan Interim Report*, November 9, at www.dec.ny.gov, accessed October 29, 2013.
NYSDOT (New York State Department of Transportation). 2014. "NYSDOT Implementation of the Smart Growth Public Infrastructure Policy Act," at www.dot.ny.gov, accessed December 16, 2014.
NYSERDA (New York State Energy Research and Development Authority). 2002. *2002 State Energy Plan and Final Environmental Impact Statement*. Albany, NY: NYSERDA.
2009. *2009 State Energy Plan*, December, at energyplan.ny.gov, accessed June 11, 2013.
2010. *Executive Order No. 111: Statewide Annual Energy Report for State Fiscal Year 2009–2010*, at ogs.ny.gov, accessed November 14, 2013.
2011. *Responding to Climate Change in New York State*, at www.nyserda.ny.gov, accessed November 1, 2013.
2012. *Fiscal Year 2012–13 Budget and Financial Plan*, at www.nyserda.ny.gov, accessed November 1, 2013.
2014. *New York State Greenhouse Gas Inventory and Forecast: Inventory 1990–2011 and Forecast 2012–2030*, April, at www.nyserda.ny.gov, accessed June 20, 2014.
2015. *New York State Renewable Portfolio Standard: Annual Performance Report through December 31, 2014*, at www.nyserda.ny.gov, accessed June 25, 2015.
NYT (New York Times). 1975a. "Carey Halts Work on Nuclear Plants," *New York Times*, February 5.
1975b. "New York's Energy Gap," *New York Times*, June 23.
1990. "America's Neck and a New Noose," *New York Times*, August 2.
1997. "Pataki Signs Bill Supporting Home Solar Power," *New York Times*, August 20.
2002. "California's Message to George Pataki," *New York Times*, July 24.
O'Dell, John. 2012. "Will California's Zero-Emissions Mandate Alter the Car Landscape?" June 14, at www.edmunds.com, accessed February 26, 2015.
Oshitani, Shizuka. 2006. *Global Warming Policy in Japan and Britain*. Manchester, UK: Manchester University Press.
OTA (Office of Technology Assessment, U.S. Congress). 1992. *Building Energy Efficiency*. Washington, DC: U.S. Government Printing Office, May.
Paine, Carl. 2005. *How the Green Guys Won*. Master's thesis, Department of Urban Studies and Planning, Massachusetts Institute of Technology, June.
Pataki, George. 2002. "Pataki's State of the State Speech," *New York Times*, January 9.
Paterson, David. 2009. "Executive Order 24, Establishing a Goal to Reduce Greenhouse Gas Emissions Eighty Percent by the Year 2050 and Preparing a Climate Action Plan," August 6, at www.dec.ny.gov, accessed February 28, 2015.
Paterson, Mathew. 1996. *Global Warming and Global Politics*. New York: Routledge.
Pelling, Mark. 2011. *Adaptation to Climate Change*. New York: Routledge.
Penna, Caetano and Frank Geels. 2011. "Enriching the Greening of Industry Literature with Issue Life Cycle Theory and a Triple Embeddedness Framework," paper presented at the DIME Final Conference, Maastricht, Netherlands, April 6–8.

Perez-Pena, Richard. 2001. "Pataki Orders a Shift Toward Renewable Energy," *New York Times*, June 11.

2002. "Governor's Task Force Is Divided on Plans to Reduce Greenhouse Gases," *New York Times*, June 12.

Perez-Pena, Richard and Patrick McGeehan. 2002. "Assault on Wall St. Misdeeds Lifts Spitzer's U.S. Profile and Makes Enemies," *New York Times*, November 4.

Perl, Anthony and James Dunn. 2007. "Reframing Automobile Fuel Economy Policy in North America," *Transport Reviews* 27:1–35.

Petersen, John. 2011. "Plug-In Vehicles and Their Dirty Little Secret," January 6, at renewableenergyworld.com, accessed May 19, 2014.

Peterson, Iver. 1976. "Business Spur Cited," *New York Times*, February 18.

Peterson, Thomas and Adam Rose. 2006. "Reducing Conflicts between Climate Policy and Energy Policy in the US," *Energy Policy* 34:619–31.

Pettit, David. 2012. "Investing in Clean Fuels Benefits Everyone," January 19, at switchboard.nrdc.org, accessed May 20, 2014.

2013. "The Ninth Circuit Decides the California Low Carbon Fuel Standard Case," September 18, at switchboard.nrdc.org, accessed May 20, 2014.

Pew Charitable Trusts. 2009. *The Clean Energy Economy*, June, at www.pewtrusts.org, accessed January 3, 2013.

Pew Research Center. 2010. "Top Stories of 2010," at www.people-press.org, accessed December 9, 2013.

2012. "Partisan Polarization Surges in Bush, Obama Years," at www.people-press .org, accessed October 9, 2014.

Plumer, Brad. 2015. "How the Next President Could Expand Obama's Climate Policies – or Dismantle Them," *Vox*, August 7, at www.vox.com, accessed December 18, 2015.

Polakovic, Gary. 2001. "States Taking the Initiative to Fight Global Warming," *Los Angeles Times*, October 7.

2002. "Assembly Bill Targets Global Warming Trend," *Los Angeles Times*, January 26.

Polakovic, Gary and Miguel Bustillo. 2002. "Davis Signs Bill to Cut Greenhouse Gases," *Los Angeles Times*, July 23.

Pollack, Ilana et al. 2013. "Trends in Ozone, its Precursors, and Related Secondary Oxidation Products in Los Angeles, California," *Journal of Geophysical Research: Atmospheres* 118:5893–5911.

Poloni-Staudinger, Lori. 2008. "Are Consensus Democracies More Environmentally Effective?" *Environmental Politics* 17:410–30.

Portney, Kent. 2013. *Taking Sustainable Cities Seriously*, 2nd edn. Cambridge, MA: MIT Press.

Posner, Paul. 2010. "The Politics of Vertical Diffusion," in Barry Rabe, ed., *Greenhouse Governance*. Washington, DC: Brookings Institution, 73–100.

Pralle, Sarah. 2006. " 'I'm Changing the Climate, Ask Me How!' " *Political Science Quarterly* 121:397–423.

2009. "Agenda-Setting and Climate Change," *Environmental Politics* 18:781–99.

Pratt, Charles. 2001. "Re-Inventing New York's Power Plant Siting Law," *Albany Law Environmental Outlook Journal* 6 (Fall):1–8.

Probst, Katherine and Sarah Szambelan. 2009. *The Role of the States in a Federal Climate Program*. Washington, DC: Resources for the Future.

Provost, Colin. 2010. "When Is AG Short for Aspiring Governor?" *Publius* 40:597–616.

Pryor, Larry. 1973. "Reagan Under Fire for Veto of Energy Conservation Bill," *Los Angeles Times*, October 4.

Rabe, Barry. 2004. *Statehouse and Greenhouse*. Washington, DC: Brookings Institution Press.

2005. "State Competition as a Source Driving Climate Change Mitigation," *New York University Environmental Law Journal* 14:2–53.

2006. "Race to the Top," Pew Center on Global Climate Change, June, at www.c2es .org, accessed March 1, 2015.

2008. "States on Steroids," *Review of Policy Research* 25:105–28.

2009a. "Governing the Climate from Sacramento," in Stephen Goldsmith and Donald Kettl, eds., *Unlocking the Power of Networks*. Washington, DC: Brookings Institution Press, 34–61.

2009b. "Second-Generation Climate Policies in the States," in Henrik Selin and Stacey VanDeveer, eds., *Changing Climates in North American Politics*. Cambridge, MA: MIT Press, 67–85.

2010a. "The Aversion to Direct Cost Imposition," *Governance* 23:583–608.

2010b. "The 'Impossible Dream' of Carbon Taxes," in Rabe, ed., *Greenhouse Governance*. Washington, DC: Brookings Institution, 126–57.

2010c. "Can Congress Govern the Climate?" in Rabe, ed., *Greenhouse Governance*. Washington, DC: Brookings Institution, 260–85.

2011. "Contested Federalism and American Climate Policy," *Publius* 41:494–521.

2013. "Reverse Diffusion and the Durability of State Cap and Trade Policy," paper prepared for the 2013 Annual Meeting of the American Political Science Association, Chicago.

2015. "A New Era in States' Climate Policies?" in Yael Wolinsky-Nahmias, ed., *Changing Climate Politics*. Washington, DC: CQ Press, 55–81.

Rabe, Barry and Philip Mundo. 2007. "Business Influence in State-Level Environmental Policy," in Michael Kraft and Sheldon Kamieniecki, eds., *Business and Environmental Policy*. Cambridge, MA: MIT Press, 265–97.

Rahm, Dianne. 2006. "Sustainable Energy," in Rahm, ed., *Sustainable Energy and the States*. Jefferson, NC: McFarland, 5–25.

Randolph, Sean. 2014. "The United States," in John Zysman and Mark Huberty, eds., *Can Green Sustain Growth?* Stanford, CA: Stanford Business Books, 125–49.

Raymond, Leigh. 2010. "The Emerging Revolution in Emissions Trading Policy," in Barry Rabe, ed., *Greenhouse Governance*. Washington, DC: Brookings Institution, 101–25.

2014. "States Leading the Way to a New Paradigm for Climate Policy," paper presented at the 2014 Midwest Political Science Association Research Conference, Chicago.

Reagan, Ronald. 1981. "Message to the Congress Transmitting the National Energy Policy Plan," July 17, at www.presidency.ucsb.edu, accessed March 1, 2015.

Redburn, Tom and Sandra Blakeslee. 1979. "Energy War: Californians in Cross Fire," *Los Angeles Times*, December 16.

Reeves, Richard. 1974. "Carey vs. Wilson," *New York Times*, October 27.

1975. " 'How Does the Governor of California Differ from a Shoemaker?' " *New York Times*, August 24.

Reich, Kenneth. 1974a. "Brown Pledges More Aides for Coast Panels," *Los Angeles Times*, September 21.

1974b. "Push for Other Power Sources Needed," *Los Angeles Times*, September 13.

Repetto, Robert, ed. 2006. *Punctuated Equilibrium and the Dynamics of U.S. Environmental Policy*. New Haven, CT: Yale University Press.

Reuters, 2005. "Seven U.S. States Sign CO_2 Plan in Break with Bush," *Reuters News Service*, December 21.

Revkin, Andrew. 1996a. "On the Environment, Pataki Is Seen in All Shades of Green," *New York Times*, January 24.

1996b. "1.5 Billion Bond Act Sought by Pataki for Environment," *New York Times*, June 7.

1997. "Pataki Bolsters State's Pursuit of Air Polluters," *New York Times*, February 1.

1999. "Spitzer and Pataki Dueling Over Environmental Mantle," *New York Times*, October 19.

2003. "Politics Reasserts Itself in the Debate Over Climate Change and Its Hazards," *New York Times*, August 5.

RGGI. (Regional Greenhouse Gas Initiative). 2013a. "Fact Sheet: RGGI CO_2 Allowance Auctions," at www.rggi.org, accessed February 25, 2015.

2013b. "RGGI States Propose Lowering Regional CO_2 Emissions Cap 45%," press release, February 7, at www.rggi.org, accessed March 2, 2015.

Richards, Kenneth. 2000. "Framing Environmental Policy Instrument Choice," *Duke Environmental Law and Policy Forum* 10:221–85.

Ricken, Christian. 1997. *Determinanten der Effektivität der Umweltpolitik*. New York: Peter Lang.

Righter, Robert. 1994. "Wind Energy in California," *California History* 73:142–55.

1996. *Wind Energy in America*. Norman, OK: University of Oklahoma Press.

Ringquist, Evan. 1993a. "Testing Theories of State Policy-Making," *American Politics Quarterly* 21:320–42.

1993b. *Environmental Protection at the State Level*. Armonk, NY: M.E. Sharpe.

Roe, David. 1984. *Dynamos and Virgins*. New York: Random House.

Roman, Mikael and Marcus Carson. 2009. *Sea Change*. Stockholm: Commission on Sustainable Development.

Ronan, Thomas. 1974. "Samuels Backed by 6 Union Chiefs," *New York Times*, August 29.

Rood, W.B. 1980. "Accord Reached on Brown Solar Energy Legislation," *Los Angeles Times*, March 4.

Roosevelt, Margot. 2010. "Proposition 23: Backers were Outspent, Out-Organized," *Los Angeles Times*, November 2.

Rosen, Amanda. 2008. "Climate Change Policymaking in the United States," paper presented at the ECPR conference, Barcelona, August 25–27.

2009. *Emission Impossible?* Ph.D. dissertation, Department of Political Science, Ohio State University.

Rosen, Hy and Peter Slocum. 1998. *From Rocky to Pataki*. Syracuse, NY: Syracuse University Press.

Rosenbaum, Walter. 1981. "Notes from No Man's Land," in Regina Axelrod, ed., *Environment, Energy, Public Policy*. Lexington, MA: D.C. Heath, 61–80.

Rosenberg, Paul. 2007. "Governor Schwarzenegger's Global Warming Act Called Hot Air," July 27, at www.arnoldwatch.org, accessed May 2, 2014.

Rosenfeld, Arthur. 2009. "A Graph Is Worth a Thousand Gigawatt-Hours," *Innovations* 4 (Fall):57–79.
Rowlands, Ian. 2010. "The Development of Renewable Electricity Policy in the Province of Ontario," *Review of Policy Research* 24:185–207.
Royden, Amy. 2010. "U.S. Climate Change Policy Under President Clinton," *Golden Gate University Law Review* 32:415–78.
RT (Russia Today). 2014. "Ohio Becomes First State to Reverse Energy Efficiency, Renewable Fuel Mandates," at www.rt.com, accessed December 11, 2015.
Rubenstein, Maribeth. 2012. *Policy Shifts toward an Energy System Transition.* Master's thesis, Sustainable Development, Utrecht University, August.
Rudolph, Richard and Scott Ridley. 1986. *Power Struggle.* New York: Harper and Row.
Russo, Frank. 2006. "California State Assembly Passes Historic Global Warming Bill," *California Progress Report*, August 31.
2007. "The History of Schwarzenegger, AB 32, and Global Warming," *California Progress Report*, February 23.
Saad, Lydia. 2015. "U.S. Views on Climate Change Stable after Extreme Winter," March 25, at www.gallup.com, accessed September 15, 2015.
Sabatier, Paul. 1988. "An Advocacy Coalition Framework of Policy Change and the Role of Policy-Oriented Learning Therein," *Policy Sciences* 21:129–68.
Sabatier, Paul and Christopher Weible. 2007. "The Advocacy Coalition Framework," in Sabatier, ed., *Theories of the Policy Process*, 2nd edn. Boulder, CO: Westview Press, 189–220.
Salkin. Patricia. 2014. "The Executive and the Environment," *Pace Environmental Law Review* 31:706–68.
San Francisco Chronicle. 1973. "Energy Crisis Bill Gets an OK," *San Francisco Chronicle*, August 10.
2002. "Stop a Changing Climate," *San Francisco Chronicle*, January 28.
Schipper, Lee et al. 1979. "The National Energy Conservation Policy Act," *Natural Resources Journal* 19:765–85.
Schlozman, Kay, Sidney Verba, and Henry Brady. 2013. *The Unheavenly Chorus.* Princeton, NJ: Princeton University Press.
Schneider, Jordan, Tony Dutzig, and Rob Sargent. 2013. "Wind Energy for a Cleaner America II," Environment Iowa Research and Policy Center, Fall, at www.environmentiowacenter.org, accessed January 15, 2015.
Schneier, Edward and John Murtaugh. 2001. *New York Politics.* Armonk, NY: M.E. Sharpe.
Schoch, Deborah and Janet Wilson. 2006. "Governor, Blair Reach Environmental Accord," *Los Angeles Times*, August 1.
Schoen, John. 2007. "California Adds up the Cost of Wildfire Damages," at www.nbcnews.com, October 26; accessed April 14, 2014.
Schoen, Luke. 2013. "Why Is China Taking Action on Clean Energy and Climate Change?" World Resources Institute, at www.chinafaqs.org, accessed November 28, 2014.
Schor, Elana. 2015. "Republicans Outfox Democrats on Climate Votes," *Politico*, January 21, at www.politico.com, accessed January 26, 2015.
Schrepfer, Susan. 1983. *Fight to Save the Redwoods.* Madison: University of Wisconsin Press.

Schreurs, Miranda. 2002. *Environmental Politics in Japan, Germany, and the United States*. New York: Cambridge University Press.

Schreurs, Miranda and Yves Tiberghien. 2007. "Multi-Level Reinforcement," *Global Environmental Politics* 7 (4):19–46.

Schroeder, Larry, David Sjoquist, and Paula Stephan. 1986. *Understanding Regression Analysis*. Newbury Park, CA: Sage.

Schulte, Daniel. 2006. *The Media and a State-Level Climate Change Policy*. Master's thesis, Department of Urban and Environmental Policy and Planning, Tufts University, July.

Schulz, Max. 2008. "NY Unplugged?" Empire Center for New York State Policy, March, at www.empirecenter.org., accessed November 13, 2013.

Schwarzenegger, Arnold. 2003. "Governor Schwarzenegger's Action Plan for California's Environment," at www.e2.org, accessed May 1, 2014.

2005. "Executive Order S-3-05," at www.gov.ca.gov, accessed May 1, 2014.

2012. *Total Recall*. New York: Simon and Schuster.

Sciortino, Michael, et al. 2011. *The 2011 State Energy Efficiency Scorecard*. Washington, DC: American Council for an Energy-Efficient Economy.

Scruggs, Lyle. 1999. "Institutions and Environmental Performance in Seventeen Western Democracies," *British Journal of Political Science* 29:1–31.

2003. *Sustaining Abundance*. New York: Cambridge University Press.

Scruggs, Lyle and Salil Benegal. 2012. "Declining Public Concern About Climate Change," *Global Environmental Change* 22:505–15.

Sekar, Samantha and Brent Sohngen. 2014. *The Effects of Renewable Portfolio Standards on Carbon Intensity in the United States*. Washington, DC: Resources for the Future.

Selin, Henrik and Stacy VanDeveer. 2005. "Canadian-U.S. Environmental Cooperation," *American Review of Canadian Studies* 35:353–78.

2007. "Political Science and Prediction," *Review of Policy Research* 24:1–27.

2009. "Climate Leadership in Northeast North America," in idem., eds., *Changing Climates in North American Politics*. Cambridge, MA: MIT Press, 111–35.

2011. "US Climate Change Politics and Policymaking," *WIREs Climate Change* 2:121–27.

2012. "Federalism, Multilevel Governance, and Climate Change Politics Across the Atlantic," in Paul Steinberg and VanDeveer, eds., *Comparative Environmental Politics*. Cambridge, MA: MIT Press, 341–68.

Sellers, Chris. 2012. *Crabgrass Crucible*. Chapel Hill: University of North Carolina Press.

Senate Rules Committee (California). 2002. "AB 1493 Bill Analysis," July 1, 2002, at leginfo.ca.gov, accessed April 28, 2014.

2006. "AB 32 Bill Analysis," August 30, 2006, at leginfo.ca.gov, accessed May 2, 2014.

Shaman, Diana. 1979. "Energy Audits Find Many Houses in Red," *New York Times*, July 22.

Sharma, Rohit. 2002. "*Niagara Mohawk Power Corp. v. FERC*," *Energy Law Journal* 23:157–77.

Shwom, Rachael. 2011. "A Middle-Range Theorization of Energy Politics," *Environmental Politics* 20:705–26.

Siaroff, Alan. 1999. "Corporatism in 24 Industrial Democracies," *European Journal of Political Research* 36:175–205.

Simon, Christopher. 2007. *Alternative Energy*. Lanham, MD: Rowman and Littlefield.

Simon, Richard. 2002. "Senate Strikes a Blow in the Energy Battle," *Los Angeles Times*, March 22.

Skelton, George. 1978a. "Brown's Record: a Lot of Gray Areas," *Los Angeles Times*, October 18.

1978b. "Brown Chides U.S. on Solar Power Lag," *Los Angeles Times*, April 26.

Skjaerseth, Jon, Guri Bang, and Miranda Schreurs. 2013. "Explaining Growing Climate Policy Differences between the European Union and the United States," *Global Environmental Politics* 13 (November):61–80.

Skocpol, Theda. 2013. "Naming the Problem," paper prepared for a symposium on The Politics of America's Fight Against Global Warming, Harvard University, February 14.

Smith, Alison. 2013. *The Climate Bonus: Co-Benefits of Climate Policy*. New York: Routledge.

Smith, Eric. 2002. *Energy, Environment, and Public Opinion*. Lanham, MD: Rowman and Littlefield.

Soifer, Hillel. 2012. "The Causal Logic of Critical Junctures," *Comparative Political Studies* 45:1572–97.

Sollen, Robert. 1998. *An Ocean of Oil*. Juneau, AK: Denali Press.

Sovacool, Benjamin. 2008. *The Dirty Energy Dilemma*. Westport, CT: Praeger.

2009. "Rejecting Renewables," *Energy Policy* 37:4500–4513.

Spross, Jeff, Ryan Koronowski, and Tiffany Germain. 2013. "The Anti-Science Climate Denier Caucus: 113th Congress Edition," June 26, at thinkprogress.org, accessed February 25, 2015.

Stall, Bill. 1977. "Brown OKs $3,000 Tax Credit on Home Solar Units," *Los Angeles Times*, September 27.

Stavins, Robert, Judson Jaffe, and Todd Schatzki. 2007. "Too Good to be True?" NBER Working Paper 13587, November, at www.nber.org, accessed February 8, 2012.

Stein, Ted. 1980. *The Path Not Taken*. Washington, DC: Common Cause.

Stern, Frank and Nicole Wobus. 2008. "Renewable Energy Credit Prices," prepared for the New York State Energy Research and Development Authority by Summit Blue Consulting, November 14, at www.puc.nh.gov, accessed May 21, 2010.

Stevens, William. 1991. "At Meeting on Global Warming, U.S. Stands Alone," *New York Times*, September 10.

Stout, David. 2001. "Bush Defends His Policies on First Day of Europe Visit," *New York Times*, June 12.

Sudarshan, Anant. 2011. "Why Is Per Capita Residential Energy Consumption in California So Low?" Department of Management Science and Engineering, Stanford University, March 5.

Sweeney, James. 2002. *The California Electricity Crisis*. Stanford, CA: Hoover Institution Press.

Swidler, Joseph. 2002. *Power and the Public Interest*. Knoxville: University of Tennessee Press.

Swift, Art. 2014. "Americans Again Pick Environment Over Economic Growth," at www.gallup.com, accessed February 26, 2015.

Swisher, Randall and Kevin Porter. 2006. "Renewable Policy Lessons from the US," in Karl Mallon, *Renewable Energy Policy and Politics*. Sterling, VA: Earthscan, 185–98.

Szarka, Joseph. 2007. *Wind Power in Europe*. New York: Palgrave.

Talberg, Anita and Kai Swoboda. 2013. "Emissions Trading Schemes Around the World," Parliament of Australia, June 6, at www.aph.gov.au, accessed May 12, 2014.

Talbot, Allan. 1972. *Power along the Hudson*. New York: Dutton.

Tang, Zhenghong et al. 2010. "Moving from Agenda to Action," *Journal of Environmental Planning and Management* 53:41–62.

Taylor, Mac. 2014. "The 2014–15 Budget: Cap-and-Trade Auction Revenue Expenditure Plan," Legislative Analyst's Office, at www.lao.ca.gov, accessed January 6, 2015.

Taylor, Margaret. 2008. "Beyond Technology-Push and Demand-Pull," *Energy Economics* 30:2829–54.

Teske, Paul, 2004. *Regulation in the States*. Washington, DC: Brookings Institution.

Thomas, Clive and Ronald Hrebenar. 1999. "Interest Group Power in the Fifty States," *Comparative State Politics* 20 (4):3–16.

Thomson, Vivian. 2014. *Sophisticated Interdependence in Climate Policy*. London: Anthem Press.

Tilly, Charles. 1984. *Big Structures, Large Processes, Huge Comparisons*. New York: Russell Sage.

Time. 2005. "Are We Making Hurricanes Worse?" *Time*, October 3, cover story.

Timney, Mary. 2004. *Power for the People*. Armonk, NY: M.E. Sharpe.

Tjernshaugen, Andreas and Ho-Ching Lee. 2007. "Shaming and Framing," in Paul Harris, ed., *Europe and Global Climate Change*. Northampton, MA: Edward Elgar, 185–210.

Trinkle, Daniel. 1995. "Cars, Congress, and Clean Air for the Northeast," *Boston College Environmental Affairs Law Review* 23:169–201.

True, James, Bryan Jones, and Frank Baumgartner. 2007. "Punctuated-Equilibrium Theory," in Paul Sabatier, ed., *Theories of the Policy Process*, 2nd ed,. Boulder, CO: Westview Press, 155–88.

Tucker, Harvey. 1982. "It's About Time," *American Journal of Political Science* 26:176–96.

UBA (Umweltbundesamt). 2007. *Environmental Data for Germany*. Berlin: UBA.

UCS (Union of Concerned Scientists). 2004. "Climate Change in California: Choosing Our Future," at www.calcleancars.org, accessed April 18, 2014.

UN (United Nations) 1992. "United Nations Framework Convention on Climate Change," at unfccc.int/key_documents/the_convention/items/2853.php, accessed January 2, 2014.

Ungar, Lowell et al. 2014. *Government Works*. Washington, DC: American Council for an Energy-Efficient Economy, September.

Urpelainen, Johannes. 2009. "Explaining the Schwarzenegger Phenomenon," *Global Environmental Politics* 9:82–105.

U.S. Census (U.S. Bureau of the Census). 1970. *Statistical Abstract of the United States: 1970*, 91st edn. Washingtown, DC: idem.

USDOE (U.S. Department of Energy). 2010. *Public Benefit Funds*, March.

2014a. "Previous Determinations," at www.energycodes.gov, accessed December 15, 2014.

2014b. "Status of State Energy Code Adoption," www.energycodes.gov, accessed December 15, 2014.

USDOS (U.S. Department of State). 1997. *Climate Action Report*. Washington, DC: U.S. Government Printing Office, July.

2002. *U.S. Climate Action Report: 2002*. Washington, DC: U.S. Government Printing Office, May.

2010. *U.S. Climate Action Report 2010*. Washington, DC: Global Publishing Services, June.

2014. *U.S. Climate Action Report 2014*, at www.state.gov, accessed June 27, 2014.

USEIA (U.S. Energy Information Agency). 2013. *State Energy Consumption Estimates: 1960 Through 2011*. Washington, DC: USEIA, June

2015. *Electric Power Monthly*. Washington, DC: USEIA, February.

USEPA (U.S. Environmental Protection Agency). 1991. "Standards of Performance for New Stationary Sources and Guidelines for Control of Existing Sources: Municipal Solid Waste Landfills," proposed regulations, May 30, at www.epa.gov, accessed March 2, 2015.

1996. "Standards of Performance for New Stationary Sources and Guidelines for Control of Existing Sources: Municipal Solid Waste Landfills," *Federal Register* 61 (49):9905–44.

2001. *The Power of Partnerships: Energy Star and Other Voluntary Programs, 2000 Annual Report*. Washington, DC: USEPA, July.

2010. "Light-Duty Vehicle Greenhouse Gas Emission Standards and Corporate Average Fuel Economy Standards: Final Rule," *Federal Register* 75 (88):25324–25728.

2011. *Energy Star and Other Climate Protection Partnerships: 2010 Annual Report*. Washington, DC: USEPA, December.

2012a. "EPA and NHTSA Set Standards to Reduce Greenhouse Gases and Improve Fuel Economy for Model Years 2017–2025 Cars and Light Trucks," August, at www.epa.gov, accessed May 20, 2014.

2012b. "2017 and Later Model Year Light-Duty Vehicle Greenhouse Gas Emissions and Corporate Average Fuel Economy Standards," *Federal Register* 77 (199):62624–63200.

2013. "EPA Proposes 2014 Renewable Fuel Standards, 2015 Biomass-Based Diesel Volume," November, at www.epa.gov, accessed August 28, 2014.

2014. "Carbon Pollution Emission Guidelines for Existing Stationary Sources: Electric Utility Generating Units," *Federal Register* 79 (117):34830–34958.

2015a. *Inventory of U.S. Greenhouse Gas Emissions and Sinks: 1990–2013*. Washington, DC: USEPA.

2015b. "Carbon Pollution Emission Guidelines for Existing Stationary Sources: Electric Utility Generating Units," Final Rule, August 3, 2015, advance online publication, forthcoming in *Federal Register*, accessed at www.epa.gov, October 30, 2015.

van Est, Rinie. 1999. *Winds of Change*. Utrecht, Netherlands: International Books.

Vasi, Ion. 2009. "Social Movements and Industry Development," *Mobilization* 14:315–36.

2011. *Winds of Change*. New York: Oxford University Press.

Vecsey, George. 1974. "Wilson Asserts Carey 'Is Deserting' Inflation Fight," *New York Times*, September 13.

Verleger, Philip. 1979. "The U.S. Petroleum Crisis of 1979," *Brookings Papers on Economic Activity* 2:463–75.

Vezirgiannidou, Sevasti-Eleni. 2013. "Climate and Energy Policy in the United States," *Environmental Politics* 22:593–609.

Vietor, Richard. 1984. *Energy Policy in America Since 1945*. New York: Cambridge University Press.

Vogel, David. 1986. *National Styles of Regulation*. Ithaca, NY: Cornell University Press.

1995. *Trading Up*. Cambridge, MA: Harvard University Press.

Vogel, Nancy. 2002. "Davis Signs Bill Boosting Clean Energy," *Los Angeles Times*, September 13.

Voorhees, Josh. 2009. "White House Rolls Out Details of Auto Fuel Economy, Emissions Standard," *New York Times*, September 15.

Walker, Jack. 1969. "The Diffusion of Innovations among the American States," *American Political Science Review* 63:880–99.

Walsh, Edward. 1981. "Resource Mobilization and Citizen Protest in Communities Around Three Mile Island," *Social Problems* 29 (October):1–21.

Walz, Rainer. 2007. "The Role of Regulation for Sustainable Infrastructure Innovations," *International Journal of Public Policy* 2:57–88.

Warren, Charles. 1973. Telegram sent to the editors of ten newspapers, Warren-Alquist Act public comments file, 131–32, at www.energy.ca.gov, accessed March 6, 2014.

1982. "From the California Assembly to the Council on Environmental Quality, 1962–1979," interview conducted by Sarah Sharp, Regional Oral History Office, Bancroft Library, University of California at Berkeley.

Warren, Jenifer. 2001. "State Joins Challenge to Bush on Air-Conditioner Standards," *Los Angeles Times*, June 20.

Watanabe, Rie. 2009. *A Comparative Analysis on Climate Policy Change Processes Between Germany and Japan*. Ph.D. dissertation, Free University of Berlin, 2009.

Weale, Albert. 1992. *The New Politics of Pollution*. New York: Manchester University Press.

Weart, Spencer. 2013. "The Discovery of Global Warming," February, at www.aip.org, accessed March 11, 2013.

Weil, Benjamin. 2009. *Renewable Energy in California*. Ph.D. dissertation, Environmental Studies, University of California, Santa Cruz, June.

Weiner, Edward 2008. *Urban Transportation Planning in the United States*, 3rd edn. New York: Springer.

Weiss, Kenneth and Miguel Bustillo. 2003. "Schwarzenegger Becomes the GOP's Green Candidate," *Los Angeles Times*, September 7.

Wellock, Thomas. 1998. *Critical Masses*. Madison: University of Wisconsin Press.

White House. 2011. "President Obama Announces Historic 54.5 mpg Fuel Efficiency Standard," press release, July 29, at www.whitehouse.gov, accessed December 15, 2014.

2015. "State Department Submits President Obama's Ambitious 2025 Target to Cut U.S. Climate Pollution by 26-28 Percent from 2005 Levels," press release, March 31, at www.whitehouse.gov, accessed June 25, 2015.

Wikle, Thomas. 1995. "Geographical Patterns of Membership in U.S. Environmental Organizations," *Professional Geographer* 47:41–48.

Wilson, John. 1985. "Efficiency Standards in California's Energy Policy," in Stephen Sawyer and John Armstrong, eds., *State Energy Policy*. Boulder, CO: Westview Press, 147–70.

Wilson, Lindsay. 2013. "The 'Electric Cars Aren't Green' Myth Debunked," at shrinkthatfootprint.com, accessed May 19, 2014.

Wines, Michael. 2013. "Climate Pact Is Signed by 3 States and a Partner," *New York Times*, October 29.

Wing, Ian Sue and Richard Eckaus. 2007. "The Decline in U.S. Energy Intensity," *Energy Policy* 35:5267–86

Wiser, Ryan, Mark Bolinger, and Galen Barbose. 2007. "Using the Federal Production Tax Credit to Build a Durable Market for Wind Power in the United States," *Electricity Journal* 20:77–88.

Xu, Beina. 2014. "China's Environmental Crisis," Council on Foreign Relations, updated February 5, at www.cfr.org, accessed November 28, 2014.

Yaccobucci, Brent and Robert Bamberger. 2007. *Automobile and Light Truck Fuel Economy, CRS Report for Congress*, updated January 17. Washington, DC: Congressional Research Service.

Yaccobucci, Brent and Larry Parker. 2008. *Climate Change: Federal Laws and Policies Related to Greenhouse Gas Reductions.* CRS Report for Congress, December 8. Washington, DC: Congressional Research Service.

Yeh, Sonia and Daniel Sperling. 2010. "Role of Low Carbon Fuel Standard in Reducing U.S. Transportation Emissions," in Sperling and James Cannon, eds., *Climate and Transportation Solutions*. Davis, CA: Institute of Transportation Studies, 48–61.

Zahariadis, Nikolaos. 2007. "The Multiple Streams Framework," in Paul Sabatier, ed., *Theories of the Policy Process*, 2nd edn. Boulder, CO: Westview Press, 65–92.

Zahran, Sammy et al. 2007. "Ecological Development and Global Climate Change," *Society and Natural Resources* 20:37–55.

Zehner, Ozzie. 2012. *Green Illusions*. Lincoln: University of Nebraska Press.

ZEV (Zero-Emission Vehicle Program Implementation Task Force). 2014. *Multi-State ZEV Action Plan*, May, at www.nescaum.org, accessed December 16, 2014.

Zierold, John. 1984. "Environmental Lobbyist in California's Capital. 1965–1984," interview conducted by Ann Lage, Regional Oral History Office, Bancroft Library, University of California at Berkeley, for the Sierra Club History Series.

Zuckerman, Seth. 2002. "Put Californians in the Driver's Seat to Reduce Global Warming," *San Francisco Chronicle*, May 6.

Index

CPSIA information can be obtained
at www.ICGtesting.com
Printed in the USA
LVOW10*0137230118
563547LV00006B/199/P

9 781107 074392